Boundary-Value Problems

Boundary-Value Problems

Ladis D. Kovach
Naval Postgraduate School

Addison-Wesley Publishing Company
Reading, Massachusetts • Menlo Park, California
London • Amsterdam • Don Mills, Ontario • Sydney

ISBN: 0-201-11728-2
ABCDEFGHIJ-DO-8987654

To the Memory of
My Parents

Contents

Preface

The author's objective in writing *Boundary-Value Problems* is to present methods of solving second-order linear partial differential equations that arise in applications. For the most part, the methods consist of separation of variables and (Laplace and Fourier) transform methods. Since these methods lead to *ordinary* differential equations, it is assumed that the student has the necessary background in this subject.

Chapter 1 provides a *ready reference* to topics in ordinary differential equations and may be consulted as needed. It is not intended that this reference chapter be included in a course in boundary-value problems, although the section on uniform convergence may be particularly helpful.

In Chapter 2 the difference between initial-value problems and boundary-value problems in *ordinary* differential equations is pointed out. This leads naturally into the prolific Sturm–Liouville theory, including the representation of a function by a series of orthonormal functions. Convergence and completeness are also presented here.

Separation of variables is introduced in Chapter 3 in connection with Laplace's equation and the heat equation. D'Alembert's solution of the vibrating infinite string is included and extended to a finite string with boundary conditions. The chapter concludes with the canonical forms of elliptic, parabolic, and hyperbolic equations.

Fourier series is the main topic in Chapter 4. In addition to cosine, sine, and exponential series, there is a section on applications and one on convergence. A heuristic argument extends Fourier series to Fourier integrals in Chapter 5. This extension then leads naturally to the Fourier transform and its applications.

All of the foregoing is brought together in Chapter 6, which is devoted to the solution of boundary-value problems expressed in rectangular coordinates. Included is the treatment of nonhomogeneous equations and nonhomogeneous boundary conditions. Fourier and Laplace transform methods of solution follow and there is also a section on the important, but often neglected, topic of the verification of solutions.

Chapter 7 contains the necessary variations to boundary-value problems phrased in polar, cylindrical, and spherical coordinates. This leads into discussions of Bessel functions and Legendre polynomials and their properties. A final chapter is devoted to numerical methods for solving boundary-value problems in both rectangular and polar coordinates.

There is a conscious effort to *separate* theory, applications, and numerical methods. This is done not only to preserve the mainstream of the development without distracting ramifications but also to provide the instructor with a maximum amount of flexibility. Since some sections are independent of the main theme, they may be included or omitted depending on the available time and the makeup of the class.

Additional flexibility can be found in the exercises. There are more than 1400 exercises at the ends of sections and these are divided into three categories. The first group helps to clarify the textual material and fills in details that have of necessity been omitted. The next, and most numerous, group consists of variations of the examples in the text and sufficient additional exercises to provide the practice most students need for a complete understanding of the material. A third group consists of more challenging exercises and those that extend the theory. Some suggestions for outside reading are also included in this group. Exercises that are of a computational nature are indicated by an asterisk (*).

It is a firm conviction of the author that the inclusion of historical sidelights is helpful to the reader and provides a welcome change of pace as well. Unfortunately, historical "facts" are sometimes controversial and provide fuel for hot debates on various topics, such as who was the originator of a particular method and the like. Unless otherwise indicated, we have used Florian Cajori's, *A History of Mathematics*, 3d ed. (New York: Chelsea, 1980) as a reference. This classic, first published in 1893, has endured to serve as an excellent source book.

Much of the material in this text has been used in classrooms during the past fifteen years. Some sections have appeared in the author's *Advanced Engineering Mathematics* (Reading, Mass.: Addison-Wesley, 1982). Preliminary versions of the present text were read by Ronald Guenther, Oregon State University; Roman Voronka, New Jersey Institute of Technology; and Euel Kennedy, California Polytechnic State University, and their incisive suggestions have been incorporated. The contributions of these reviewers is hereby gratefully acknowledged. Thanks are also due to Judy Caswell who did all the typing and to Wayne Yuhasz for his encouragement and editorial guidance. The staff at Addison-Wesley has been cooperative and helpful throughout the project. Finally, the many thousands of students who have offered valuable suggestions have provided the most important resource an author can have.

Monterey, California **L.D.K.**
July 1983

1 Review of Ordinary Differential Equations

1.1 LINEAR HOMOGENEOUS EQUATIONS WITH CONSTANT COEFFICIENTS

We present here a brief review of some of the types of ordinary differential equations a student should have met prior to studying boundary-value problems. Succeeding chapters will lean heavily on the material in this first chapter. Since practically all our references to ordinary differential equations will be to *second-order* linear equations, we will confine our discussion to these. Examples will be used freely, and theory will be kept to a minimum.

A common type of problem, which fortunately is also one of the simplest to solve, is the following **initial-value problem**:

$$\begin{aligned} y'' + ay' + by &= 0, \\ y(0) = c, \quad y'(0) &= d. \end{aligned} \tag{1.1-1}$$

Here a, b, c, and d are real numbers, and the primes denote differentiation with respect to some independent variable, say, x. The homogeneous differential equation in (1.1–1) has a general solution that is a linear combination of *two* **linearly independent functions**, each of which satisfies the given equation. Since there are *two* initial conditions, a unique solution can always be found. We illustrate with some examples.

EXAMPLE 1.1–1 Solve the initial-value problem

$$y'' - y' - 6y = 0, \qquad y(0) = 3, \qquad y'(0) = 4.$$

Solution The substitution $y = \exp(rx)$ results in the **characteristic equation**

$$r^2 - r - 6 = 0$$

and the **general solution**

$$y = c_1 \exp(-2x) + c_2 \exp(3x).$$

Substituting the initial values into the general solution and its derivative produces the system of linear equations

$$\begin{aligned} c_1 + c_2 &= 3, \\ -2c_1 + 3c_2 &= 4. \end{aligned} \qquad \textbf{(1.1-2)}$$

We find $c_1 = 1$, $c_2 = 2$, so the **particular solution** sought is

$$y = \exp(-2x) + 2\exp(3x). \quad ■$$

We observe that the two functions $y_1(x) = \exp(-2x)$ and $y_2(x) = \exp(3x)$ are **linearly independent** on every interval because their **Wronskian***

$$\begin{vmatrix} y_1(x) & y_2(x) \\ y_1'(x) & y_2'(x) \end{vmatrix} = \begin{vmatrix} \exp(-2x) & \exp(3x) \\ -2\exp(-2x) & 3\exp(3x) \end{vmatrix} = \exp x \begin{vmatrix} 1 & 1 \\ -2 & 3 \end{vmatrix}$$
$$= 5 \exp x$$

is everywhere different from zero. It is precisely for this reason that the system (1.1–2) has a unique solution, which, in turn, leads to the unique solution of the initial-value problem.

EXAMPLE 1.1–2 Solve the initial-value problem

$$y'' + 2y' + y = 0, \qquad y(0) = 3, \qquad y'(0) = -5.$$

Solution The characteristic equation

$$r^2 + 2r + 1 = (r + 1)^2 = 0$$

has two equal roots, and it can be shown that $y_1(x) = \exp(-x)$ and $y_2(x) = x\exp(-x)$ are two linearly independent solutions of the given differential equation. Hence the general solution is

$$y = c_1 \exp(-x) + c_2 x \exp(-x),$$

so that the initial conditions yield the system

$$\begin{aligned} c_1 &= 3 \\ -c_1 + c_2 &= -5, \end{aligned}$$

and the particular solution (Fig. 1.1–1)

$$y = 3\exp(-x) - 2x\exp(-x). \quad ■$$

*The Wronskian is a special determinant, named in honor of Hoëné Wronski (1778–1853), a Polish mathematician and philosopher. The word was coined in 1881 by Sir Thomas Muir (1844–1934), an English mathematician and educator, who also originated the word "radian."

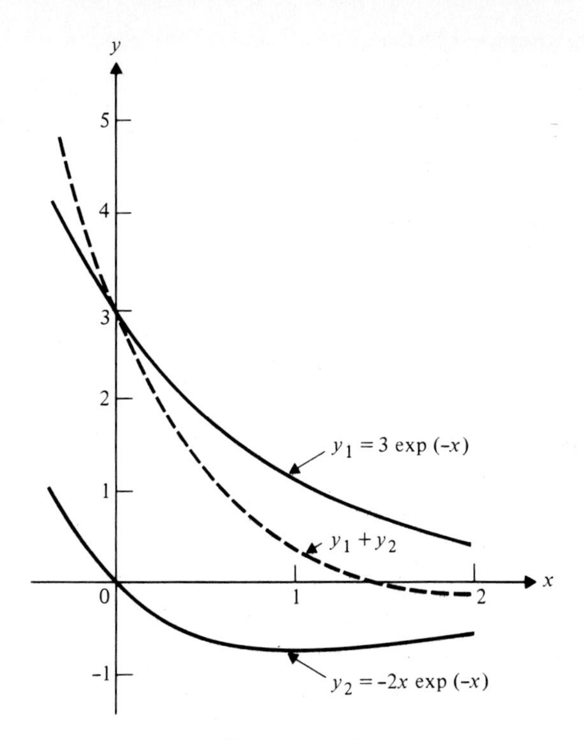

Figure 1.1-1
Graph of 3 exp (− x) − 2x exp (− x).

In this last example, knowing that $y_1 = \exp(-x)$ is a solution, we can assume that $y_2(x) = u(x) \exp(-x)$ is a second solution, substitute it into the given differential equation, and solve the resulting differential equation for $u(x)$. This elegant method of finding a second linearly independent solution of a differential equation when one solution is known is called the method of **reduction of order**. The name comes from the fact that the resulting differential equation for $u(x)$ can be treated as a *first-order* equation. The reduction of order method is not limited to differential equations with constant coefficients but can be applied to any **linear** equation, that is, to an equation of the form

$$y'' + a(x)y' + b(x)y = 0. \tag{1.1-3}$$

EXAMPLE 1.1-3 Solve the initial-value problem

$$y'' - 4y' + 13y = 0, \qquad y(0) = y'(0) = 3.$$

Solution In this example the roots of the characteristic equation

$$r^2 - 4r + 13 = 0$$

are the complex numbers $2 \pm 3i$. Hence the general solution is

$$y = \exp(2x)(c_1 \cos 3x + c_2 \sin 3x)$$

so that the initial conditions result in the system

$$\begin{aligned} c_1 &= 3, \\ 2c_1 + 3c_2 &= 3 \end{aligned}$$

and the particular solution

$$y = \exp(2x)(3 \cos 3x - \sin 3x). \qquad \blacksquare$$

In the last three examples we have examined the three possible cases that can arise in the study of second-order linear homogeneous equations with constant coefficients. In each case the characteristic equation is a quadratic equation, the roots of which must fall into one of the three cases illustrated in Fig. 1.1-2.

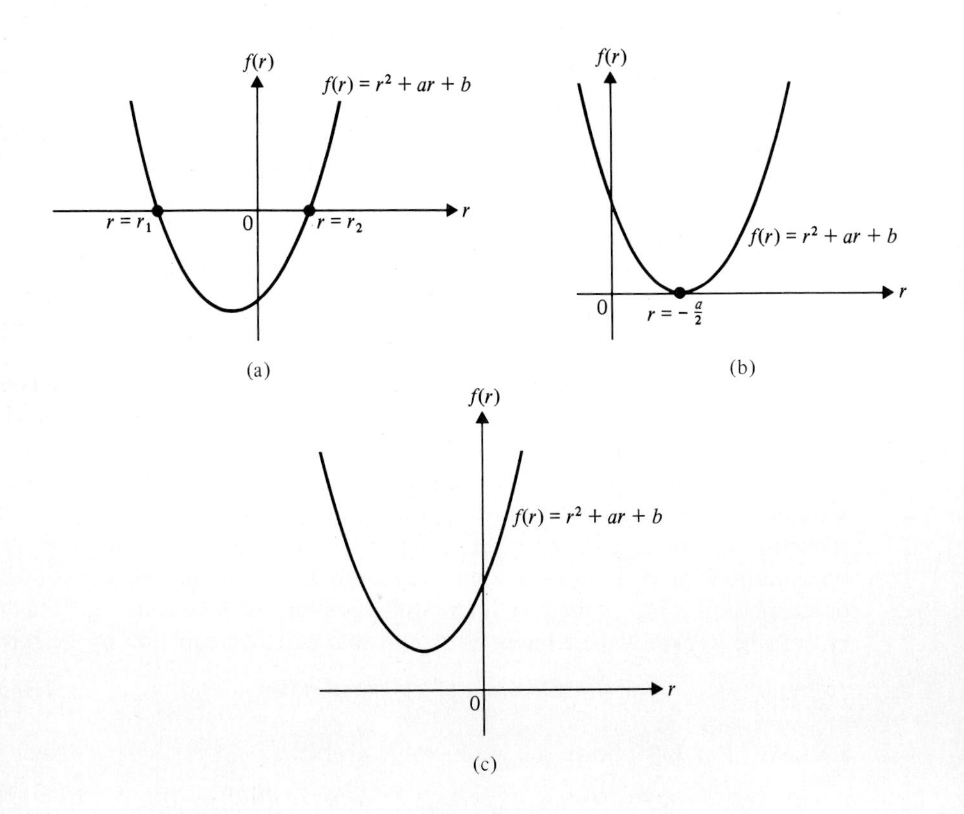

Figure 1.1-2
Graphs of characteristic equations. (a) Real and unequal roots. (b) Real and equal roots. (c) Complex roots.

In the next section we will consider nonhomogeneous equations of a special kind.

Key Words and Phrases

initial-value problem	linearly independent solutions
characteristic equation	Wronskian
general solution	reduction of order
particular solution	linear differential equation

1.1 Exercises*

• **1.** Verify that in order for $y = \exp(rx)$ to satisfy $y'' - y' - 6y = 0$ it is necessary that r satisfy the characteristic equation $r^2 - r - 6 = 0$. (Example 1.1–1)

2. Obtain the system (1.1–2), and then solve it for c_1 and c_2.

3. Verify that the Wronskian of $y_1(x) = \exp(-2x)$ and $y_2(x) = \exp(3x)$, written $W[y_1(x), y_2(x)]$, has the value $5 \exp x$.

4. Verify that the substitution of $y = \exp(rx)$ into $y'' + 2y' + y = 0$ leads to the characteristic equation $(r + 1)^2 = 0$. (Example 1.1–2)

5. Verify that $y_1(x) = \exp(-x)$ and $y_2(x) = x \exp(-x)$ both satisfy $y'' + 2y' + y = 0$.

6. Compute the Wronskian, $W[y_1(x), y_2(x)]$, of the functions in Exercise 5.

7. Obtain the particular solution in Example 1.1–2.

8. Obtain the roots of the characteristic equation of $y'' - 4y' + 13y = 0$.

9. Verify that $y_1(x) = \exp(2x) \cos 3x$ and $y_2(x) = \exp(2x) \sin 3x$ both satisfy $y'' - 4y' + 13y = 0$.

10. Compute the Wronskian of the functions $y_1(x)$ and $y_2(x)$ of Exercise 9.

11. Obtain the particular solution in Example 1.1–3.

•• **12.** Find the general solution to each of the following equations.
(a) $y'' - 3y' + 2y = 0$
(b) $y'' - 6y' + 9y = 0$
(c) $y'' - 6y' + 25y = 0$

13. Find the general solution of

$$y'' - 3y' = 0.$$

---

*Exercises at the end of each section are divided into three parts. Those in the first group are related directly to the text in that they help to clarify the textual material or supply computational steps that have been omitted. In the second group are exercises that provide the necessary practice so that the student may become more familiar with the methods and techniques presented. The last group includes exercises of a more theoretical or challenging nature.

14. Solve each of the following equations completely.

(a) $y'' + 2y' + 2y = 0$

(b) $y'' + y' + 2y = 0$

(c) $8y'' + 4y' + y = 0, \quad y(0) = 0, \quad y'(0) = 1$

(d) $x'' + 4x = 0, \quad x(\pi/4) = 1, \quad x'(\pi/4) = 3$

(*Hint:* The prime denotes differentiation with respect to some independent variable, say, t.)

15. Solve each of the following initial-value problems.

(a) $y'' + 3y' + 2y = 0, \quad y(0) = 0, \quad y'(0) = 2$

(b) $y'' + 9y = 0, \quad y(0) = 2, \quad y'(0) = 9$

(c) $y'' - 4y' + 4y = 0, \quad y(0) = 3, \quad y'(0) = -6$

16. Find the general solutions in each of the following.

(a) $\ddot{y} - 4y = 0,$ the dot representing $\dfrac{d}{dt}$

(b) $u'' + 6u' + 9u = 0,$ with $u = u(v)$

(c) $y'' + 5y' = 0,$ with $y = y(r)$

(d) $y'' - 2y' - 35y = 0,$ with $y = y(x)$

17. Solve each of the following initial-value problems.

(a) $x'' + x' - 3x = 0, \quad x(0) = 0, \quad x'(0) = 1$

(b) $u'' + 5u' + 6u = 0, \quad u(0) = 1, \quad u'(0) = 2$

(c) $\ddot{\theta} + 2\pi\dot{\theta} + \pi^2\theta = 0, \quad \theta(1) = 1, \quad \dot{\theta}(1) = 1/\pi$

(d) $4y'' + 20y' + 25y = 0, \quad y(0) = 1, \quad y'(0) = 2$

••• **18.** Show, by computing their Wronskian, that the functions $\exp(\alpha x)\cos(\beta x)$ and $\exp(\alpha x)\sin(\beta x)$ are linearly independent for all x.

19. **(a)** Given that $y_1 = \exp(-x)$ is a solution of $y'' + 2y' + y = 0$, assume that $y_2(x) = u(x)\exp(-x)$ is a solution and thus obtain $u''(x) = 0$.

(b) Find $u(x)$ and $y_2(x)$ in part (a). Observe that $y_2(x)$ is actually the general solution of the given differential equation.

20. Consider the linear, homogeneous differential equation

$$y'' + a(x)y' + b(x)y = 0,$$

and assume that a solution $y_1(x)$ has been found.

(a) Let $y_2(x) = u(x)y_1(x)$ be a second solution, and obtain the following equation for $u(x)$:

$$y_1(x)u''(x) + [2y_1'(x) + a(x)y_1(x)]u'(x) = 0.$$

(b) Apply the method in part (a) to the equation

$$x^2y'' + xy' - 4y = 0,$$

to find a second solution $y_2(x)$, given that $y_1(x) = x^2$. (*Note:* The given differential equation must be divided by x^2 in order to use the result of part (a).)

21. Prove that a common factor may be removed from any row and/or column of a determinant. (Compare the paragraph following Example 1.1-1.)

22. Prove that the value of a determinant is unchanged if any row (or column) is multiplied by a constant and the result is added to corresponding elements in any other row (or column).
23. Prove that interchanging any two rows (or columns) of a determinant changes its sign.
24. Prove that the value of a determinant having zeros below (or above) the main diagonal is the product of the diagonal terms.
25. Use the properties in Exercises 21–24 to evaluate the determinant

$$\begin{vmatrix} 3 & 6 & 9 \\ 2 & 4 & -6 \\ -4 & 12 & -24 \end{vmatrix}.$$

1.2 NONHOMOGENEOUS EQUATIONS WITH CONSTANT COEFFICIENTS

To consider nonhomogeneous equations in general would lead us too far afield. Hence we will deal with second-order linear differential equations with *constant* coefficients, that is, equations of the form

$$y'' + ay' + by = f(x), \tag{1.2–1}$$

where a and b are constants.

It can be shown that the general solution of Eq. (1.2–1) is the sum

$$y(x) = y_c(x) + y_p(x),$$

where $y_c(x)$, called the **complementary solution**, is the general solution of the **reduced equation**

$$y'' + ay' + by = 0 \tag{1.2–2}$$

and $y_p(x)$, called a **particular integral** (or particular solution) is *any* solution of Eq. (1.2–1). We saw how to find $y_c(x)$ in Section 1.1, so the present section is concerned mainly with finding $y_p(x)$.

What appears to be a simple problem is really not that simple unless we put some restrictions on $f(x)$ in Eq. (1.2–1). Accordingly, consider functions belonging to the set

$$\{P_n(x) \exp(\alpha x) \sin(\beta x),\ P_n(x) \exp(\alpha x) \cos(\beta x)\}, \tag{1.2–3}$$

where $P_n(x)$ is a polynomial in x of degree n, that is,

$$P_n(x) = c_0 + c_1x + c_2x^2 + \cdots + c_nx^n.$$

Although it may seem restrictive to limit $f(x)$ to functions of the types shown in (1.2–3), a large percentage of elementary problems fall into this category. This is indeed fortunate, since the class of functions (1.2–3) has a most useful property, namely, that the class is *closed* under differentiation. This means that if a function of the type shown in (1.2–3) is differentiated, the result will be a function (or possibly a sum of functions) of the same type. We illustrate with an example.

EXAMPLE 1.2–1 Find the first and second derivatives of the function

$$f(x) = (x^2 - 2x + 3) \exp (2x) \sin (3x).$$

Solution Recall from calculus that to differentiate a triple product, it is necessary to differentiate each factor, multiply the result by the other two factors, and add the three products. Hence

$$\begin{aligned} f'(x) &= (2x - 2) \exp (2x) \sin (3x) \\ &\quad + (x^2 - 2x + 3)\, 2 \exp (2x) \sin (3x) \\ &\quad + (x^2 - 2x + 3) \exp (2x)\, 3 \cos (3x) \\ &= (2x^2 - 2x + 4) \exp (2x) \sin (3x) \\ &\quad + (3x^2 - 6x + 9) \exp (2x) \cos (3x), \end{aligned}$$

and similarly,

$$\begin{aligned} f''(x) &= (-5x^2 + 18x - 21) \exp (2x) \sin (3x) \\ &\quad + (12x^2 - 12x + 24) \exp (2x) \cos (3x). \quad \blacksquare \end{aligned}$$

Because of the unusual property of the functions (1.2–3) illustrated in Example 1.2–1, we can find a particular integral of Eq. (1.2–1) by a method called the **method of undetermined coefficients**. This method consists of assuming a particular *form* of the solution, substituting into the differential equation, and then reconciling the coefficients of like terms.

EXAMPLE 1.2–2 Find a particular integral of the differential equation

$$y'' + y' - 6y = (4x + 5) \exp x.$$

Solution Since the function on the right is a polynomial of degree one multiplied by exp x, we assume a particular integral $y_p(x)$ of the form

$$y_p(x) = (Ax + B) \exp x.$$

Then

$$\begin{aligned} y_p'(x) &= (Ax + B) \exp x + A \exp x \\ y_p''(x) &= (Ax + B) \exp x + 2A \exp x, \end{aligned}$$

and substituting into the given differential equation, we find

$$(-4Ax + 3A - 4B) \exp x = (4x + 5) \exp x.$$

When we equate coefficients of like terms, we have $A = -1$ and $B = -2$, so the required particular integral is

$$y_p(x) = -(x + 2) \exp x. \quad \blacksquare$$

We point out that the complementary solution in Example 1.2-2 is

$$y_c(x) = c_1 \exp(-3x) + c_2 \exp(2x)$$

and that this solution has no terms in common with the assumed $y_p(x)$. When the two *do* have terms in common, this eventuality must be taken into account as we will show later.

EXAMPLE 1.2-3 Find a particular integral of the differential equation

$$y'' + y' - 6y = (50x + 40) \sin x.$$

Solution Since differentiating $\sin x$ results in $\cos x$, we choose a particular integral of the form

$$y_p(x) = (Ax + B) \sin x + (Cx + D) \cos x,$$

which includes *both* $\sin x$ and $\cos x$ terms. When we substitute this and its two derivatives into the given differential equation, the result is

$$\sin x[(-7A - C)x + A - 7B - 2C - D] \\ + \cos x[(A - 7C)x + 2A + B + C - 7D] = (50x + 40) \sin x.$$

When we equate the coefficients of like terms, we obtain a system of four linear equations in four unknowns, which can be solved by elimination or by matrix methods (for example, Gaussian elimination). We find $A = -7$, $B = -6$, $C = -1$, $D = -3$, and

$$y_p(x) = -(7x + 6) \sin x - (x + 3) \cos x. \quad \blacksquare$$

In the next example we show the procedure that must be followed when a term in $y_p(x)$ duplicates a term in $y_c(x)$.

EXAMPLE 1.2-4 Find a particular integral of the differential equation

$$y'' - 2y' + y = 12x \exp x.$$

Solution We first note that the complementary solution is

$$y_c(x) = (c_1 + c_2 x) \exp x.$$

Since the right-hand member of the differential equation is the product of a polynomial of degree one and an exponential function, we would ordinarily assume a particular integral of the form $(Ax + B) \exp x$. This cannot possibly be a solution of the given equation, however, since it is the solution of $y'' - 2y' + y = 0$. When this situation occurs, we multiply the function $(Ax + B)$

$r^2 - 2r + 1 = 0$

$y = Ae^{+x} + Bxe^{x}$

$(r-1)(r-1) = \emptyset$

$\exp x$ by x^s, where s is the *smallest* positive integer that will prevent duplication of terms in $y_c(x)$ and $y_p(x)$. In the present case, $s = 2$; hence we assume

$$y_p(x) = x^2(Ax + B) \exp x.$$

Differentiating this last expression twice and substituting into the given differential equation yields

$$(6Ax + 2B) \exp x = 12x \exp x,$$

so that $A = 2$, $B = 0$, and

$$y_p(x) = 2x^3 \exp x.$$

The general solution is then

$$y(x) = (c_1 + c_2x + 2x^3) \exp x. \quad \blacksquare$$

When the function $f(x)$ in (1.2–1) becomes unwieldy, that is, when it consists of a sum (or difference) of a large number of terms, a simplification can be made. We use the **Principle of Superposition**, which states that if $y = \phi_1(x)$ is a solution of $y'' + ay' + by = f_1(x)$ and $y = \phi_2(x)$ is a solution of $y'' + ay' + by = f_2(x)$, then $y = \phi_1(x) + \phi_2(x)$ is a solution of $y'' + ay' + by = f_1(x) + f_2(x)$. Hence we can solve a more difficult problem by adding the appropriate solutions of simpler ones. The exercises provide opportunities to use this technique.

We note that in Example 1.2–4 we found $B = 0$. This shows that we could have assumed a particular integral of the form $y_p(x) = Ax^3 \exp(x)$ and obtained the correct result more directly. Unfortunately, shortcuts of this kind cannot always be ascertained a priori; hence it is advisable to include all the terms of a polynomial in a particular integral.

Key Words and Phrases

complementary solution
reduced equation
particular integral
method of undetermined coefficients
Principle of Superposition

1.2 Exercises

• **1.** Show that if $y_c(x)$ is the general solution of $y'' + ay' + by = 0$ and $y_p(x)$ is any solution of $y'' + ay' + by = f(x)$, then the general solution of the latter is $y_c(x) + y_p(x)$. Here a and b are constants, and $f(x)$ is an arbitrary function of x. (*Hint*: A second-order differential equation must contain two arbitrary constants in its general solution.)

2. Verify that $y = -(x + 2) \exp x$ is a particular integral of $y'' + y' - 6y = (4x + 5) \exp x$.

3. Verify that $y = -(7x + 6) \sin x - (x + 3) \cos x$ is a particular integral of $y'' + y' - 6y = (50x + 40) \sin x$.

4. Fill in the details required to obtain the particular integral in Example 1.2–4.

5. Use the Principle of Superposition to solve the equation

$$y'' - 2y' + y = x^2 - 2x + 3 \sin x.$$

•• 6. In each of the following, indicate the proper form of $y_p(x)$, assuming that there are no duplications of terms in $y_c(x)$. Refer to Eq. (1.2–1).
(a) $f(x) = \exp(x) \cos x + \exp(2x) \cos 2x$
(b) $f(x) = x^3 \exp x$
(c) $f(x) = 3 + 2 \cos x$

7. Solve each of the following to obtain the general solution.
(a) $y'' - 2y' - 3y = 2 \exp x - 3 \exp(2x)$
(b) $y'' - 2y' + y = 3 \sin 2x$
(c) $y'' + 3y' - 4y = 3 \exp x$
(d) $y'' - 2y' + y = 3 \exp x$
(e) $y'' + 4y' + 4y = (2 + x) \exp(-2x)$
(f) $y'' - 4y = 4 \exp(2x)$
(g) $y'' + 4y' + 4y = 4x^2 - 8x$
(h) $y'' + 2y' + y = \sin x + \cos 2x$

8. Find a particular integral of

$$y'' + 3y' + 2y = 2 \exp(3x).$$

9. Find a particular integral of

$$y'' + 4y = 3 \sin 2x.$$

10. Solve the initial-value problem

$$y'' - 2y' + y = x \exp x, \qquad y(0) = 3, \qquad y'(0) = 5.$$

11. Find the general solution of each of the following equations.
(a) $y'' + y = \cos x + 3 \sin 2x$
(b) $y'' - 5y' + 6y = \cosh x$
(c) $y'' + 2y' + y = \cos^2 x = \frac{1}{2}(1 + \cos 2x)$

12. Solve each of the following initial-value problems.
(a) $y'' + 2y' + 5y = 10 \cos x, \quad y(0) = 5, \quad y'(0) = 6$
(b) $y'' - 7y' + 10y = 100x, \quad y(0) = 0, \quad y'(0) = 5$
(c) $y'' - 2y' + y = x^2 - 1, \quad y(0) = 2, \quad y'(0) = 1$

••• 13. The method of undetermined coefficients is limited to the functions shown in (1.2–3). There is a more general method known as **variation of parameters**, which we now illustrate using the equation

$$y'' + y = \tan x.$$

(a) Obtain the complementary solution

$$y_c(x) = c_1 \cos x + c_2 \sin x.$$

(b) Assume a particular solution of the form

$$y_p(x) = u(x) \cos x + v(x) \sin x.$$

(Note that the constants or parameters c_1 and c_2 have been replaced by functions $u(x)$ and $v(x)$. Our objective will be to obtain two equations in $u'(x)$ and $v'(x)$ that can then be solved simultaneously.) Differentiate to obtain

$$y_p'(x) = -u(x) \sin x + v(x) \cos x$$

with

$$u'(x) \cos x + v'(x) \sin x = 0.$$

Observe that this last condition simplifies $y_p'(x)$, $y_p''(x)$ and provides a second equation in $u'(x)$ and $v'(x)$.

(c) Differentiate $y_p'(x)$ in part (b) and substitute into the given differential equation to obtain

$$-u'(x) \sin x + v'(x) \cos x = \tan x.$$

(d) Solve the system

$$\begin{aligned} -u'(x) \sin x + v'(x) \cos x &= \tan x \\ u'(x) \cos x + v'(x) \sin x &= 0 \end{aligned}$$

for $u'(x)$ and $v'(x)$ by Cramer's rule or by elimination.

(e) Integrate $u'(x)$ and $v'(x)$ to find $u(x)$ and $v(x)$.

(f) Find $y_p(x)$ and thus obtain the general solution. Note that success in using the method of variation of parameters is contingent on being able to obtain $u(x)$ and $v(x)$ from $u'(x)$ and $v'(x)$.

14. Use the method of Exercise 13 to obtain the general solutions of each of the following equations.

(a) $y'' - y' = \sec^2 x - \tan x$

(b) $y'' - 2y' + y = \exp(x)/(1 - x)^2$

(c) $y'' + y = \sec x \tan x$

(d) $y'' + y = \sec x$

(e) $y'' - 2y' + y = e^x/x^2$

(f) $y'' + 4y = \cot 2x$

15. Solve the initial-value problem

$$y'' - 2y' + y = e^x/(1 - x)^2, \qquad y(0) = 2, \qquad y'(0) = 6.$$

16. Verify that $y_1(x) = x$ and $y_2(x) = 1/x$ are solutions of

$$x^3y'' + x^2y' - xy = 0.$$

Then use this information and the method of variation of parameters to find the general solution of

$$x^3y'' + x^2y' - xy = x/(1 + x).$$

17. (a) Solve the equation

$$y'' - y = x \sin x$$

by the method of undetermined coefficients.

(b) Solve the equation in part (a) by using the method of variation of parameters.

18. Consider

$$y'' + ay' + by = f(x),$$

where a and b are constants with $b \neq 0$ and $f(x)$ is a polynomial of degree n. Show that this equation always has a solution that is a polynomial of degree n.

19. In Exercise 18, show that the solution is a polynomial of degree $n + 1$ in the case where $b = 0$.

20. If $u(x)$, $v(x)$, and $w(x)$ are differentiable functions of x, use the formula for differentiating a product,

$$\frac{d(uv)}{dx} = u\frac{dv}{dx} + v\frac{du}{dx},$$

to find $d(uvw)/dx$.

21. Show that reconciling the coefficients of like terms in the method of undetermined coefficients depends on the forming of a linearly independent set by the functions involved. (*Hint*: Recall that if $f_i(x)$, $i = 1, 2, \ldots, n$, are n linearly independent functions over the reals, then

$$\sum_{i=1}^{n} c_i f_i(x) = 0$$

holds only if each $c_i = 0$.)

1.3 CAUCHY-EULER EQUATIONS

In the two preceding sections we discussed second-order linear ordinary differential equations with *constant* coefficients. While equations of this type will occur frequently throughout the remaining chapters, we will also have occasion to solve other linear differential equations, that is, equations of the form

$$a_0(x)y'' + a_1(x)y' + a_2(x)y = f(x). \qquad \textbf{(1.3-1)}$$

One type of linear differential equation with variable coefficients that can be reduced to a form already considered is the **Cauchy-Euler equation**, which has the normal form

$$x^2y'' + axy' + by = f(x), \qquad x > 0, \qquad \textbf{(1.3-2)}$$

where a and b are constants. This equation is also called a *Cauchy** equation, an *Euler*† equation, and an *equidimensional* equation. The last term comes from the fact that the physical dimension of x in the left-hand member of Eq. (1.3-2) is immaterial, since replacing x by cx, where c is a nonzero constant,

*Augustin-Louis Cauchy (1789–1857), a French mathematician.
†Leonhard Euler (1707–1783), a Swiss mathematician.

leaves the dimension of the left-hand member unchanged. We will meet a form of this equation later in our study of boundary-value problems having circular symmetry. We assume throughout this section that $x \neq 0$. For the most part we assume that $x > 0$, although we also deal with the case $x < 0$ later. We begin with an example to illustrate the method of solution.

EXAMPLE 1.3–1 Find the complementary solution of the equation

$$x^2y'' + 2xy' - 2y = x^2 \exp(-x), \qquad x > 0.$$

Solution This is a Cauchy–Euler equation, and we make the following substitutions in the reduced equation

$$y_c(x) = x^m, \qquad y_c' = mx^{m-1}, \qquad y_c'' = m(m-1)x^{m-2},$$

so the homogeneous equation becomes

$$[m(m-1) + 2m - 2]x^m = 0.$$

Because of the restriction $x \neq 0$, we must have

$$m^2 + m - 2 = 0,$$

which has roots $m_1 = -2$ and $m_2 = 1$. Thus

$$y_c(x) = c_1x^{-2} + c_2x. \qquad \blacksquare$$

We remark that the substitution, $y_c(x) = x^m$, did not come from thin air. It was dictated by the *form* of the left-hand member of the differential equation, which in turn ensured that each term of the equation would contain the common factor x^m.

It should be pointed out also that if we were interested in obtaining the *general* solution of the equation in Example 1.3–1, we could use the complementary solution above and the method of **variation of parameters**. (See Exercise 13 in Section 1.2.)

EXAMPLE 1.3–2 Obtain the complementary solution of

$$x^2y'' + 3xy' + y = x^3, \qquad x > 0.$$

Solution This time the substitution $y_c = x^m$ leads to

$$m^2 + 2m + 1 = 0,$$

which has a double root $m = -1$. Hence we have *one* solution of the homogeneous equation, namely,

$$y_1(x) = x^{-1}.$$

One might "guess" that a second linearly independent solution could be obtained by multiplying $y_1(x)$ by x. This procedure, however, is limited to the case of equations with *constant* coefficients and is thus not applicable here. It is easy to check that $y = 1$ is *not* a solution. In this case we can use the method of **reduction of order** to find a second solution.

We set $y_2(x) = u(x)/x$ and compute two derivatives. Thus

$$y_2'(x) = \frac{u'x - u}{x^2},$$

$$y_2''(x) = \frac{x^2(xu'' - 2u') + 2ux}{x^4},$$

and substitution into the homogeneous equation results in

$$xu'' + u' = 0,$$

which can be solved* by setting $u' = v$ and **separating the variables**. Then

$$v = \frac{du}{dx} = \frac{c_2}{x},$$

and†

$$u = c_2 \log x.$$

Hence

$$y_2(x) = \frac{c_2}{x} \log x,$$

and the complementary solution is

$$y_c(x) = \frac{c_1}{x} + \frac{c_2}{x} \log x.$$

We shall see later that the function $\log x$ occurs in the case of repeated roots. ■

EXAMPLE 1.3–3 Find the complementary solution of

$$x^2y'' + xy' + y = \cos x, \qquad x > 0.$$

Solution In this example we have, after substituting $y_c(x) = x^m$,

$$m^2 + 1 = 0$$

so that the solutions are x^i and x^{-i}. Hence the complementary solution is

$$y_c(x) = C_1x^i + C_2x^{-i}. \qquad \textbf{(1.3–3)}$$

A more useful form can be obtained, however, by replacing C_1 and C_2 by $(c_1 - ic_2)/2$ and $(c_1 + ic_2)/2$, respectively, and noting that

$$x^i = \exp(i \log x) = \cos(\log x) + i \sin(\log x).$$

*An alternative method is to note that $xu'' + u' = d(xu') = 0$, leading to $xu' = c_2$.

†We will consistently use $\log x$ for the *natural logarithm* of x.

$x^i = e^{i \log x} = e^{\log x^i} = x^i$

With these changes the complementary solution (1.3–3) can be written

$$y_c(x) = c_1 \cos(\log x) + c_2 \sin(\log x). \quad \blacksquare$$

We shall summarize the various cases that occur when solving the *homogeneous* Cauchy–Euler equation

$$x^2y'' + axy' + by = 0. \tag{1.3–4}$$

Substitution of $y_c(x) = x^m$ and its derivatives into Eq. (1.3–4) leads to the equation

$$m(m - 1) + am + b = 0$$

or

$$m^2 + (a - 1)m + b = 0. \tag{1.3–5}$$

This is called the **auxiliary equation** of the homogeneous Cauchy–Euler equation (1.3–4).

Case I. $(a - 1)^2 - 4b > 0$. The roots of Eq. (1.3–5) are real and unequal, say, m_1 and m_2. Then

$$y_c(x) = c_1x^{m_1} + c_2x^{m_2}, \tag{1.3–6}$$

and since the Wronskian

$$\begin{vmatrix} x^{m_1} & x^{m_2} \\ m_1x^{m_1-1} & m_2x^{m_2-1} \end{vmatrix} = (m_2 - m_1)x^{m_1+m_2-1} \neq 0,$$

showing that x^{m_1} and x^{m_2} are linearly independent.*

Case II. $(a - 1)^2 - 4b = 0$. The roots of Eq. (1.3–5) are real and equal, say, $m_1 = m_2 = m$. Then

$$y_1(x) = x^m$$

is one solution of Eq. (1.3–4). A second solution can be found by the method of reduction of order. Let

$$y_2(x) = x^m u(x)$$

be a second solution, differentiate twice, and substitute into Eq. (1.3–4). Then

$$u[m(m - 1) + am + b]x^m + u'(2m + a)x^{m+1} + u''x^{m+2} = 0.$$

*Note that the assumption $x > 0$ is essential here.

But the coefficient of u vanishes because x^m is a solution of Eq. (1.3–4) and $2m + a = 1$ from Eq. (1.3–5). Thus

$$xu'' + u' = 0,$$

which is satisfied by $u = \log x$ and

$$y_2(x) = x^m \log x.$$

In this case the complementary solution is

$$y_c(x) = x^m(c_1 + c_2 \log x). \tag{1.3–7}$$

Case III. $(a - 1)^2 - 4b < 0$. The roots of Eq. (1.3–5) are complex conjugates, say, $m_1 = \alpha + \beta i$ and $m_2 = \alpha - \beta i$. Then two linearly independent solutions of the homogeneous equation are

$$y_1(x) = x^{\alpha+\beta i} = x^\alpha x^{\beta i} = x^\alpha \exp(i\beta \log x)$$

and

$$y_2(x) = x^{\alpha-\beta i} = x^\alpha x^{-\beta i} = x^\alpha \exp(-i\beta \log x).$$

Using Euler's formula* to transform the exponential gives us

$$y_1(x) = x^\alpha[\cos(\beta \log x) + i \sin(\beta \log x)]$$

and

$$y_2(x) = x^\alpha[\cos(\beta \log x) - i \sin(\beta \log x)].$$

Hence the complementary solution becomes

$$y_c(x) = x^\alpha[c_1 \cos(\beta \log x) + c_2 \sin(\beta \log x)]. \tag{1.3–8}$$

If the general solution to Eq. (1.3–2) is required, it is necessary to add a particular solution to the appropriate complementary solution. A particular solution can be found by the method of variation of parameters, although difficulties may be encountered, as was mentioned in Exercise 13 of Section 1.2. Note that the method of undetermined coefficients is not applicable here, since the Cauchy–Euler differential equation does not have constant coefficients.

There is an alternative method for solving Eq. (1.3–4). Since $x > 0$, we can make the substitution

$$u = \log x.$$

This leads to $x = \exp u$ and, using the chain rule, to

$$\frac{dy}{dx} = \frac{dy}{du}\frac{du}{dx} = \frac{1}{x}\frac{dy}{du}.$$

*$\exp(i\theta) = \cos\theta + i \sin\theta$.

We also have

$$\begin{aligned}\frac{d^2y}{dx^2} &= \frac{d}{dx}\left(\frac{dy}{dx}\right) = \frac{d}{dx}\left(\frac{1}{x}\frac{dy}{du}\right)\\ &= \frac{1}{x}\frac{d}{du}\left(\frac{dy}{du}\right)\frac{du}{dx} - \frac{1}{x^2}\frac{dy}{du}\\ &= \frac{1}{x^2}\left(\frac{d^2y}{du^2} - \frac{dy}{du}\right).\end{aligned}$$

This method has the advantage that Eq. (1.3–4) is transformed into

$$\frac{d^2y}{du^2} + (a - 1)\frac{dy}{du} + by = f(\exp u).$$

Thus the differential equation has constant coefficients, and the methods of Section 1.1 are available for finding the complementary solution of the Cauchy–Eulcr cquation. In fact, if $f(\exp u)$ has the proper form, then the method of undetermined coefficients may lead to the general solution of the nonhomogeneous equation quite easily.

We have considered exclusively the case where $x > 0$. If solutions are desired for values of x satisfying $x < 0$, then x may be replaced by $-x$ in the differential equation and in the solution.

Key Words and Phrases

Cauchy-Euler equation
variation of parameters
reduction of order
separating the variables
auxiliary equation

1.3 Exercises

In the following exercises, assume that the independent variable is positive unless otherwise stated.

• **1.** Use the substitution $u = \log x$ to solve each of the following equations.
   (a) $x^2y'' + 2xy' - 2y = 0$ (Compare Example 1.3–1)
   (b) $x^2y'' + 3xy' + y = 0$ (Compare Example 1.3–2)
   (c) $x^2y'' + xy' + y = 0$ (Compare Example 1.3–3)

2. Obtain the general solution of the equation

$$x^2y'' + 3xy' + y = x^3.$$

(*Hint*: Use the substitution of Exercise 1 and the result of Example 1.3–2.)

3. Use Euler's formula

$$\exp(i\theta) = \cos\theta + i\sin\theta$$

to fill in the details in Case III.

•• 4. Solve the initial-value problem

$$x^2y'' - 2y = 0, \qquad y(1) = 6, \qquad y'(1) = 3.$$

5. Find the general solution of

$$x^2y'' + 5xy' - 5y = 0.$$

6. Find the general solution of

$$t^2y'' + 5ty' + 5y = 0.$$

7. Find the general solution of

$$r^2u'' + 3ru' + u = 0.$$

8. Obtain the general solution for

$$x^2y'' + xy' - 9y = x^2 - 2x.$$

9. Find the general solution of

$$x^2y'' + xy' + 4y = \log x.$$

10. Solve the initial-value problem

$$x^2y'' - xy' + 2y = 5 - 4x, \qquad y(1) = 0, \qquad y'(1) = 0.$$

11. Solve each of the following initial-value problems.
(a) $x^2y'' + xy' = 0, \quad y(1) = 1, \quad y'(1) = 2$
(b) $x^2y'' - 2y = 0, \quad y(1) = 0, \quad y'(1) = 1$

••• 12. Use the method of reduction of order to find a second solution for each of the following differential equations, given one solution as shown.
(a) $x^2y'' - xy' + y = 0, \quad y_1(x) = x$
(b) $xy'' + 3y' = 0, \quad y_1(x) = 2$
(c) $x^2y'' + xy' - 4y = 0, \quad y_1(x) = x^2$
(d) $x^2y'' - xy' + y = 0, \quad y_1(x) = x\log x^2$

13. Show that the products xy' and x^2y'' remain unchanged if x is replaced by cx, where c is a nonzero constant.

14. Show that the substitution $x = \exp u$ transforms the equation

$$x^2y'' + axy' + by = 0,$$

where a and b are constants, into

$$\frac{d^2y}{du^2} + (a - 1)\frac{dy}{du} + by = 0.$$

15. Solve each of the following initial-value problems.
(a) $4x^2y'' - 4xy' + 3y = 0, \quad y(1) = 0, \quad y'(1) = 1$
(b) $x^2y'' + 5xy' + 4y = 0, \quad y(1) = 1, \quad y'(1) = 3$

16. Obtain the general solution of the equation

$$x^2y'' + axy' = 0,$$

where a is a constant.

1.4 INFINITE SERIES

Since we will need to solve second-order linear equations with variable coefficients that are not of Cauchy–Euler type (Eq. 1.3–2), we must explore other methods of solution. One powerful method is the power series method. In using this method we assume that the solution to a given differential equation can be expressed as a power series. Inasmuch as we will need certain facts about infinite series and their convergence, we digress in this section to review some aspects of these.

Each of the following is an example of a **series of constants**:

$$1 + 2 + 3 + 4 + \cdots + n + \cdots, \tag{1.4–1}$$

$$1 - 1 + 1 - 1 + 1 - + \cdots + (-1)^{n+1} + \cdots, \tag{1.4–2}$$

$$0 + 0 + 0 + \cdots + 0 + \cdots, \tag{1.4–3}$$

$$1 + \frac{1}{2^p} + \frac{1}{3^p} + \cdots + \frac{1}{n^p} + \cdots, \tag{1.4–4}$$

$$a(1 + r + r^2 + \cdots + r^n + \cdots), \tag{1.4–5}$$

$$1 - \frac{1}{3} + \frac{1}{5} - + \cdots + \frac{(-1)^{n+1}}{2n - 1} + \cdots. \tag{1.4–6}$$

The series in (1.4–1) is **divergent** because the **partial sums**

$$S_1 = 1, \quad S_2 = 1 + 2, \quad S_3 = 1 + 2 + 3, \quad S_4 = 1 + 2 + 3 + 4, \quad \ldots$$

form a **sequence**

$$\{S_1, S_2, S_3, \ldots\} = \{1, 3, 6, 10 \ldots\},$$

which has *no* **limit point.*** On the other hand, the series in (1.4–2) is divergent because its sequence of partial sums has *two* limit points, $+1$ and 0. The series in (1.4–3) is a trivial example of a **convergent** series, since its sequence of partial sums has a unique limit point, namely zero.

---

*A point is called a *limit point* of a sequence if an infinite number of terms in the sequence are within a distance of ϵ of the point, where ϵ is an arbitrarily small positive number. A limit point need not be unique and need not be an element of the sequence. For example, 1 is the unique limit point of the sequence

$$\left\{\frac{1}{2}, \frac{3}{4}, \frac{7}{8}, \frac{15}{16}, \ldots\right\}$$

By a more sophisticated test (the **integral test**) it can be shown that the series of (1.4–4) is convergent for $p > 1$ and divergent for $p \leq 1$. When $p = 1$, the series is called a **harmonic series**. The series of (1.4–5) is a **geometric series** with first term a and **common ratio** r. It can be shown (by the ratio test) that (1.4–5) converges if $|r| < 1$ and diverges if $|r| \geq 1$ and $a \neq 0$. The sum of (1.4–5) can be written in **closed form** as

$$a \sum_{n=0}^{\infty} r^n = \frac{a}{1 - r}, \qquad |r| < 1.$$

Finally, the series of (1.4–6) is an example of an **alternating series** that can be proved (using a theorem of Leibniz*) to be convergent because the following two conditions hold:

1. The absolute value of each term is less than or equal to the absolute value of its predecessor.
2. The limiting value of the nth term is zero as $n \to \infty$.

We remark that it is one thing to determine whether a given series converges, but quite another to determine what it converges to. It is not obvious, for example, that the sum of the series in (1.4–6) is $\pi/4$, although we will obtain this result and some others in the exercises for Section 4.3. (See Exercises 23 and 25 of that section.)

Of greater interest to us than a series of constants will be **power series** of the form

$$a_0 + a_1(x - x_0) + a_2(x - x_0)^2 + \cdots + a_n(x - x_0)^n + \cdots. \tag{1.4–7}$$

Such a series is called a power series in $x - x_0$. A power series *always* converges. For example, (1.4–7) converges for $x = x_0$, but we will be interested in convergence on an *interval* such as $(x_0 - R, x_0 + R)$. We call R $(R > 0)$ the **radius of convergence** of the power series and determine its value by using the **ratio test**† as shown in the next example.

EXAMPLE 1.4–1 Find the radius of convergence of the series

$$(x - 1) - \frac{(x - 1)^2}{2} + \frac{(x - 1)^3}{3} - \frac{(x - 1)^4}{4} + \cdots.$$

---

*Gottfried Wilhelm Leibniz (1646–1716), the co-inventor (with Sir Isaac Newton) of the calculus, who proved the theorem in 1705.

†Also called D'Alembert's ratio test after Jean-le-Rond D'Alembert (1717–1783), a French mathematician who made important contributions in analytical mechanics.

Solution It will be convenient to write the series using summation notation,

$$\sum_{n=1}^{\infty} \frac{(-1)^{n+1}(x-1)^n}{n}.$$

According to the ratio test, a series converges whenever

$$\lim_{n\to\infty} \left| \frac{u_{n+1}}{u_n} \right| < 1,$$

where u_n represents the nth term of the series. In the present case,

$$\lim_{n\to\infty} \left| \frac{u_{n+1}}{u_n} \right| = \lim_{n\to\infty} \left| \frac{(-1)^{n+2}(x-1)^{n+1}}{n+1} \cdot \frac{n}{(-1)^{n+1}(x-1)^n} \right|$$
$$= |x-1| \lim_{n\to\infty} \frac{n}{n+1} = |x-1| < 1.$$

Hence $-1 < x - 1 < 1$ or $0 < x < 2$, showing that the *radius* of convergence is 1. In many problems it is necessary to examine the endpoints of the **interval of convergence** as well, and this must be done separately. It can be shown (Exercise 3) that the interval of convergence here is $0 < x \leq 2$. ■

Obtaining a power series representation of a function is an important mathematical technique that has many applications. Recall that the **Maclaurin* series** for a function $f(x)$ is given by

$$f(x) = f(0) + f'(0)x + \frac{f''(0)}{2!}x^2 + \frac{f'''(0)}{3!}x^3 + \cdots. \quad \textbf{(1.4-8)}$$

Following are some familiar Maclaurin series expansions:

$$\exp x = 1 + x + \frac{x^2}{2!} + \frac{x^3}{3!} + \cdots + \frac{x^n}{n!} + \cdots, \quad \textbf{(1.4-9)}$$

$$\sin x = x - \frac{x^3}{3!} + \frac{x^5}{5!} - + \cdots + \frac{(-1)^{n+1}x^{2n-1}}{(2n-1)!} + \cdots, \quad \textbf{(1.4-10)}$$

$$\cos x = 1 - \frac{x^2}{2!} + \frac{x^4}{4!} - + \cdots + \frac{(-1)^{n+1}x^{2n-2}}{(2n-2)!} + \cdots, \quad \textbf{(1.4-11)}$$

$$\log(1+x) = x - \frac{x^2}{2} + \frac{x^3}{3} - + \cdots + \frac{(-1)^{n+1}x^n}{n} + \cdots. \quad \textbf{(1.4-12)}$$

*After Colin Maclaurin (1698–1746), a Scottish mathematician.

The Maclaurin series for $\exp x$, $\sin x$, and $\cos x$ converge for all finite values of x, while the series (1.4–12) has an interval of convergence given by $-1 < x \leq 1$. All four of the series are particular cases of Eq. (1.4–8). Using summation notation, we can write*

$$\exp x = \sum_{n=0} \frac{x^n}{n!}$$

$$\sin x = \sum_{n=1} \frac{(-1)^{n+1} x^{2n-1}}{(2n-1)!}$$

$$\cos x = \sum_{n=1} \frac{(-1)^{n+1} x^{2n-2}}{(2n-2)!}$$

$$\log(1+x) = \sum_{n=1} \frac{(-1)^{n+1} x^n}{n}$$

where we have used the convention $0! = 1$. Observe that n is a **dummy index** and may be replaced by something else if this is desirable. For example, replacing n by $m - 1$ in Eq. (1.4–12) produces

$$\log(1+x) = \sum_{m=0} \frac{(-1)^m x^{m-1}}{m-1}.$$

We shall make use of this flexibility of the dummy index in the next section.

A Maclaurin series representation of a function can be thought of as an approximation of the function in the neighborhood of $x = 0$ as shown in Fig. 1.4–1. If a series expansion about some point other than $x = 0$ is required, then we can use a **Taylor**† **series**

$$f(x) = f(a) + f'(a)(x-a) + \frac{f''(a)}{2!}(x-a)^2 + \frac{f'''(a)}{3!}(x-a)^3 + \cdots. \qquad \textbf{(1.4–13)}$$

Note that Maclaurin's series is a special case of Taylor's series, the case where $a = 0$.

It would appear from Eqs. (1.4–8) and (1.4–13) that any function that has an infinite number of derivatives that are defined at $x = a$ can be represented by a Taylor series in the neighborhood of $x = a$. This is not entirely

*We shall omit the upper values of n on summations henceforth if they are $n = \infty$.

†After Brook Taylor (1685–1731), a British mathematician who discovered it in 1712. Historically, Taylor's series predated Maclaurin's series.

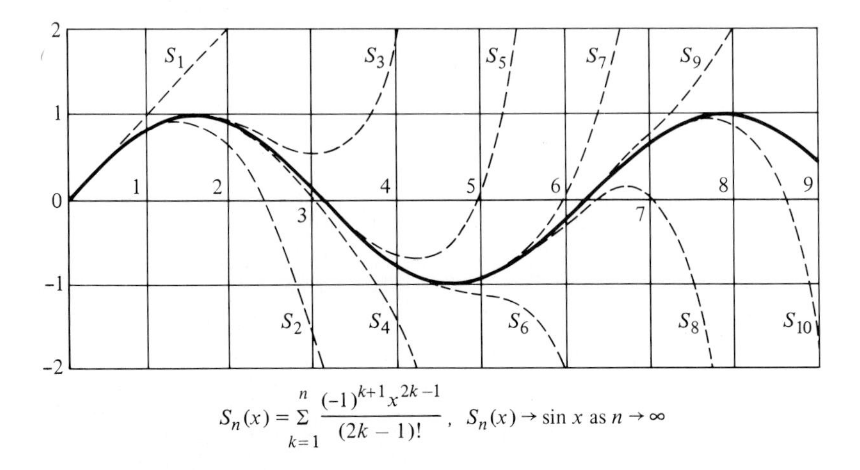

Figure 1.4-1
Approximations of partial sums to sin x. (From H. M. Kammerer, "Sine and Cosine Approximation Curves," *MAA Monthly* 43, p. 293.)

true. The conjecture in the last statement represents an oversimplification of the facts.* We will discuss this and related topics further in the next section.

Key Words and Phrases

series of constants
divergent
partial sums
sequence
limit point
convergent
integral test
harmonic series
geometric series
common ratio
closed form
alternating series
power series
radius of convergence
ratio test
interval of convergence
Maclaurin series
dummy index
Taylor series

1.4 Exercises

• **1.** Write the sequence of partial sums for each of the following series.
(a) $1 - 1 + 1 - 1 + - \cdots + (-1)^{n+1} + \cdots$
(b) $0 + 0 + 0 + 0 + \cdots$

*A small caveat is necessary here, since there are some (pathological) functions that have derivatives of all orders at a point yet cannot be represented by a Taylor series there.

(c) $1 - \frac{1}{3} + \frac{1}{5} - \frac{1}{7} + - \cdots$
(d) $1 + \frac{1}{3} + \frac{1}{9} + \frac{1}{27} + \cdots$

2. What is the sum of the series in Exercise 1(d)?

3. Show that the series of Example 1.4–1 converges when $x = 2$ and diverges when $x = 0$.

4. Use the ratio test to show that the Maclaurin series for $\exp x$, $\sin x$, and $\cos x$ converge for all x.

5. Show that the interval of convergence of the Maclaurin series for $\log(1 + x)$ is $-1 < x \leq 1$. See Eq. (1.4–12).

6. Verify that each of the following summations is correct.

(a) $$\sin x = \sum_{n=0} \frac{(-1)^n x^{2n+1}}{(2n + 1)!}$$

(b) $$\cos x = \sum_{n=0} \frac{(-1)^n x^{2n}}{(2n)!}$$

(c) $$\exp(-x) = \sum_{n=1} \frac{(-1)^{n-1} x^{n-1}}{(n - 1)!}$$

(d) $$\cosh x = \frac{e^x + e^{-x}}{2} = \sum_{n=0} \frac{x^{2n}}{(2n)!}$$

(e) $$\sinh x = \frac{e^x - e^{-x}}{2} = \sum_{n=1} \frac{x^{2n-1}}{(2n - 1)!}$$

*7. Use Maclaurin's series to compute $\sin \frac{\pi}{4}$ and $\cos \frac{\pi}{4}$ to four decimals. (*Hint*: Use the fact that the error has the same sign as the first neglected term but has a smaller absolute value.)

•• 8. Consider the sequence

$$\left\{\frac{1}{2}, \frac{3}{4}, \frac{7}{8}, \frac{15}{16}, \cdots\right\}.$$

(a) Write the nth term of the sequence.
(b) Show that 1 is the limit point of the sequence.

9. Apply the ratio test to the series (1.4–5) to show that the series converges for $|r| < 1$.

10. (a) Identify the series

$$\frac{1}{10} + \frac{1}{100} + \frac{1}{1000} + \cdots.$$

(b) Find the sum of the series.

---

*Calculator problem.

11. Find the radius of convergence of each of the following power series.

(a) $(x - 1) + \dfrac{(x - 1)^3}{3} + \dfrac{(x - 1)^5}{5} + \cdots$

(b) $1 + \dfrac{x^2}{2!} + \dfrac{x^4}{4!} + \dfrac{x^6}{6!} + \cdots$

(c) $1 + \dfrac{(x + 3)}{2} + \dfrac{(x + 3)^2}{3} + \dfrac{(x + 3)^3}{4} + \cdots$

(d) $x + \dfrac{2!x^2}{2^2} + \dfrac{3!x^3}{3^3} + \dfrac{4!x^4}{4^4} + \cdots$ (*Hint*: Use the limit definition of e.)

(e) $1 + \dfrac{(x + 2)}{3} + \dfrac{(x + 2)^2}{2 \cdot 3^2} + \dfrac{(x + 2)^3}{3 \cdot 3^3} + \cdots$

(f) $1 + \dfrac{(x - 1)^2}{2!} + \dfrac{(x - 1)^4}{4!} + \dfrac{(x - 1)^6}{6!} + \cdots$

12. Find the interval of convergence of each of the following power series. If the interval is finite, investigate the convergence of the series at the endpoints of the interval.

(a) $\displaystyle\sum_{n=0} \frac{x^n}{2^n}$

(b) $\displaystyle\sum_{n=0} \frac{x^{2n+1}}{n!}$

(c) $\displaystyle\sum_{n=1} \frac{(x - 1)^n}{2n}$

(d) $\displaystyle\sum_{n=1} \frac{x^n}{n^2}$

(e) $\displaystyle\sum_{n=1} \frac{x^n}{n^n}$

13. Verify that each of the following series is convergent.

(a) $\displaystyle\sum_{n=1} \frac{n}{(n + 1)^3}$

(b) $\displaystyle\sum_{n=1} \frac{1}{n!}$

(c) $\displaystyle\sum_{n=1} n\left(\frac{1}{2}\right)^n$

(d) $\displaystyle\sum_{n=2} \frac{\sqrt{n + 3}}{(n - 1)^2}$

14. Verify that each of the following series is divergent.

(a) $\sum_{n=2} \frac{1}{\sqrt{n} \log n}$

(b) $\sum_{n=1} \frac{2^n}{n^2}$

(c) $\sum_{n=1} \frac{(n-1)^n}{n!}$

(d) $\sum_{n=1} \frac{n}{(n+1)^2}$

15. Determine the values of p for which the following series converges and diverges.

$$\sum_{n=2} \frac{1}{n^p \log n},$$

where p is a positive integer.

••• **16.** Use the integral test to determine for what values of p (p is a real number) the series

$$\sum_{n=1} \frac{1}{n^p}$$

converges.

17. Consider the series

$$\frac{1}{1 \cdot 3} + \frac{1}{2 \cdot 4} + \frac{1}{3 \cdot 5} + \cdots .$$

(a) Write S_n, the sum of the first n terms in closed form. (*Hint*: Decompose the nth term of the series into partial fractions.)

(b) Obtain the sum of the series.

18. Generalize Exercise 17 for

$$\sum_{n=1} \frac{1}{n(n+p)}$$

where p is a positive integer.

1.5 SERIES SOLUTIONS

Before we give an example of how a power series solution of a linear differential equation can be obtained, we need the results of two theorems. These are presented without proofs in order to preserve the continuity.

THEOREM 1.5-1
A power series $\sum_{n=0} a_n(x - x_0)^n$ *and its derivative* $\sum_{n=1} na_n(x - x_0)^{n-1}$ *have the same radius of convergence.*

THEOREM 1.5-2
Let a function $f(x)$ *be represented by a power series* $\sum_{n=0} a_n(x - x_0)^n$ *in the interior of its interval of convergence. Then the function is differentiable there, and its derivative is given by*

$$f'(x) = \sum_{n=1} na_n(x - x_0)^{n-1}.$$

We are now ready to look at a simple differential equation with a view to solving it by using series. To begin with, we will take x_0 to be zero. Later we will indicate why this is not always possible.

EXAMPLE 1.5-1 Find a solution of the equation $y'' - xy = 0$.

Solution We *assume* that there is a solution of the form

$$y = \sum_{n=0} a_n x^n.$$

Then

$$y' = \sum_{n=1} na_n x^{n-1}$$

and

$$y'' = \sum_{n=2} n(n - 1)a_n x^{n-2}.$$

Substituting these values into the given equation produces

$$\sum_{n=2} n(n - 1)a_n x^{n-2} - \sum_{n=0} a_n x^{n+1} = 0.$$

In order to collect terms it would be convenient to have x^n in both summations. This can be accomplished by realizing that n is a *dummy index* of summation and can be replaced by any other letter just as we change variables in definite integrals. Accordingly, we replace *each* n in the first sum by $n + 2$ and each n in the second sum by $n - 1$. Then

$$\sum_{n=0} (n + 2)(n + 1)a_{n+2}x^n - \sum_{n=1} a_{n-1}x^n = 0.$$

Next we combine the two sums into one with n going from 1 to ∞, adding any terms that are left out of this sum. Thus

$$\sum_{n=1} [(n + 2)(n + 1)a_{n+2} - a_{n-1}]x^n + 2a_2 = 0,$$

which is a linear combination of $1, x, x^2, \ldots$. Since the set of functions

$$\{1, x, x^2, x^3, \ldots\}$$

is a linearly independent set, a linear combination of these functions can be zero if and only if *each* coefficient is zero. Hence

$$2a_2 = 0,$$

and, in general,

$$(n + 2)(n + 1)a_{n+2} - a_{n-1} = 0.$$

From the first of these, $a_2 = 0$, and from the second we obtain the **recursion formula**

$$a_{n+2} = \frac{a_{n-1}}{(n + 2)(n + 1)}, \qquad n = 1, 2, \ldots .$$

- For $n = 1$ we have $a_3 = \dfrac{a_0}{3 \cdot 2}$, so that a_0 can be arbitrary.
- For $n = 2$ we have $a_4 = \dfrac{a_1}{4 \cdot 3}$, so that a_1 can be arbitrary.
- For $n = 3$ we have $a_5 = \dfrac{a_2}{5 \cdot 4} = 0$; consequently, $a_2, a_5, a_8, \ldots$ are all zero.
- For $n = 4$ we have $a_6 = \dfrac{a_0}{3 \cdot 2 \cdot 6 \cdot 5}$.
- For $n = 5$ we have $a_7 = \dfrac{a_4}{7 \cdot 6} = \dfrac{a_1}{4 \cdot 3 \cdot 7 \cdot 6}$, and so on.

The solution to the given differential equation is

$$y = a_0 + a_1 x + \frac{a_0}{6}x^3 + \frac{a_1}{12}x^4 + \frac{a_0}{180}x^6 + \frac{a_1}{504}x^7 + \cdots .$$

This last equation can also be written as

$$y = a_0\left(1 + \frac{x^3}{6} + \frac{x^6}{180} + \cdots\right) + a_1\left(x + \frac{x^4}{12} + \frac{x^7}{504} + \cdots\right),$$

which shows the two arbitrary constants we expect to find in the solution of a second-order differential equation. It can be shown (Exercise 1) that both series converge for $-\infty < x < \infty$. ■

While many series can be written in **closed form**, for example,

$$e^x = 1 + x + \frac{x^2}{2!} + \frac{x^3}{3!} + \cdots, \tag{1.5-1}$$

$$\sin x = x - \frac{x^3}{3!} + \frac{x^5}{5!} - \frac{x^7}{7!} + - \cdots, \tag{1.5-2}$$

$$\cos x = 1 - \frac{x^2}{2!} + \frac{x^4}{4!} - \frac{x^6}{6!} + - \cdots, \tag{1.5-3}$$

$$\log(1 + x) = x - \frac{x^2}{2} + \frac{x^3}{3} - \frac{x^4}{4} + - \cdots, \tag{1.5-4}$$

this is not always possible. If a function can be represented in an open interval containing x_0 by a *convergent* series of the form $\sum_{n=0} a_n(x - x_0)^n$, then the function is said to be **analytic** at $x = x_0$. The functions in Eqs. (1.5-1) through (1.5-4) are all analytic at $x = 0$. If a function is analytic at every point where it is defined, it is called an **analytic function**. All polynomials are analytic, and so are rational functions except where their denominators vanish.

Now let us look at another example of a series solution of a differential equation.

EXAMPLE 1.5-2 Solve the equation

$$(x - 1)y'' - xy' + y = 0.$$

Solution As before, assume

$$y = \sum_0 a_n x^n, \qquad y' = \sum_1 n a_n x^{n-1}, \qquad y'' = \sum_2 n(n - 1)a_n x^{n-2},$$

and substitute into the given differential equation. Then

$$\sum_2 n(n - 1)a_n x^{n-1} - \sum_2 n(n - 1)a_n x^{n-2} - \sum_1 n a_n x^n + \sum_0 a_n x^n = 0.$$

Replace n by $n + 1$ in the first sum and replace n by $n + 2$ in the second sum so that we have

$$\sum_1 (n + 1)n a_{n+1} x^n - \sum_0 (n + 2)(n + 1)a_{n+2} x^n$$

$$- \sum_1 n a_n x^n + \sum_0 a_n x^n = 0$$

or

$$\sum_1 [n(n + 1)a_{n+1} - (n + 1)(n + 2)a_{n+2} - n a_n + a_n]x^n - 2a_2 + a_0 = 0.$$

Equating the coefficients of various powers of x to zero gives us the following:

$$
\begin{aligned}
&a_0 \text{ is arbitrary,}\\
&a_1 \text{ is arbitrary,}\\
a_2 &= \frac{1}{2}a_0,\\
a_{n+2} &= \frac{n(n+1)a_{n+1} + (1-n)a_n}{(n+1)(n+2)}, \qquad n = 1, 2, \ldots,\\
a_3 &= \frac{2a_2}{2\cdot 3} = \frac{a_2}{3} = \frac{a_0}{3\cdot 2},\\
a_4 &= \frac{6a_3 - a_2}{3\cdot 4} = \frac{a_3}{2} - \frac{a_2}{12} = \frac{a_0}{12} - \frac{a_0}{24} = \frac{a_0}{4!},
\end{aligned}
$$

etc. Hence

$$y = a_1x + a_0\left(1 + \frac{x^2}{2!} + \frac{x^3}{3!} + \frac{x^4}{4!} + \cdots\right),$$

and it can be shown (Exercise 2) that x and e^x are two linearly independent solutions of the given equation. Here the solution can be written in closed form in contrast to the solution of Example 1.5–1 (Exercise 3). ■

Unfortunately, the series method of solving ordinary differential equations is not as simple as the last two examples seem to indicate. Consider the equation

$$2x^2y'' + 5xy' + y = 0.$$

We leave it as an exercise (Exercise 4) to show that the series method with $x_0 = 0$ will produce only the trivial solution $y = 0$. Yet the given equation is a Cauchy–Euler equation, and both $x^{-1/2}$ and $1/x$ are solutions (Exercise 5). The answer to the apparent mystery lies in the fact that the individual solutions of a Cauchy–Euler equation are not linearly independent on any interval that includes the origin. Recall that in Section 1.3 we solved Cauchy–Euler equations assuming that $x > 0$ *or* $x < 0$.

Consider the most general second-order, linear, homogeneous ordinary differential equation,

$$y'' + P(x)y' + Q(x)y = 0. \qquad \textbf{(1.5–5)}$$

Those values of x, call them x_0, at which *both* $P(x)$ and $Q(x)$ are analytic are called **ordinary points** of Eq. (1.5–5). If either $P(x_0)$ or $Q(x_0)$ is not analytic, then x_0 is a **singular point** of Eq. (1.5–5). If, however, x_0 is a singular point of Eq. (1.5–5) but *both* $(x - x_0)P(x)$ and $(x - x_0)^2Q(x)$ are analytic at $x = x_0$, then x_0 is a **regular singular point** of Eq. (1.5–5). All other singular points are called **irregular singular points**.

EXAMPLE 1.5-3 Classify the singular points of the equation

$$(1 - x^2)y'' - 2xy' + n(n + 1)y = 0,$$

where $n = 0, 1, 2, \ldots$.

The only singular points are $x_0 = \pm 1$. If $x_0 = -1$, then

$$\frac{(x + 1)(-2x)}{1 - x^2} = \frac{2x}{x - 1} \quad \text{and} \quad \frac{(x + 1)^2 n(n + 1)}{1 - x^2} = \frac{n(n + 1)(x + 1)}{x - 1}.$$

Since both of these rational functions are analytic at $x = -1$, the latter is a regular singular point. Similarly, for $x_0 = 1$ we have

$$\frac{(x - 1)(-2x)}{1 - x^2} = \frac{2x}{x + 1} \quad \text{and} \quad \frac{(x - 1)^2 n(n + 1)}{1 - x^2} = \frac{n(n + 1)(1 - x)}{1 + x},$$

so that $x_0 = 1$ is also a regular singular point. ■

The point of all this is contained in an 1865 theorem due to Fuchs.* Fuchs' theorem states that it is always possible to obtain *at least one* power series solution to a linear differential equation provided that the assumed series solution is about an ordinary point or, at worst, a regular singular point.

The work of Fuchs was extended by Frobenius,† who in 1874 suggested that instead of assuming a series solution of the form $\sum_0 a_n x^n$, one should use the form $\sum_0 a_n x^{n+r}$. The use of this form to solve linear, ordinary differential equations is known today as the **method of Frobenius**. We illustrate with an example using the Cauchy-Euler equation referred to above.

EXAMPLE 1.5-4 Solve the equation

$$2x^2y'' + 5xy' + y = 0$$

by the method of Frobenius.

Solution We have

$$y = \sum_0 a_n x^{n+r},$$

$$y' = \sum_0 (n + r)a_n x^{n+r-1},$$

$$y'' = \sum_0 (n + r)(n + r - 1)a_n x^{n+r-2},$$

*Lazarus Fuchs (1833–1902), a German mathematician.
†Georg Frobenius (1849–1917), a German mathematician.

and on substituting into the given equation we have

$$\sum_{0}^{\infty} [2(n + r)(n + r - 1) + 5(n + r) + 1]a_n x^{n+r} = 0.$$

Since the coefficient of x^{n+r} must be zero for $n = 0, 1, 2, \ldots$, we have for $n = 0$,

$$(2r^2 + 3r + 1)a_0 = 0.$$

Choosing a_0 to be arbitrary, that is, nonzero, produces

$$2r^2 + 3r + 1 = 0,$$

which is called the **indicial equation.** Its roots are -1 and $-1/2$. In general,

$$a_n(2n^2 + 4nr + 3n) = 0, \qquad n = 1, 2, \ldots,$$

which can be satisfied only by taking $a_n = 0$, $n = 1, 2, \ldots$. Hence we are left with the two possibilities,

$$y_1(x) = a_0 x^{-1} \qquad \text{and} \qquad y_2(x) = b_0 x^{-1/2}.$$

Note that the two constants are arbitrary, since each root of the indicial equation leads to an infinite series. In this example, however, each series consists of a single term. ■

When solving *second-order* linear differential equations by the method of Frobenius, the indicial equation is a quadratic equation, and three possibilities exist. We list these together with their consequences here.

1. If the roots of the indicial equation are *equal*, then only *one* solution can be obtained.
2. If the roots of the indicial equation differ by a number that is not an integer, then two linearly independent solutions may be obtained.
3. If the roots of the indicial equation differ by an integer, then the larger integer of the two will yield a solution, whereas the smaller may or may not yield a solution.

It should be mentioned that the theory behind the method of Frobenius is by no means simple. A good discussion of the various cases that may arise (although the case where the indicial equation has complex roots is omitted) can be found in Albert L. Rabenstein, *Elementary Differential Equations with Linear Algebra*, 3d ed. (New York: Academic Press, 1982), pp. 391 ff.

We conclude this section by solving two important differential equations that will appear later in the text in connection with certain types of boundary-value problems.

EXAMPLE 1.5-5 Obtain a solution of the differential equation

$$\frac{d^2y}{dx^2} + \frac{1}{x}\frac{dy}{dx} + \left(1 - \frac{n^2}{x^2}\right)y = 0, \qquad n = 0, 1, 2, \ldots . \tag{1.5-6}$$

This equation is known as **Bessel's differential equation**. It was originally obtained by Friedrich Wilhelm Bessel (1784–1846), a German mathematician, in the course of his studies of planetary motion. Since then, this equation has appeared in problems of heat conduction, electromagnetic theory, and acoustics that are expressed in *cylindrical coordinates*.

Solution Since the coefficients are not constant, we seek a series solution. Multiplying Eq. (1.5-6) by x^2, we obtain

$$x^2y'' + xy' + (x^2 - n^2)y = 0. \tag{1.5-7}$$

We note that $x = 0$ is a regular singular point; hence we use the method of Frobenius. Assume that

$$y = \sum_{m=0} a_m x^{m+r},$$

$$y' = \sum_{m=0} a_m(m + r)x^{m+r-1},$$

$$y'' = \sum_{m=0} a_m(m + r)(m + r - 1)x^{m+r-2}$$

and substitute into Eq. (1.5-7). Then

$$\sum_{m=0} a_m(m + r)(m + r - 1)x^{m+r} + \sum_{m=0} a_m(m + r)x^{m+r} + \sum_{m=0} a_m x^{m+r+2} - n^2 \sum_{m=0} a_m x^{m+r} = 0.$$

If we replace m by $m - 2$ in the third series, the last equation can be written as

$$\sum_{m=2} [a_m(m + r)(m + r - 1) + a_m(m + r) + a_{m-2} - n^2a_m]x^{m+r} + a_0r(r - 1)x^r + a_0rx^r - n^2a_0x^r + a_1r(r + 1)x^{r+1} + a_1(r + 1)x^{r+1} - n^2a_1x^{r+1} = 0.$$

Simplifying, we get

$$\sum_{m=2} [a_m((m + r)^2 - n^2) + a_{m-2}]x^{m+r} + a_0(r^2 - n^2)x^r + a_1(r^2 + 2r + 1 - n^2)x^{r+1} = 0.$$

The coefficient of x^r must be zero; hence if we assume $a_0 \neq 0$, then we obtain $r = \pm n$. We choose the positive sign, since n was defined as a nonnegative integer in Eq. (1.5-6). Since the coefficient of x^{r+1} must also be zero, we may choose $a_1 = 0$. Then the recursion formula is obtained by setting the coefficient of x^{m+r} equal to zero. Thus

$$a_m = \frac{-a_{m-2}}{m(m+2n)}, \qquad m = 2, 3, \ldots .$$

The first few coefficients can be computed from this formula. They are as follows:

$$m = 2: \qquad a_2 = \frac{-a_0}{2^2(n+1)},$$

$$m = 4: \qquad a_4 = \frac{-a_2}{2^3(n+2)} = \frac{a_0}{2^4 \cdot 2(n+1)(n+2)},$$

$$m = 6: \qquad a_6 = \frac{-a_4}{2^2 \cdot 3(n+3)} = \frac{a_0}{2^6 \cdot 3!(n+1)(n+2)(n+3)}.$$

In general we have

$$a_{2m} = \frac{(-1)^m a_0}{2^{2m} m!(n+1)(n+2)\cdots(n+m)}, \qquad m = 1, 2, \ldots,$$

and a solution to Eq. (1.5–6) can be written as

$$y_n(x) = a_0 \sum_{m=0} \frac{(-1)^m x^{2m+n}}{2^{2m} m!(n+1)(n+2)\cdots(n+m)}$$

$$= 2^n n! a_0 \sum_{m=0} \frac{(-1)^m}{m!(m+n)!} \left(\frac{x}{2}\right)^{2m+n}. \qquad \blacksquare$$

The **Bessel function of the first kind of order *n*** is defined by giving a_0 the value $1/2^n n!$. We have

$$J_n(x) = \sum_{m=0} \frac{(-1)^m}{m!(m+n)!} \left(\frac{x}{2}\right)^{2m+n}, \qquad n = 0, 1, 2, \ldots, \qquad \textbf{(1.5–8)}$$

a solution of Bessel's differential equation. We will consider this function in greater detail in Chapter 7.

EXAMPLE 1.5–6 Obtain a solution of the equation

$$(1 - x^2)y'' - 2xy' + n(n+1)y = 0, \qquad \textbf{(1.5–9)}$$

where n is a constant. This equation is known as **Legendre's differential equation.***

*After Adrien Marie Legendre (1752–1833), a French mathematician who is known mainly for his work in number theory, elliptic functions, and calculus of variations.

Solution Since $x = \pm 1$ are regular singular points (see Example 1.5–3), we may assume a power series about $x = 0$, which is an ordinary point. Accordingly, put

$$y = \sum_{m=0} a_m x^m, \qquad y' = \sum_{m=1} a_m m x^{m-1}, \qquad y'' = \sum_{m=2} a_m m(m-1)x^{m-2}.$$

Substituting these values into Eq. (1.5–9) produces

$$\sum_{m=2} a_m m(m-1)x^{m-2} - \sum_{m=2} a_m m(m-1)x^m - 2\sum_{m=1} a_m m x^m + n(n+1)\sum_{m=0} a_m x^m = 0.$$

Replacing m by $m+2$ in the first sum, we get

$$\sum_{m=0} a_{m+2}(m+2)(m+1)x^m - \sum_{m=2} a_m(m-1)x^m - 2\sum_{m=1} a_m m x^m + n(n+1)\sum_{m=0} a_m x^m = 0,$$

or

$$\sum_{m=2} [a_{m+2}(m+2)(m+1) - a_m m(m-1) - 2a_m m + a_m n(n+1)]x^m + 2a_2 + 6a_3 x - 2a_1 x + n(n+1)a_0 + n(n+1)a_1 x = 0.$$

Setting the coefficient of each power of x equal to zero in the above, we have

$$2a_2 + n(n+1)a_0 = 0, \qquad a_2 = \frac{-n(n+1)a_0}{2}, \qquad a_0 \text{ arbitrary;}$$

$$6a_3 - 2a_1 + n(n+1)a_1 = 0, \qquad a_3 = \frac{[2 - n(n+1)]a_1}{6}, \qquad a_1 \text{ arbitrary.}$$

In general, we can say

$$a_{m+2}(m+2)(m+1) - [m(m-1) + 2m - n(n+1)]a_m = 0;$$

$$a_{m+2} = \frac{m(m+1) - n(n+1)}{(m+2)(m+1)} a_m;$$

$$a_{m+2} = \frac{(m-n)(m+n+1)}{(m+2)(m+1)} a_m, \qquad m = 0, 1, 2, \ldots . \quad \textbf{(1.5–10)}$$

Equation (1.5–10) is the recurrence relation from which the coefficients can be found.

Computing the first few coefficients gives us

$$a_2 = \frac{-n(n+1)}{1\cdot 2} a_0,$$

$$a_4 = \frac{(2-n)(n+3)}{4\cdot 3} a_2 = \frac{n(n-2)(n+1)(n+3)}{4!} a_0,$$

$$a_6 = \frac{(4-n)(n+5)}{6\cdot 5} a_4 = \frac{-n(n-2)(n-4)(n+1)(n+3)(n+5)}{6!} a_0,$$

$$a_3 = \frac{(1-n)(n+2)}{3\cdot 2} a_1 = \frac{-(n-1)(n+2)}{3!} a_1,$$

$$a_5 = \frac{(3-n)(n+4)}{5\cdot 4} a_3 = \frac{(n-1)(n-3)(n+2)(n+4)}{5!} a_1,$$

$$a_7 = \frac{(5-n)(n+6)}{7\cdot 6} a_5$$
$$= \frac{-(n-1)(n-3)(n-5)(n+2)(n+4)(n+6)}{7!} a_1.$$

Hence a solution to Legendre's equation can be written as

$$y_n(x) = a_0\left[1 - \frac{n(n+1)}{2!}x^2 + \frac{n(n-2)(n+1)(n+3)}{4!}x^4 - \frac{n(n-2)(n-4)(n+1)(n+3)(n+5)}{6!}x^6 + - \cdots\right]$$
$$+ a_1\left[x - \frac{(n-1)(n+2)}{3!}x^3 + \frac{(n-1)(n-3)(n+2)(n+4)}{5!}x^5 - \frac{(n-1)(n-3)(n-5)(n+2)(n+4)(n+6)}{7!}x^7 + - \cdots\right]. \quad \textbf{(1.5–11)}$$

Both series converge for $-1 < x < 1$.

If $n = 0, 2, 4, \ldots$ and a_1 is chosen to be zero, then the solutions, using Eq. (1.5–11), become

$$y_0(x) = a_0,$$
$$y_2(x) = a_0(1 - 3x^2),$$
$$y_4(x) = a_0\left(1 - 10x^2 + \frac{35}{3}x^4\right), \qquad \text{etc.}$$

If we also impose the condition that $y_n(1) = 1$, then we can evaluate the a_0 to obtain

$$P_0(x) = 1,$$
$$P_2(x) = \frac{1}{2}(3x^2 - 1), \tag{1.5-12}$$
$$P_4(x) = \frac{1}{8}(35x^4 - 30x^2 + 3), \ldots .$$

These *polynomials* are called the **Legendre polynomials of even degree.**

If $n = 1, 3, 5, \ldots$ and a_0 is chosen to be zero, then the solutions, using Eq. (1.5–11), become

$$y_1(x) = a_1x,$$
$$y_3(x) = a_1\left(x - \frac{5}{3}x^3\right),$$
$$y_5(x) = a_1\left(x - \frac{14}{3}x^3 + \frac{21}{5}x^5\right), \qquad \text{etc.}$$

If we again impose the condition that $y_n(1) = 1$, then we can evaluate the a_1 to obtain

$$P_1(x) = x, \qquad P_3(x) = \frac{1}{2}(5x^3 - 3x),$$
$$P_5(x) = \frac{1}{8}(63x^5 - 70x^3 + 15x), \ldots . \tag{1.5-13}$$

These polynomials are called the **Legendre polynomials of odd degree.** ■

The **Legendre polynomials** will be of use in Section 7.3, since they arise in boundary-value problems expressed in *spherical coordinates*.

Key Words and Phrases

recursion formula
closed form
analytic
analytic function
ordinary point
singular point
regular and irregular singular point
method of Frobenius
indicial equation
Bessel's differential equation
Bessel function of the first kind of order n,
Legendre's differential equation
Legendre polynomials of even degree
Legendre polynomials of odd degree

1.5 Exercises

• **1.** **(a)** Show that one solution of the differential equation $y'' - xy = 0$ in Example 1.5–1 can be written as

$$y_1(x) = a_0 \sum_{n=0} \frac{1 \cdot 4 \cdot 7 \cdots (3n - 2)}{(3n)!} x^{3n}.$$

(b) Write the second solution in a similar form.

(c) Find the radius of convergence of the series in part (a).

(d) Observe that $x_0 = 0$ is an ordinary point and hence that both solutions are analytic.

2. Verify that $y_1(x) = x$ and $y_2(x) = e^x$ are linearly independent solutions of the differential equation $(x - 1)y'' - xy' + y = 0$.

3. **(a)** Show that the solution obtained in Example 1.5–2 is equivalent to

$$y(x) = c_1x + c_2e^x.$$

(b) For what values of x is the above solution valid?

4. Show that assuming a solution of the form $y = \sum_0 a_nx^n$ for the equation

$$2x^2y'' + 5xy' + y = 0$$

leads to the trivial solution $y = 0$.

5. Verify that $x^{-1/2}$ and $1/x$ are linearly independent solutions of the equation in Exercise 4 on every interval not containing the origin.

•• **6.** Classify the singular points of each of the following differential equations.

(a) $x^2y'' + xy' + (x^2 - n^2)y = 0, \quad n = 0, 1, 2, \ldots$

(b) $x^3y'' - xy' + y = 0$

(c) $x^2y'' + (4x - 1)y' + 2y = 0$

(d) $x^3(x - 1)^2y'' + x^4(x - 1)^3y' + y = 0$

7. Use power series to solve each of the following equations.

(a) $y'' + y = 0$

(b) $y'' - y = 0$

(c) $y' - y = x^2$

(Note that the power series method is not limited to *homogeneous* equations.)

(d) $y' - xy = 0$

(If possible, write the solution in closed form.)

(e) $(1 + x^2)y'' + 2xy' - 2y = 0$

8. Solve each of the following differential equations by the method of Frobenius.

(a) $xy'' + y' + xy = 0$

(b) $4xy'' + 2y' + y = 0$

(c) $x^2y'' + 2xy' - 2y = 0$

9. Solve the equation

$$xy'' + 2y' = 0$$

by two methods. (*Hint*: x is an integrating factor.)

10. Solve the equation

$$y'' - xy' - y = 0$$

by assuming a solution that is a power series in $(x - 1)$. In this case the coefficient x must also be written in terms of $x - 1$. This can be done by assuming that $x = A(x - 1) + B$ and determining the constants A and B.

11. Obtain a solution of

$$xy'' + y' + 4xy = 0.$$

12. Obtain a solution of

$$xy'' - 2y = 0.$$

••• **13.** The differential equation

$$y'' - xy = 0$$

is known as **Airy's equation,*** and its solutions are called **Airy functions** (Fig. 1.5-1), which have applications in the theory of diffraction.

(a) Obtain the solution in terms of a power series in x.

(b) Obtain the solution in terms of a power series in $(x - 1)$. (Compare Exercise 10.)

14. Compare the solutions of Exercise 13 with those of $y'' - y = 0$. Comment.

15. Solve the initial-value problem

$$y'' + xy = 2, \qquad y(0) = y'(0) = 1.$$

16. Solve the differential equation

$$4xy'' + 2y' + y = 0.$$

17. Find the interval of convergence of the two series in the solution of Exercise 16.

18. Solve the initial-value problem

$$y'' + y' + xy = 0, \qquad y(0) = y'(0) = 1.$$

19. Obtain the general solution of

$$2x^2y'' - xy' + (1 - x^2)y = 0.$$

20. Obtain the general solution of

$$x^2y'' + x^2y' - 2y = 0.$$

21. Illustrate Theorem 1.5-1 by differentiating the series in Exercise 11 of Section 1.4 and finding the radii of convergence of the differentiated series. (Note that this does *not* constitute a *proof* of the theorem.)

22. Illustrate Theorem 1.5-2 by differentiating the functions and series in Eqs. (1.5-1) through (1.5-4).

23. Find the general solution of

$$y'' + xy' + y = 0.$$

---

*After Sir George B. Airy (1801–1892), an English mathematician and astronomer.

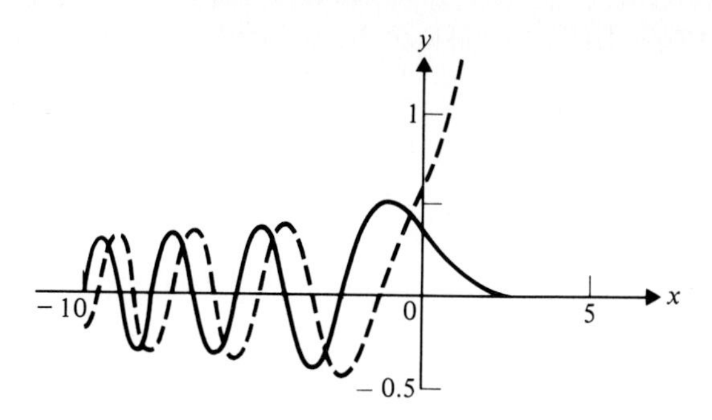

Figure 1.5-1
Airy functions.

1.6 UNIFORM CONVERGENCE

We give a brief discussion of some of the theoretical aspects in the present section, not for the sake of the theory per se but to prepare for a wide variety of applications. It will be shown, for example, that we will have to broaden our concept of convergence.

Pointwise Convergence

Recall what is meant by saying that a sequence of functions, defined for $a \leq x \leq b$,

$$\{f_1(x), f_2(x), f_3(x), \ldots, f_n(x), \ldots\},$$

converges to $f(x)$. When we write

$$f(x) = \lim_{n \to \infty} f_n(x) \qquad \text{for all } x \text{ in } [a, b],$$

we mean that the difference between $f(x)$ and $f_n(x)$ can be made arbitrarily small provided that n is taken large enough.

In more precise mathematical language we would say that given any positive number ϵ and any point x_0 in the interval $[a, b]$, we can satisfy

$$|f_n(x_0) - f(x_0)| < \epsilon \tag{1.6-1}$$

whenever $n \geq N(\epsilon, x_0)$, an integer. In other words, given $\epsilon > 0$, we can find an integer N such that the inequality (1.6-1) holds whenever $n \geq N$. The important thing to notice here is that N will usually depend on ϵ *and* x_0, so we write $N(\epsilon, x_0)$.

EXAMPLE 1.6–1 Consider the sequence

$$\{x^n\}$$

with $0 < x_0 < 1$ and take $0 < \epsilon < 1$. Then we can show that

$$\lim_{n \to \infty} x^n = 0,$$

since (1.6–1) becomes $x_0^n < \epsilon$ or $n \log x_0 < \log \epsilon$ or

$$\frac{\log \dfrac{1}{\epsilon}}{\log \dfrac{1}{x_0}} < n. \tag{1.6–2}$$

To prove that the given sequence converges, we must find an integer N large enough so that the relation (1.6–2) holds for all $n \geq N$. Thus (1.6–2) shows clearly that N depends on both ϵ and x_0. Observe from (1.6–2) that we must have $\epsilon > 0$ and $0 < x_0 < 1$. ■

The convergence illustrated in Example 1.6–1 is called **pointwise convergence**, since, given $\epsilon > 0$, the inequality (1.6–2) can be satisfied for each point x_0 of the interval (0, 1). In other words, the sequence of functions $\{x_n\}$ converges to zero at every point of the interval (0, 1).

In later chapters we will be dealing with functions that will be defined by means of an infinite series. For example, we will have

$$f(x) = \sum_{n=1} u_n(x),$$

where each $u_i(x)$ is defined on some interval I given by $a \leq x \leq b$. If the series is (pointwise) convergent for each x in I, then the sum of the series is also a function of x, say, f. The following important questions arise:

1. Under what conditions will f be continuous?
2. Under what conditions will f be differentiable?
3. Under what conditions can we write

$$\int_a^b \left[\sum_{n=1} u_n(x)\right] dx = \int_a^b f(x)\, dx \stackrel{?}{=} \sum_{n=1} \int_a^b u_n(x)\, dx.$$

4. Under what conditions can we write

$$\frac{d}{dx} \sum_{n=1} u_n(x) = f'(x) \stackrel{?}{=} \sum_{n=1} u_n'(x).$$

The answers to these questions involve something other than pointwise convergence, and we consider this topic next.

Uniform Convergence

In case a sequence of functions $\{f(x_n)\}$ converges to $f(x)$ and, for a given $\epsilon > 0$, we can establish the inequality

$$|f(x_n) - f(x)| < \epsilon$$

for all $n \geq N(\epsilon)$ *and* for each x belonging to an interval I simultaneously, then we can say that the sequence **converges uniformly** on I. In contrast to pointwise convergence the integer N in uniform convergence usually depends on ϵ but *never* on x, so we write $N(\epsilon)$.

EXAMPLE 1.6–2 Let $f_n(x) = x^n/n$, $0 \leq x \leq 1$, $\epsilon > 0$. Then $f(x) = 0$, and

$$|f_n(x) - f(x)| < \epsilon$$

becomes

$$\left|\frac{x^n}{n}\right| \leq \frac{1}{n} < \epsilon$$

provided that $1/\epsilon < n$. Here we may take N to be the smallest integer that is greater than $1/\epsilon$, a choice that is independent of x. We say that x^n/n converges to zero *uniformly on the interval* $0 \leq x \leq 1$. ■

Since convergence of an infinite series is intimately related to the convergence of a sequence, the foregoing discussion applies to infinite series as well. For example, we say that the series

$$\sum_{n=1} u_n(x)$$

converges *pointwise* to $f(x)$ if the sequence $\{S_n(x)\}$ of **partial sums** with

$$S_n(x) = \sum_{i=1}^{n} u_i(x)$$

converges pointwise to $f(x)$. In a similar fashion we define the *uniform* convergence on an interval I of an infinite series of functions.

A related idea deals with improper integrals that contain a parameter. For example, the integral

$$F(x) = \int_a^\infty f(t, x)\, dt \qquad \textbf{(1.6–3)}$$

defines $F(x)$ on I if the integral converges for each value of x in the interval I. If x_0 is in I, then pointwise convergence of the integral in Eq. (1.6–3) means that we can find a number $N(\epsilon, x_0)$ such that, for a given $\epsilon > 0$,

$$\left| \int_a^\infty f(t, x_0)\, dt - \int_a^b f(t, x_0)\, dt \right| = \left| \int_b^\infty f(t, x_0)\, dt \right| < \epsilon$$

whenever $b > N$. Ordinarily, N depends on x_0 and ϵ. On the other hand, the convergence is uniform on the interval I whenever there is a number $N(\epsilon)$ independent of x such that

$$\left| \int_b^\infty f(t, x)\, dt \right| < \epsilon$$

for all $b > N$ and *all* x on I.

EXAMPLE 1.6–3 The improper integral

$$\int_0^\infty xe^{-xt}\, dt = \begin{cases} 1, & 0 < x \leq 1, \\ 0, & x = 0, \end{cases}$$

does not converge uniformly on the interval $0 \leq x \leq 1$, although it converges at each point of the interval.

Solution To prove pointwise convergence, consider

$$\left| \int_b^\infty xe^{-xt}\, dt \right| = \int_b^\infty xe^{-xt}\, dt = e^{-xb} < \epsilon$$

whenever $b > N$. It follows that we must have $-b < (1/x) \log \epsilon$, which shows that b (and N) depends on both x and ϵ. Moreover, for $0 < \epsilon < 1$, the quantity $|(1/x) \log \epsilon|$ can be made as great as we please, so there is no possibility of satisfying the inequality $-b < (1/x) \log \epsilon$ for *all* x on $[0, 1]$ simultaneously. ■

EXAMPLE 1.6–4 The improper integral

$$\int_0^\infty e^{-xt}\, dt$$

converges uniformly to $1/x$ on the interval $1 \leq x \leq 2$. Here we have

$$\left| \int_b^\infty e^{-xt}\, dt \right| = \int_b^\infty e^{-xt}\, dt = \frac{e^{-xb}}{x} < \epsilon$$

whenever $b > N$, from which it follows that $-b < (1/x) \log x\epsilon$. Now the right-hand member of the last inequality varies between $\log \epsilon$ and $1/2 \log 2\epsilon$ as x varies between 1 and 2. Hence, taking the greater of the two values, we must satisfy $-b < 1/2 \log 2\epsilon$. In this case, b (and N) depends only on ϵ, showing that the convergence is uniform on the interval $[1, 2]$. ■

In later chapters we will solve boundary-value problems by various transform methods. In these cases the solutions will be expressed in terms of integrals, and we will be interested in knowing under what conditions such integrals can be differentiated. Stated another way, if

$$F(x) = \int_c^\infty f(t, x)\,dt,$$

then under what conditions can we obtain

$$F'(x) = \frac{\partial}{\partial x}\int_c^\infty f(x, t)\,dt \overset{?}{=} \int_c^\infty \frac{\partial f(x, t)}{\partial x}\,dt.$$

The following theorems, whose proofs we omit, will be of use later.*

THEOREM 1.6-1

If

$$F(x) = \int_c^\infty f(x, t)\,dt$$

converges uniformly on the interval $a \le x \le b$, *and if* $\partial f/\partial x$ *is continuous in* t *and* x *when* $c \le t$ *and* $a \le x \le b$, *and if*

$$\int_c^\infty \frac{\partial f(x, t)}{\partial x}\,dt$$

converges uniformly on $a \le x \le b$, *then*

$$F'(x) = \int_c^\infty \frac{\partial f(x, t)}{\partial x}\,dt.$$

THEOREM 1.6-2

Suppose $\sum u_n(x)$ *is a series of functions each differentiable on the interval* I $(a \le x \le b)$ *and such that* $\sum u_n(x_0)$ *converges for some value* x_0 *in* I. *If* $\sum u_n'(x)$ *converges uniformly on* I, *then* $\sum u_n(x)$ *converges uniformly on* I *to a function* $u(x)$, *and*

$$u'(x) = \sum u_n'(x), \qquad a \le x \le b.$$

Key Words and Phrases

pointwise convergence
uniform convergence
partial sums

*Proofs can be found in Walter Rudin, *Principles of Mathematical Analysis*, 3d ed. (New York: McGraw-Hill, 1976).

1.6 Exercises

• **1.** **(a)** In Example 1.6–1, compute N if $x_0 = 0.1$ and $\epsilon = 0.01$.
(b) In Example 1.6–1, compute N if $x_0 = 0.5$ and $\epsilon = 0.01$.

2. Evaluate

$$\int_0^\infty xe^{-xt}\,dt$$

on the interval $0 \le x \le 1$.

•• **3.** Show that

$$\int_0^\infty \frac{\sin xt \exp(at)}{t}\,dt = \arctan\frac{x}{a} \qquad \text{for } a > 0.$$

(*Hint*: Calling the improper integral $F(x)$, find $F'(x)$, and then integrate.)

4. Obtain the result

$$\int_0^\infty \frac{\sin xt}{t}\,dt = \frac{\pi}{2} \qquad \text{for } x > 0.$$

(*Hint*: Consider

$$G(a) = \int_0^\infty \frac{\sin xt \exp(-at)}{t}\,dt.$$

Use Exercise 3 to find $G(0)$. Under what condition can we find $\lim_{a\to 0^+} G(a)$?)

5. Use Exercise 4 to obtain the more complete result

$$\int_0^\infty \frac{\sin xt}{t}\,dt = \begin{cases} \dfrac{\pi}{2} & \text{for } x > 0 \\[2ex] -\dfrac{\pi}{2} & \text{for } x < 0 \\[2ex] 0 & \text{for } x = 0. \end{cases}$$

6. Graph the function defined in Exercise 5.

7. A useful test for the uniform convergence of a series of functions is the following *Weierstrass* M-test*:

Let $\sum u_n(x)$ be a series of functions defined on an interval I. Let $\sum M_n$ be a *convergent* series of nonnegative constants. If for every x on I,

$$|u_n(x)| \le M_n, \qquad n = 1, 2, \ldots,$$

then $\sum u_n(x)$ converges uniformly on I.

---

*After Karl T. Weierstrass (1815–1897), a German mathematician who pioneered rigor in analysis.

(a) Show that $\sum_{n=1} n^2 x^n$ converges uniformly on $\left[-\frac{1}{2}, \frac{1}{2}\right]$

(b) Show that $\sum_{n=1} (x \log x)^n$ converges uniformly on (0, 1].

●●● **8.** Show that the sequence

$$\left\{\frac{x}{x+n}\right\}$$

converges pointwise on $[0, \infty)$ and uniformly on [0, a], $a > 0$.

9. (a) Show that the series

$$S = \sum_{n=1} \frac{\sin n^2 x}{n^2}$$

is absolutely convergent (hence convergent for all x) by the comparison test.

(b) Show that the differentiated series

$$\sum_{n=1} \cos n^2 x$$

is divergent for all x.

(c) Explain.

10. (a) Show that the series

$$\sum_{n=1} \frac{\sin nx}{n^2}$$

converges uniformly for all x.

(b) Show that the differentiated series

$$\sum_{n=1} \frac{\cos nx}{n}$$

diverges for some values of x.

11. (a) Show that the infinite series defines the function $f(x)$:

$$f(x) = \frac{1}{1-x} = \sum_{n=0} x^n, \qquad -1 < x < 1.$$

(b) By differentiation, obtain

$$f'(x) = \frac{1}{(1-x)^2} = \sum_{n=1} nx^{n-1}, \qquad -1 < x < 1.$$

(c) Justify the procedure in part (b).

REFERENCES

Birkhoff, G., and G.-C. Rota, *Ordinary Differential Equations*, 3rd ed. New York: John Wiley, 1978.

Boyce, W. E., and R. C. DiPrima, *Elementary Differential Equations and Boundary Value Problems*, 3rd ed. New York: John Wiley, 1977.

Derrick, W. R., and S. I. Grossman, *Elementary Differential Equations with Applications*, 2d ed. Reading, Mass.: Addison-Wesley, 1980.

Hyslop, J. M., *Infinite Series*, 3rd ed. Edinburgh: Oliver and Boyd, 1947.

Leighton, W., *A First Course in Ordinary Differential Equations*, 5th ed. Belmont, Calif.: Wadsworth, 1981.

Martin, R. H., *Ordinary Differential Equations*. New York: McGraw-Hill, 1983.

Ross, S. L., *Introduction to Ordinary Differential Equations*, 3rd ed. New York: John Wiley, 1980.

Tierney, J. A., *Differential Equations*. Boston: Allyn and Bacon, 1979.

Wylie, C. R., *Differential Equations*. New York: McGraw-Hill, 1979.

Zill, D. G., *A First Course in Differential Equations with Applications*, 2d ed. Boston: Prindle, Weber and Schmidt, 1982.

2 Boundary-Value Problems in Ordinary Differential Equations

2.1 SOME PRELIMINARIES

In Chapter 1 we reviewed some aspects of ordinary differential equations. We saw, for example, that a second-order, linear differential equation contains *two* arbitrary constants in its *general solution*. In an *initial-value* problem these constants can be found from *initial conditions*, that is, from values of the dependent variable and its derivative at some **common value** of the independent variable. Without loss of generality the latter can be taken to be zero, since, if it has some other value, a simple scale change can be made. In summary, a *unique solution* can be obtained to a linear differential equation* provided that the initial value and the initial slope of the dependent variable are known.

In many problems of practical interest, however, information is given about the unknown function at two **distinct** points. For example, a function $y(x)$ may be required to satisfy a differential equation on an interval $a \leq x \leq b$. When $y(a)$ and $y(b)$ are specified, we call these given values **boundary conditions**, and the resulting problem is called a **boundary-value problem**. It may be that $y(a)$ and $y'(b)$ are specified, or $y'(a)$ and $y(b)$, or some combination of these. Thus there are many more possibilities than in an initial-value problem. We illustrate with some simple examples in which λ is a nonzero real parameter.

EXAMPLE 2.1-1 The boundary-value problem

$$y'' - \lambda^2 y = 0, \qquad 0 < x < a,$$
$$y(0) = y(a) = 0$$

*We continue to consider only *second-order* equations in this chapter.

has a **unique solution**. Here the general solution of the differential equation is

$$y(x) = c_1 \cosh \lambda x + c_2 \sinh \lambda x.$$

The boundary condition $y(0) = 0$ yields $c_1 = 0$, while the condition $y(a) = 0$ results in $c_2 \sinh \lambda a = 0$. This last equation shows that $c_2 = 0$, since the hyperbolic sine is zero only if its argument is zero, which is not the case here. Thus $y(x) = 0$ is the unique solution. ■

Since the differential equation in Example 2.1–1 is homogeneous and since both boundary conditions are homogeneous, it might appear that the zero solution is natural and inevitable. The next example shows that something quite different may happen.

EXAMPLE 2.1–2 The boundary-value problem

$$\begin{aligned} &y'' + \lambda^2 y = 0, \qquad 0 < x < a, \\ &y(0) = y(a) = 0 \end{aligned} \tag{2.1–1}$$

has **infinitely many solutions**. This time the general solution is

$$y(x) = c_1 \cos \lambda x + c_2 \sin \lambda x,$$

and again the condition $y(0) = 0$ leads to $c_1 = 0$. The condition $y(a) = 0$, however, yields $c_2 \sin \lambda a = 0$, which is satisfied by $c_2 = 0$ but also by $\lambda a = n\pi$, $n = 1, 2, 3, \ldots$. In other words, if the parameter λ has any of the values given by

$$\lambda_n = \frac{n\pi}{a}, \qquad n = 1, 2, \ldots, \tag{2.1–2}$$

then the solution of the problem is

$$y_n(x) = \sin \frac{n\pi}{a} x, \qquad n = 1, 2, \ldots \tag{2.1–3}$$

or any constant multiple* of this. ■

Recall from linear algebra that if T is a linear transformation (an $n \times n$ matrix) that transforms a vector $\mathbf{u}$ of an n-dimensional vector space into a vector $\mathbf{w}$ in the same space, we can write $T\mathbf{u} = \mathbf{w}$. If, however, for some vector $\mathbf{v}$ in the space (not the zero vector $\mathbf{0}$) it happens that $T\mathbf{v} = \lambda\mathbf{v}$, then we call λ an eigenvalue of T and $\mathbf{v}$ an eigenvector belonging to λ.

If we define

$$L \equiv \frac{d^2}{dx^2},$$

then it is easy to show that L is a **linear differential operator** that transforms

*Note that since $\sin(-\alpha) = -\sin \alpha$, $n = 1, 2, \ldots$ gives all solutions.

twice-differentiable functions into functions. In the case illustrated in Example 2.1–2 we have*

$$Ly = -\lambda^2 y; \qquad \textbf{(2.1–4)}$$

hence it is natural to call the functions $y_n(x)$ in Eq. (2.1–3) **eigenfunctions** and the corresponding values of $\lambda_n^2 = n^2\pi^2/a^2$ **eigenvalues**. We observe that eigenfunctions are not unique since if $y(x)$ is an eigenfunction, then so is $cy(x)$ for any constant c.

EXAMPLE 2.1–3 Find the eigenvalues and the corresponding eigenfunctions of the boundary-value problem

$$y'' + \lambda^2 y = 0, \qquad 0 < x < \pi,$$
$$y'(0) = y'(\pi) = 0.$$

Solution From the general solution of the differential equation, namely,

$$y(x) = c_1 \cos \lambda x + c_2 \sin \lambda x,$$

we find

$$y'(x) = -c_1\lambda \sin \lambda x + c_2\lambda \cos \lambda x.$$

The condition $y'(0) = 0$ leads to $c_2 = 0$, while the condition $y'(\pi) = 0$ gives

$$-c_1\lambda \sin \lambda\pi = 0.$$

Although the last equation is satisfied by $c_1 = 0$, this value of the constant results in $y = 0$, called the **trivial solution** of the given problem. In contrast with the situation in Example 2.1–1, where $y = 0$ was the unique solution, here we can choose $\lambda = n$, $n = 0, 1, 2, \ldots$, to satisfy the boundary conditions. Hence we have the eigenvalues

$$0, 1, 4, 9, \ldots, n^2, \ldots$$

and the corresponding eigenfunctions

$$1, \cos x, \cos 2x, \cos 3x, \ldots, \cos nx, \ldots. \qquad ■$$

The next example will show that the eigenvalues need not belong to a **denumerable set.**†

EXAMPLE 2.1–4 Solve the boundary-value problem with a nonhomogeneous boundary condition

$$\ddot{y} + \lambda^2 y = 0, \qquad 0 < t < 1,$$
$$y(0) = 0, \qquad y(1) = 1,$$

where the dot represents d/dt.

---

*Equation (2.1–4) is the differential equation of a **simple harmonic oscillator**.

†A set of numbers S is said to be denumerable if the elements of S can be put into a one-to-one correspondence with the positive integers 1, 2, 3, 4, ...

Solution From the general solution

$$y(t) = c_1 \cos \lambda t + c_2 \sin \lambda t$$

we have $c_1 = 0$ by virtue of $y(0) = 0$. The condition $y(1) = 1$ then results in

$$c_2 \sin \lambda = 1$$

and the eigenfunctions

$$y_\lambda(t) = \frac{\sin \lambda t}{\sin \lambda}.$$

Thus λ may be any real number as long as $\lambda \neq \pm n\pi$, $n = 0, 1, 2, \ldots$. ■

It is interesting to examine the last example from a geometrical viewpoint. The differential equation is the equation of a **simple harmonic oscillator**, and we have $y = 0$ at $t = 0$. If $\lambda = 2\pi$, say, then y will be *zero* again when $t = 1/2, 1, 3/2, \ldots$; so it is impossible to satisfy the second boundary condition.

The next example shows that the form of the boundary condition may be less simple than those considered previously.

EXAMPLE 2.1-5 Find the eigenvalues and eigenfunctions of the boundary-value problem

$$\begin{aligned} &y'' + \lambda^2 y = 0, \qquad 0 < x < \pi, \\ &y'(0) + 2y'(\pi) = 0, \qquad y(\pi) = 0. \end{aligned}$$

Solution Starting with

$$y = c_1 \cos \lambda x + c_2 \sin \lambda x$$

and

$$y' = -c_1\lambda \sin \lambda x + c_2\lambda \cos \lambda x,$$

we apply the two boundary conditions to obtain

$$\begin{cases} c_2\lambda - 2c_1\lambda \sin \lambda\pi + 2c_2\lambda \cos \lambda\pi = 0, \\ c_1 \cos \lambda\pi + c_2 \sin \lambda\pi = 0. \end{cases}$$

This system is easily solved for c_1 and c_2 by solving the second equation for c_1 and substituting this value into the first equation. The result is

$$\cos \lambda\pi = -2 \tag{2.1-5}$$

if we rule out the trivial solution $y = 0$. Hence Eq. (2.1-5) can be satisfied only by taking λ to be a complex number. Thus the problem of Example 2.1-5 has only complex eigenvalues and, of course, complex eigenfunctions. ■

It should be obvious that two-point boundary-value problems may have a great many forms. If we need to learn how to solve such problems, it would be well to narrow the field to those that occur frequently. This we do in the remainder of this chapter.

Key Words and Phrases

boundary conditions
unique solution
infinitely many solutions
linear differential operator
eigenfunction
eigenvalue
trivial solution
denumerable set
simple harmonic oscillator

2.1 Exercises

Unless otherwise stated, assume that λ is real and $\lambda \neq 0$.

• **1.** Verify that the functions

$$y_n(x) = \sin \frac{n\pi}{a} x, \qquad n = 1, 2, 3, \ldots,$$

satisfy the differential equation and boundary conditions in Eqs. (2.1–1).

2. Show that

$$L \equiv \frac{d^2}{dx^2}$$

is a linear operator, that is, that it has the properties:
(i) $L(cf) = cLf$
(ii) $L(f + g) = Lf + Lg$,
where f and g are twice-differentiable functions and c is a constant.

3. Carry out the necessary details required to arrive at Eq. (2.1–5) in Example 2.1–5. State any assumptions that must be made.

•• **4.** Find the eigenvalues and corresponding eigenfunctions in each of the following boundary-value problems.
(a) $y'' - \lambda^2 y = 0, \quad 0 < x < a,$
$y'(0) = y'(a) = 0$
(b) $y'' - \lambda^2 y = 0, \quad 0 < x < a,$
$y(0) = 0, \quad y(a) = 1$
(c) $y'' + \lambda^2 y = 0, \quad 0 < x < a,$
$y(0) = 0, \quad y'(a) = 0$
(d) $y'' + \lambda^2 y = 0, \quad 0 < x < a,$
$y(0) = 1, \quad y'(a) = 0$

5. In Exercise 4, determine whether zero is an eigenvalue; if it is, find the corresponding eigenfunction. (Only nonzero eigenfunctions are permitted.)

6. Find the eigenfunctions of the following boundary-value problem.

$$y'' + \lambda^2 y = 0, \qquad 0 < x < 2\pi,$$
$$y(0) - y(2\pi) = 0, \qquad y'(0) - y'(2\pi) = 0$$

7. Find the solution corresponding to $\lambda = 0$ in Example 2.1–4.

8. Obtain the eigenvalues and corresponding eigenfunctions of the following problem.

$$y'' + \lambda^2 y = 0, \qquad -1 < x < 1,$$
$$y(-1) = y(1) = 0$$

9. Obtain the eigenvalues and eigenfunctions of the problem

$$y'' + y' + (\lambda + 1)y = 0, \qquad 0 < x < \pi,$$
$$y(0) = y(\pi) = 0$$

10. Show that the problem

$$y'' + (1 + \lambda)y' + \lambda y = 0, \qquad 0 < x < 1,$$
$$y'(0) = y(1) = 0$$

has only the trivial solution unless λ is complex. (*Note*: Examine the three cases $\lambda < 0$, $\lambda = 0$, and $\lambda > 0$.)

11. Show that the only real eigenvalue of the problem

$$x^2y'' - \lambda xy' + \lambda y = 0, \qquad 1 < x < 2,$$
$$y(1) = 0, \qquad y(2) - y'(2) = 0,$$

is zero, and find the corresponding eigenfunction. (*Hint*: The differential equation is a Cauchy–Euler equation.)

••• **12.** **(a)** Show that $y_n = \cos 2nx$, $n = 0, 1, 2, \ldots$, are solutions of

$$y'' + \lambda^2 y = 0, \qquad 0 < x < \pi,$$
$$y(0) - y(\pi) = 0, \qquad y'(0) + y'(\pi) = 0.$$

(b) If $\lambda^2 \neq 4n^2$, show that the problem of part (a) has an infinite number of solutions corresponding to every complex value of λ^2.

13. Show that the problem

$$y'' + \lambda^2 y = 0, \qquad 0 < x < \pi,$$
$$2y(0) - y(\pi) = 0, \qquad 2y'(0) + y(\pi) = 0,$$

possesses only the trivial solution.

14. Show that each of the following boundary-value problems has a unique solution.

(a) $y'' + y = x, \qquad 0 < x < \pi/2,$
$y(0) = 2, \qquad y(\pi/2) = 1$

(b) $y'' + 4y = 0, \qquad 0 < x < \pi,$
$y(0) - 2y'(0) = -2, \qquad y(\pi) + 3y'(\pi) = 3$

(c) $y'' - 3y' + 2y = 0, \qquad 0 < x < 1,$
$y(0) = y'(1) = 0$

15. Prove that the problem

$$y'' + 9y = 0, \qquad 0 < x < \pi,$$
$$y(0) = 1, \qquad y(\pi) = -1$$

does not have a unique solution.

16. Prove that the problem

$$y'' + y = x, \qquad 0 < x < \pi,$$
$$y(0) = 2, \qquad y(\pi) = 1$$

has no solution.

2.2 STURM-LIOUVILLE PROBLEMS

In Section 2.1 we saw some of the many forms that two-point boundary-value problems can take. In some cases the solutions were functions of a real variable, in others they were functions of a complex variable, and occasionally the solution was the trivial one. Moreover, the set of solution functions was denumerable at times and nondenumerable at other times. In short, the results were quite varied, although most of the time we considered a particularly simple form of a second-order, homogeneous differential equation with constant coefficients.

In the remainder of this chapter we limit our study to second-order, linear differential equations, that is, to equations of the form

$$y'' + p(x)y' + q(x)y = r(x). \tag{2.2-1}$$

Boundary conditions will also be restricted in form to

$$\begin{aligned} a_1 y(a) + a_2 y'(a) &= a_3, \\ b_1 y(b) + b_2 y'(b) &= b_3, \end{aligned} \tag{2.2-2}$$

where the a_i and b_i are real constants with $a_3 \neq 0$, $b_3 \neq 0$. An equation of the form (2.2–1) with $a \leq x \leq b$ together with boundary-conditions of the form (2.2–2) will constitute a **two-point nonhomogeneous boundary-value problem.** We will assume that $p(x)$, $q(x)$, and $r(x)$ are continuous on $[a, b]$, which assures the existence of a solution of the differential equation.

Whenever the differential equation *and* the boundary conditions are homogeneous, we have a **two-point homogeneous boundary-value problem,**

$$y'' + p(x)y' + q(x)y = 0, \tag{2.2-3}$$

$$\begin{cases} a_1 y(a) + a_2 y'(a) = 0, \\ b_1 y(b) + b_2 y'(b) = 0. \end{cases} \tag{2.2-4}$$

When we solve boundary-value problems in partial differential equations by the method of separation of variables (Section 3.2), we will introduce a **separation parameter** into the ordinary differential equation, and Eq. (2.2–3) will be generalized to

$$y'' + p(x, \lambda)y' + q(x, \lambda)y = 0. \tag{2.2-5}$$

It is this last type of differential equation that we considered in Section 2.1. Note, however, that not all the boundary conditions there were of the type in

Eqs. (2.2–4). The boundary conditions in Example 2.1–5 were not of the form (2.2–4), for example.

A homogeneous boundary-value problem that has the form

$$\frac{d}{dx}\left[r(x)\frac{dy}{dx}\right] + [q(x) + \lambda w(x)]y = 0, \qquad -\infty < a \le x \le b < \infty, \tag{2.2–6}$$

$$\begin{cases} a_1 y(a) + a_2 y'(a) = 0, \\ b_1 y(b) + b_2 y'(b) = 0. \end{cases} \tag{2.2–7}$$

is called a **Sturm–Liouville system** if the above constants and functions have certain properties that will be described later.

The system is named after Jacques C. F. Sturm (1803–1855), a Swiss who taught at the Sorbonne, and Joseph Liouville (1809–1882), Professor at the Collège de France. These two mathematicians, together with Augustin-Louis Cauchy (1789–1857), a colleague of Liouville's, developed most of the extensive theory relating to these systems.

A **Sturm–Liouville problem** consists of finding the values of λ and the corresponding values of y that satisfy the system. Systems described by Eqs. (2.2–6) and (2.2–7) occur in a wide variety of applications, as we shall see in later chapters.

In Eq. (2.2–6) we require that $r(x)$ be a real function that is continuous and has a continuous derivative over the interval of interest $a \le x \le b$. We also require that $q(x)$ be real and continuous and that $w(x)$ be continuous and nonnegative on the same interval. It may happen that $w(x) = 0$ at isolated points in the interval. We further require that λ be a constant (independent of x) and that a_1 and a_2 be not both zero and that b_1 and b_2 be not both zero in Eqs. (2.2–7). If $r(x) > 0$ for $a \le x \le b$, then the Sturm–Liouville problem is called **regular.**

To solve a regular Sturm–Liouville problem means to find values of λ (called **eigenvalues** or characteristic values) and to find the corresponding nontrivial functions y (called **eigenfunctions** or characteristic functions).

Next we give some examples of the foregoing. In each case we identify the terms in the examples and compare them to the terms in Eqs. (2.2–6) and (2.2–7).

EXAMPLE 2.2–1 Solve the following system:

$$y'' + \lambda y = 0, \qquad 0 \le x \le \pi,$$
$$y(0) = 0, \qquad y(\pi) = 0.$$

Solution Here $r(x) = 1$, $q(x) = 0$, $w(x) = 1$, $a = 0$, $b = \pi$, $a_1 = b_1 = 1$, and $a_2 = b_2 = 0$. The solution to the differential equation is

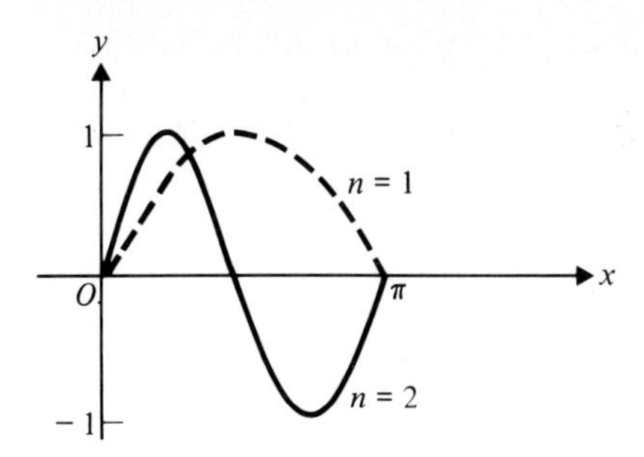

Figure 2.2-1
Eigenfunctions (Example 2.2-1).

$$y = c_1 \cos(\sqrt{\lambda}x) + c_2 \sin(\sqrt{\lambda}x)$$

with $\lambda > 0$. If $\lambda \leq 0$, then the system has only the trivial solution $y = 0$ (Exercise 1). This is not of interest, since *every* Sturm–Liouville system has a trivial solution. Note that we admit zero as an eigenvalue but not as an eigenfunction.

The condition $y(0) = 0$ implies that $c_1 = 0$; hence the updated solution becomes

$$y = c_2 \sin(\sqrt{\lambda}x).$$

The second condition $y(\pi) = 0$ implies that either $c_2 = 0$ (which would lead to the trivial solution) or $\sqrt{\lambda}\pi = n\pi$, that is, $\lambda = n^2$, $n = 1, 2, 3, \ldots$. Thus the eigenvalues of the system are $\lambda_1 = 1$, $\lambda_2 = 4$, $\lambda_3 = 9, \ldots$. The corresponding eigenfunctions are

$$y_1(x) = \sin x, \qquad y_2(x) = \sin 2x, \qquad y_3(x) = \sin 3x, \ldots,$$

and, in general (Figure 2.2-1),

$$y_n(x) = \sin nx, \qquad n = 1, 2, 3, \ldots,$$

where the arbitrary constants have been set equal to one, since eigenfunctions are unique only to within a multiplicative constant. ■

EXAMPLE 2.2-2 Solve the following system:

$$y'' + \lambda y = 0, \qquad 0 \leq x \leq \pi,$$
$$y'(0) = 0, \qquad y'(\pi) = 0.$$

Solution Now we have $r(x) = w(x) = 1$, $q(x) = 0$, $a = 0$, $b = \pi$, $a_1 = b_1 = 0$, and $a_2 = b_2 = 1$. The solution to the differential equation is again

$$y = c_1 \cos(\sqrt{\lambda}x) + c_2 \sin(\sqrt{\lambda}x), \qquad \lambda \geq 0.$$

From this we have

$$y' = -c_1\sqrt{\lambda} \sin(\sqrt{\lambda}x) + c_2\sqrt{\lambda}\cos(\sqrt{\lambda}x).$$

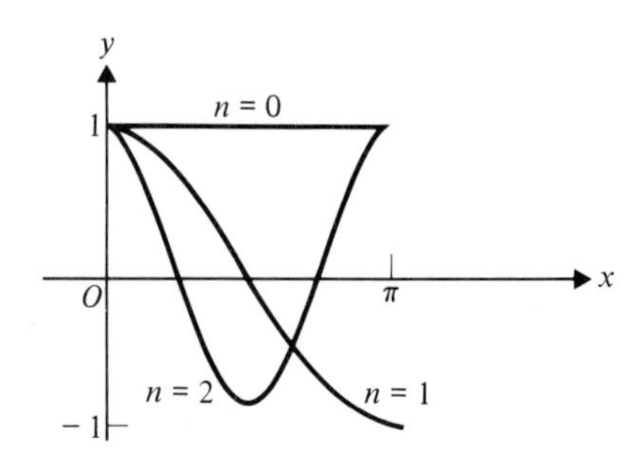

Figure 2.2-2
Eigenfunctions (Example 2.2-2).

The first condition $y'(0) = 0$ implies that $c_2 = 0$ or $\lambda = 0$. If we take the case $c_2 = 0$, the solution becomes

$$y = c_1 \cos(\sqrt{\lambda}x) \qquad \text{and} \qquad y' = -c_1\sqrt{\lambda}\sin(\sqrt{\lambda}x).$$

The second condition $y'(\pi) = 0$ implies that $\sqrt{\lambda} = n$, $\lambda = n^2$, $n = 1, 2, \ldots$.

The case $\lambda = 0$ must be considered separately, since it leads to a *different* problem, namely,

$$y'' = 0, \qquad 0 \le x \le \pi, \qquad y'(0) = 0, \qquad y'(\pi) = 0.$$

This problem has the solution $y =$ constant, which we can take to be one.

Hence the eigenvalues are $\lambda_0 = 0$, $\lambda_1 = 1$, $\lambda_2 = 4$, $\lambda_3 = 9, \ldots$, and the corresponding eigenfunctions are $y_0 = 1$, $y_1 = \cos x$, $y_2 = \cos 2x$, $y_3 = \cos 3x, \ldots$, where again we have set the arbitrary constants equal to unity (see Fig. 2.2-2). It can be shown (Exercise 2) that $\lambda < 0$ leads to the trivial solution. ■

EXAMPLE 2.2-3 Solve the following system:

$$y'' + \lambda y = 0, \qquad 0 \le x \le 1,$$
$$y(0) + y'(0) = 0, \qquad y(1) = 0.$$

Solution Here we have $r(x) = w(x) = 1$, $q(x) = 0$, $a = 0$, $b = 1$, $a_1 = a_2 = b_1 = 1$, and $b_2 = 0$. If $\lambda < 0$, we have the trivial solution (Exercise 3). If $\lambda = 0$, then $y = k_1 + k_2x$, and the boundary conditions applied to this function show that an eigenfunction belonging to the eigenvalue $\lambda = 0$ is $1 - x$.

If $\lambda > 0$, we have as the solution of the differential equation

$$y = c_1\cos(\sqrt{\lambda}x) + c_2\sin(\sqrt{\lambda}x).$$

The condition $y(0) + y'(0) = 0$ implies that $c_1 + c_2\sqrt{\lambda} = 0$, that is, $c_1 = -c_2\sqrt{\lambda}$; then the condition $y(1) = 0$ implies that $\sqrt{\lambda} = \tan\sqrt{\lambda}$. Thus the eigenvalues are the squares of the solutions of the transcendental equation $t = \tan t$. This equation cannot be solved by algebraic methods, so we graph the curves $u = t$ and $u = \tan t$ and observe the values of t where the two

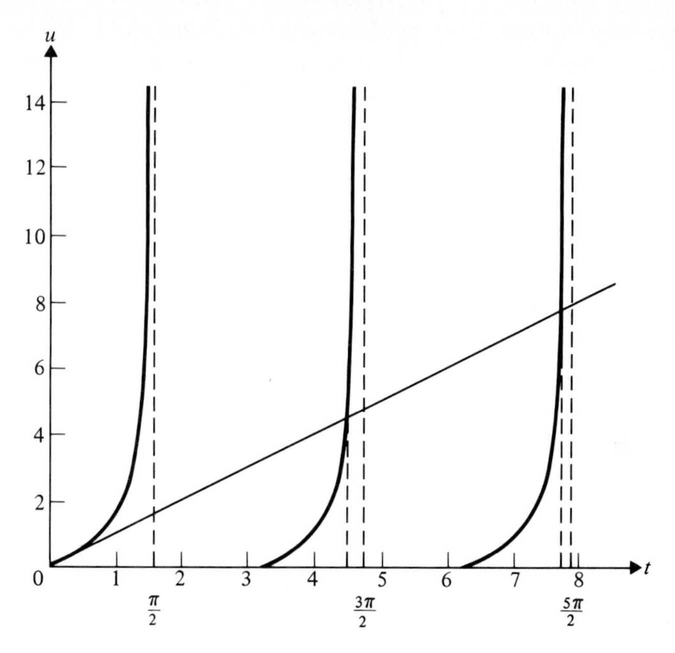

Figure 2.2-3
Simultaneous solutions of $u = t$ and $u = \tan t$.

curves intersect. Figure 2.2-3 shows the first two eigenvalues $\lambda_1 \doteq (4.5)^2$ and $\lambda_2 \doteq (7.7)^2$, together with the eigenvalue $\lambda = 0$ that has already been found.

We can see from Fig. 2.2-3 that there are an infinite number of eigenvalues and that they are approaching the square of the odd multiples of $\pi/2$. In other words,

$$\lambda_n \doteq \frac{(2n + 1)^2\pi^2}{4},$$

the approximation getting better as n increases through the positive integers. A computer is useful in obtaining the eigenvalues. For example, the following computational scheme shows how to obtain λ_1 to three-decimal accuracy with the aid of a hand-held calculator.

t	$\tan t$	
4.5	4.637	
4.4	3.096	
4.45	3.723	
4.49	4.422	
4.495	4.527	
4.4935	4.495	
4.4934	4.493	Hence $\lambda_1 \doteq 20.187$.

The values of tan t are a guide to what should be chosen for the next value of t. The eigenvalues in this problem can be found in some published tables.

The eigenfunctions corresponding to the eigenvalues λ_n, $n = 1, 2, \ldots,$ are

$$y_n(x) = \sin(\sqrt{\lambda_n}x) - \sqrt{\lambda_n}\cos(\sqrt{\lambda_n}x). \quad \blacksquare$$

Up to this point, all the examples presented involved a very simple second-order differential equation. We can show, however, that *every* second-order, linear, homogeneous differential equation can be transformed into the form shown in Eq. (2.2–6). Consider

$$A(x)y'' + B(x)y' + C(x)y = 0, \tag{2.2–8}$$

with $A'(x) \neq B(x)$. Note that if this restriction is *not* imposed, then Eq. (2.2–8) will have the required form. If we multiply Eq. (2.2–8) by

$$\frac{1}{A(x)}\exp\left(\int^{x}\frac{B(t)\,dt}{A(t)}\right) = \mu(x)/A(x),$$

we can write the result as

$$\frac{d}{dx}\left[\mu(x)\frac{dy}{dx}\right] + \frac{C(x)}{A(x)}\mu(x)y = 0,$$

which is the form required for a Sturm–Liouville system. The parameter λ is, of course, an essential part of the system.

We can simplify the notation by defining a **linear differential operator** L,

$$L \equiv \frac{d}{dx}\left[r(x)\frac{d}{dx}\right] + q(x), \tag{2.2–9}$$

that is,

$$Ly = \frac{d}{dx}\left[r(x)\frac{dy}{dx}\right] + q(x)y = (ry')' + qy,$$

the primes denoting differentiation with respect to x. With this notation, Eq. (2.2–6) can be written

$$Ly = -\lambda wy. \tag{2.2–10}$$

So far, we have been dealing with *regular* problems. A Sturm–Liouville problem is said to be **singular** if (1) $r(a) = 0$ and $a_1 = a_2 = 0$, or (2) $r(b) = 0$ and $b_1 = b_2 = 0$. Singular problems also arise when $r(x)$ or $w(x)$ vanishes at $x = a$ or $x = b$, when $q(x)$ is discontinuous at these points, or when the interval $a \leq x \leq b$ is unbounded.

Singular Sturm–Liouville problems will be treated in Section 2.7. In the next section we will study a most useful property, namely, the orthogonality of eigenfunctions.

Key Words and Phrases

two-point nonhomogeneous boundary-value problem
two-point homogeneous boundary-value problem
separation parameter
Sturm–Liouville system
Sturm–Liouville problem
regular and singular Sturm–Liouville problems
eigenvalue
eigenfunction
linear differential operator

2.2 Exercises

• **1.** Verify that the problem of Example 2.2–1 has only the trivial solution if $\lambda \leq 0$.

2. Show that $\lambda < 0$ in Example 2.2–2 leads to the trivial solution.

3. Verify that $\lambda < 0$ leads to the trivial solution in the problem of Example 2.2–3.

4. In Example 2.2–3, obtain the eigenfunction belonging to the eigenvalue zero.

***5.** Compute λ_2, the second nonzero eigenvalue in Example 2.2–3, to three decimals.

•• **6.** Obtain the eigenvalues and corresponding eigenfunctions of the regular Sturm–Liouville system

$$y'' + \lambda y = 0, \qquad y'(0) = 0, \qquad y(\pi) = 0.$$

In Exercises 7–11, find the eigenvalues and corresponding eigenfunctions of each system.

7. $y'' + \lambda y = 0, \quad y'(-\pi) = 0, \quad y'(\pi) = 0$

8. $y'' + \lambda y = 0, \quad y(0) = 0, \quad y'(\pi) = 0$

9. $y'' + (1 + \lambda)y = 0, \quad y(0) = 0, \quad y(\pi) = 0$

10. $y'' + 2y' + (1 - \lambda)y = 0, \quad y(0) = 0, \quad y(1) = 0$

11. $y'' + 2y' + (1 - \lambda)y = 0, \quad y'(0) = 0, \quad y'(\pi) = 0$

12. Transform the differential equation of Exercise 11 into the form (2.2–6). (*Hint*: Find $\mu(x)$ as shown in the text.)

13. Show that the problem

$$y'' - 4\lambda y' - 4\lambda^2 y = 0, \qquad y(0) = 0, \qquad y(1) + y'(1) = 0,$$

has only one eigenvalue, and find the corresponding eigenfunction.

14. Solve the *nonhomogeneous* problem

$$y'' = 1, \qquad y(-1) = 0, \qquad y(1) - 2y'(1) = 0.$$

••• **15.** Show that the differential operator L defined in Eq. (2.2–9) is a linear operator.

16. Laguerre's equation,

$$xy'' + (1 - x)y' + \lambda y = 0,$$

is of importance in **quantum mechanics**.

(a) Transform the equation into the form (2.2–6).

(b) What is the weight function $w(x)$?

17. Hermite's differential equation,

$$y'' - 2xy' + 2\lambda y = 0,$$

arises in the quantum mechanical theory of the **linear oscillator**.

(a) Transform the equation into the form (2.2–6).

(b) What is the weight function $w(x)$?

18. Tchebycheff's differential equation,

$$(1 - x^2)y'' - xy' + n^2y = 0,$$

is found in mathematical physics.

(a) Transform the equation into the form (2.2–6).

(b) What is the weight function $w(x)$?

19. Let $y_1(x)$ and $y_2(x)$ be linearly independent solutions of Eq. (2.2–6). Prove that the Sturm–Liouville problem, Eqs. (2.2–6) and (2.2–7), has a nontrivial solution if and only if the determinant

$$\begin{vmatrix} a_1y_1(a) + a_2y_1'(a) & b_1y_1(b) + b_2y_1'(b) \\ a_1y_2(a) + a_2y_2'(a) & b_1y_2(b) + b_2y_2'(b) \end{vmatrix}$$

is zero. (*Hint*: Recall that a system of linear homogeneous equations for c_1 and c_2

$$\begin{cases} \alpha c_1 + \beta c_2 = 0 \\ \gamma c_1 + \delta c_2 = 0 \end{cases}$$

has a nontrivial solution if and only if the determinant

$$\begin{vmatrix} \alpha & \beta \\ \gamma & \delta \end{vmatrix} = 0.)$$

2.3 ORTHOGONALITY OF EIGENFUNCTIONS

In Section 2.2 we introduced a linear differential operator

$$L \equiv \frac{d}{dx}\left[r(x)\frac{d}{dx}\right] + q(x), \tag{2.3–1}$$

that is,

$$Ly = \frac{d}{dx}\left[r(x)\frac{dy}{dx}\right] + q(x)y = (ry')' + qy,$$

the primes denoting differentiation with respect to x. By using this notation the Sturm–Liouville equation (2.2–6) can be written as

$$Ly = -\lambda wy. \tag{2.3–2}$$

In this section we investigate some properties of the eigenfunctions y of the differential operator L in Eq. (2.3–2). First, however, we need a few definitions.

DEFINITION 2.3–1
The set of functions,

$$\{\phi_i(x),\ i = 1, 2, 3, \ldots\},$$

each of which is continuous on [a, b], is **orthogonal** *on [a, b] with* **weight function*** *w(x) if*

$$\int_a^b \phi_n(x)\phi_m(x)w(x)\,dx = 0, \qquad n \neq m,$$

and

$$\int_a^b [\phi_n(x)]^2 w(x)\,dx \neq 0.$$

Orthogonality of *functions* as given in the above definition is a generalization of orthogonality of *vectors.* Note that the *sum* of products in the scalar (or dot) multiplication of vectors has been replaced by the *integral* of products.

That the eigenfunctions of Example 2.2–1 form an *orthogonal set* on the interval $0 \leq x \leq \pi$ with weight function $w(x) = 1$ is shown partly by the fact that

$$\int_0^\pi \sin nx \sin mx\,dx = \left.\frac{\sin (n - m)x}{2(n - m)} - \frac{\sin (n + m)x}{2(n + m)}\right|_0^\pi = 0,$$

if $m \neq n$ (see also Exercise 1). The orthogonality of these eigenfunctions is of prime importance in Fourier series expansions, as we shall see in Chapter 4.

Similarly, the orthogonality of the eigenfunctions of Example 2.2–2 requires (see also Exercise 2)

$$\int_0^\pi \cos nx \cos mx\,dx = \left.\frac{\sin (n - m)x}{2(n - m)} + \frac{\sin (n + m)x}{2(n + m)}\right|_0^\pi = 0,$$

if $n \neq m$. Hence the set

$$\{1, \cos x, \cos 2x, \cos 3x, \ldots\}$$

is an orthogonal set on the interval $0 \leq x \leq \pi$ with weight function $w(x) = 1$. This property will also be useful in Fourier series expansions.

A set of functions having an additional property, given in the following definition, will be of even greater value in our work.

DEFINITION 2.3–2
The set of functions,

$$\{\phi_i(x),\ i = 1, 2, 3, \ldots\},$$

each of which is continuous on [a, b], is **orthonormal** *on [a, b] with*

*Also called a *density function.*

weight function w(x) if

$$\int_a^b \phi_n(x)\phi_m(x)w(x)\,dx = 0, \qquad n \neq m,$$

and

$$\int_a^b [\phi_n(x)]^2 w(x)\,dx = 1.$$

Thus orthonormal functions have the same properties as orthogonal functions, but, in addition, they have been *normalized*. Again we have an analogy to *unit* (or normalized) vectors. The two relations in Definition 2.3–2 can be expressed more simply by using a notational symbol called the **Kronecker delta*** defined by

$$\delta_{mn} = \begin{cases} 0, & \text{if } m \neq n, \\ 1, & \text{if } m = n. \end{cases} \tag{2.3–2}$$

Using this symbol, the functions in an orthonormal set have the property

$$\int_a^b \phi_n(x)\phi_m(x)w(x)\,dx = \delta_{mn}. \tag{2.3–3}$$

It is a simple matter to construct an orthonormal set from a given orthogonal set. For example, the set

$$\left\{ \frac{1}{\sqrt{\pi}}, \frac{\cos x}{\sqrt{\pi/2}}, \frac{\cos 2x}{\sqrt{\pi/2}}, \frac{\cos 3x}{\sqrt{\pi/2}}, \ldots \right\}$$

is an orthonormal set of eigenfunctions of the problem in Example 2.2–2. The orthonormal set is formed (see Exercise 14) by dividing each function ϕ of the corresponding orthogonal set by the **norm** of that function, defined as

$$\|\phi\| = \left(\int_a^b [\phi(x)]^2 w(x)\,dx \right)^{1/2}. \tag{2.3–4}$$

Note here again the analogy to vectors, where we obtain a unit vector by dividing a vector by its length, that is, by the square root of the *sum* of the squares of its components.

We next investigate under what conditions the eigenfunctions of the differential operator L defined in Eq. (2.3–1) form an orthogonal set. If y_1 and y_2 are functions that are twice differentiable on $[a, b]$, then

$$\begin{aligned} y_1Ly_2 - y_2Ly_1 &= y_1(ry_2')' - y_2(ry_1')' \\ &= y_1(ry_2'' + r'y_2') - y_2(ry_1'' + r'y_1') \\ &= r'(y_1y_2' - y_2y_1') + r(y_1y_2'' - y_2y_1'') \\ &= [r(y_1y_2' - y_2y_1')]'. \end{aligned}$$

The last result is known as **Lagrange's identity.**†

---

*After Leopold Kronecker (1823–1891), a German mathematician.

†After Joseph L. Lagrange (1736–1813), a mathematician of French extraction educated in Italy, who made many contributions to applied mathematics.

On the other hand, if λ_1 is an eigenvalue belonging to y_1 while λ_2 is an eigenvalue belonging to y_2, then, from Eq. (2.3–2),

$$y_1 L y_2 - y_2 L y_1 = (\lambda_1 - \lambda_2) w y_1 y_2 .$$

Thus, equating the two quantities, we get

$$(\lambda_1 - \lambda_2) w y_1 y_2 = [r(y_1 y_2' - y_2 y_1')]',$$

and integrating from a to b gives us

$$(\lambda_1 - \lambda_2) \int_a^b w(x) y_1(x) y_2(x)\, dx = r(y_1 y_2' - y_2 y_1') \Big|_a^b . \qquad \textbf{(2.3–5)}$$

We now see that if $\lambda_1 \neq \lambda_2$, then $y_1(x)$ and $y_2(x)$ are orthogonal on the interval $[a, b]$ with weight function $w(x)$, provided that the boundary conditions are such that the right-hand side of Eq. (2.3–5) vanishes. Thus for a regular Sturm–Liouville problem where $r(x) > 0$ on $[a, b]$ and (see Section 2.2)

$$\begin{aligned} a_1 y(a) + a_2 y'(a) &= 0, \\ b_1 y(b) + b_2 y'(b) &= 0, \end{aligned}$$

we have

$$y_1 y_2' - y_2 y_1' \Big|_a^b = 0, \qquad \textbf{(2.3–6)}$$

as can easily be shown (Exercise 5). We have proved the following theorem.

THEOREM 2.3-1

In a regular Sturm–Liouville problem,

$$\frac{d}{dx}\left[r(x)\frac{dy}{dx}\right] + [q(x) + \lambda w(x)]y = 0, \qquad -\infty < a \leq x \leq b < \infty,$$

$$\begin{aligned} a_1 y(a) + a_2 y'(a) &= 0, \\ b_1 y(b) + b_2 y'(b) &= 0, \end{aligned} \qquad \textbf{(2.3–7)}$$

eigenfunctions belonging to different eigenvalues are orthogonal on $[a, b]$ with weight function $w(x)$.

It follows from Theorem 2.3–1 that the eigenfunctions of a regular Sturm–Liouville problem can be used to form an othonormal set of functions. These functions, in turn, are useful in approximating other functions, as we will show in Section 2.4. In Section 2.7 we will see that under certain other conditions, Eq. (2.3–6) can be satisfied so that obtaining orthogonal eigenfunctions will not be limited to regular Sturm–Liouville problems.

Key Words and Phrases

orthogonal set of functions
weight function
orthonormal set of functions
Kronecker delta
norm of a function
Lagrange's identity

2.3 Exercises

• **1.** Show that

$$\int_0^\pi \sin nx \sin mx \, dx = \frac{\pi}{2}$$

if $m = n$.

2. **(a)** Show that

$$\int_0^\pi \cos nx \cos mx \, dx = \frac{\pi}{2}$$

if $m = n$.

(b) Show that

$$\int_0^\pi \cos^2 nx \, dx \neq 0, \qquad n = 0, 1, 2, \ldots .$$

3. Verify that the norm of the function $f(x) = 1$ on $[0, \pi]$ is $\sqrt{\pi}$.

4. Obtain an orthonormal set from the orthogonal set

$$\{\sin nx, n = 1, 2, \ldots\}$$

of Example 2.2-1.

5. Verify Eq. (2.3-6).

•• **6.** State the orthogonality relation for the eigenfunctions of the problem

$$y'' + \lambda y = 0, \qquad 0 \le x \le 1,$$
$$y(0) + y'(0) = 0, \qquad y(1) = 0.$$

(Compare Example 2.2-3.)

***7.** Verify the orthogonality of the eigenfunctions of Exercise 6, using $\lambda_1 = 4.493$ and $\lambda_2 = 7.725$.

8. Obtain the orthonormal set of eigenfunctions for the problem

$$y'' + \lambda y = 0, \qquad y'(0) = 0, \qquad y(\pi) = 0.$$

(*Hint*: See Exercise 6 in Section 2.2.)

9. Obtain the orthonormal set of eigenfunctions for each of the following problems. Note that in some cases, zero may be an eigenvalue.

(a) $y'' + \lambda y = 0, \quad y(0) = 0, \quad y'(\pi) = 0$

(b) $y'' + (1 + \lambda) y = 0, \quad y(0) = 0, \quad y(\pi) = 0$
(*Hint*: See Exercise 9 in Section 2.2.)

(c) $y'' + \lambda y = 0, \quad y'(0) = 0, \quad y'(c) = 0$

(d) $y'' + \lambda y = 0, \quad y'(0) = 0, \quad y(c) = 0$

10. **(a)** Verify that the eigenfunctions of the problem

$$y'' + 2y' + (1 - \lambda)y = 0, \qquad y(0) = 0, \qquad y(1) = 0,$$

are $y_n(x) = \sin n\pi x \exp(-x)$, $n = 1, 2, 3, \ldots .$

(b) Transform the problem of part (a) so that it has the form (2.3-7). (*Hint*: See Section 2.2.)

(c) Verify that the eigenfunctions of the problem in part (a) are orthogonal on [0, 1] using the proper weight function.

---

*Calculator problem.

••• **11.** **(a)** Verify that the eigenfunctions of the problem

$$y'' + 2y' + (1 - \lambda)y = 0, \qquad y'(0) = 0, \qquad y'(\pi) = 0,$$

are $y_0 = 1$ and $y_n(x) = (n \cos nx + \sin nx) \exp(-x)$, $n = 1, 2, \ldots$.

(b) Obtain an orthonormal set of eigenfunctions for the problem in part (a).

12. **(a)** Verify that the eigenfunctions of the problem

$$y'' + \lambda y = 0, \qquad y'(-\pi) = y'(\pi) = 0$$

are $y_n(x) = \cos \frac{n}{2}(x + \pi), \qquad n = 0, 1, 2, \ldots$.

(b) Show that the eigenfunctions in part (a) are orthogonal on $[-\pi, \pi]$. (*Hint*: Use the trigonometric identity

$$\sin A \cos B = \frac{1}{2}\sin(A + B) + \frac{1}{2}\sin(A - B).)$$

(c) Obtain the orthonormal set of eigenfunctions for the problem in part (a).

13. Show that Lagrange's identity leads to

$$\int_a^b y_1 L y_2 \, dx = \int_a^b y_2 L y_1 \, dx$$

for functions y_1 and y_2 satisfying Eqs. (2.2–6) and (2.2–7). This shows that L is a *symmetric* operator on the space of all real-valued twice continuously differentiable functions on $[a, b]$ that satisfy the boundary conditions (2.2–7).

14. If

$$\{\phi_n(x), n = 1, 2, \ldots\}$$

is an orthogonal set on $[a, b]$ with weight function $w(x)$, verify that

$$\left\{ \frac{\phi_n(x)}{\|\phi_n(x)\|}, n = 1, 2, \ldots \right\}$$

is an orthonormal set on $[a, b]$ with weight function $w(x)$.

2.4 REPRESENTATION BY A SERIES OF ORTHONORMAL FUNCTIONS

The set of vectors

$$S = \{\mathbf{e}_1, \mathbf{e}_2, \ldots, \mathbf{e}_n\}$$

of $\mathbb{R}^n$ where

$$\mathbf{e}_i = (e_{i1}, e_{i2}, \ldots, e_{in}), \qquad \text{with } e_{ij} = \delta_{ij}, \tag{2.4–1}$$

forms a **natural basis** for $\mathbb{R}^n$. From the above definition we see that the vectors in S are mutually orthogonal, that is, the **inner product*** of any pair, denoted by $(\mathbf{e}_i, \mathbf{e}_j)$, is given by

$$(\mathbf{e}_i, \mathbf{e}_j) = \delta_{ij}. \tag{2.4-2}$$

An important consequence of the foregoing is that an *arbitrary* vector $\mathbf{v}$ of $\mathbb{R}^n$ can be uniquely expressed as

$$\mathbf{v} = \sum_{i=1}^{n} c_i \mathbf{e}_i, \tag{2.4-3}$$

where the c_i are scalars. Moreover, the c_i can be easily computed from Eq. (2.4-3). To find c_k $(1 \le k \le n)$, for example, we form the inner product of both members of Eq. (2.4-3) with $\mathbf{e}_k$. Then

$$(\mathbf{v}, \mathbf{e}_k) = \sum_{i=1}^{n} (c_i \mathbf{e}_i, \mathbf{e}_k) = \sum_{i=1}^{n} c_i (\mathbf{e}_i, \mathbf{e}_k) = c_k,$$

so that Eq. (2.4-3) can be written as

$$\mathbf{v} = \sum_{i=1}^{n} (\mathbf{v}, \mathbf{e}_i)\mathbf{e}_i. \tag{2.4-4}$$

We are tacitly assuming a knowledge of the properties of the so-called **Euclidean inner product*** in $\mathbb{R}^n$ in the above. (See Exercise 2.)

Now define the inner product of two functions f and g with respect to the weight function w as

$$(f, g) = \int_a^b f(x)g(x)w(x)\,dx. \tag{2.4-5}$$

In order that this definition be useful it is necessary that† $w(x) > 0$ on (a, b), and it will simplify matters if f and g are both piecewise continuous on (a, b). A function f is **piecewise continuous** on (a, b) if this interval can be partitioned into a finite number of subintervals such that:

1. f is continuous on each open subinterval and
2. f has finite left- and right-hand limits at each point of subdivision.

We give some examples to clarify the above definition.

---

*Also called the *dot product* and the *scalar product*. The notation $<\mathbf{e}_i, \mathbf{e}_j>$ is also used by some authors.

†If $w(x) = 0$ at *isolated* points of (a, b), then this will not cause any problems.

EXAMPLE 2.4–1 The function f defined as

$$f(t) = \begin{cases} 2t, & \text{for } 0 < t < 1, \\ 1, & \text{for } 1 \leq t < 2, \\ -1, & \text{for } 2 \leq t < 3, \end{cases}$$

is piecewise continuous on (0, 3). Figure 2.4–1 shows the graph of this function. At $t = 1$ the left-hand limit is found to be

$$\lim_{t \to 1^-} f(t) = 2,$$

whereas the right-hand limit is given by

$$\lim_{t \to 1^+} f(t) = 1. \quad \blacksquare$$

We remark that continuous functions belong to the class of piecewise continuous functions and that points of continuity are characterized by the fact that the left- and right-hand limits are *equal* and equal to the functional values at the points. The type of discontinuity at $t = 1$ and $t = 2$ exhibited by the function of Example 2.4–1 is called a (finite) **jump discontinuity**.

Another example of a piecewise continuous function is shown in Fig. 2.4–2. Note that this function satisfies the definition of a piecewise continuous function, although $g(1)$ does not exist and $g(2)$ differs from the left- and right-hand limits of $g(t)$ at $t = 2$. (See Exercise 3.)

Observe that the inner product of functions defined in Eq. (2.4–5) has all the properties ascribed to the inner product of vectors. For example, if f, g, and h are piecewise continuous on (a, b) and c is a constant, then

$$\begin{aligned} &(a) \quad (f, g) = (g, f), \\ &(b) \quad (f, g + h) = (f, g) + (f, h), \\ &(c) \quad (f, cg) = (cf, g) = c(f, g). \end{aligned} \tag{2.4–6}$$

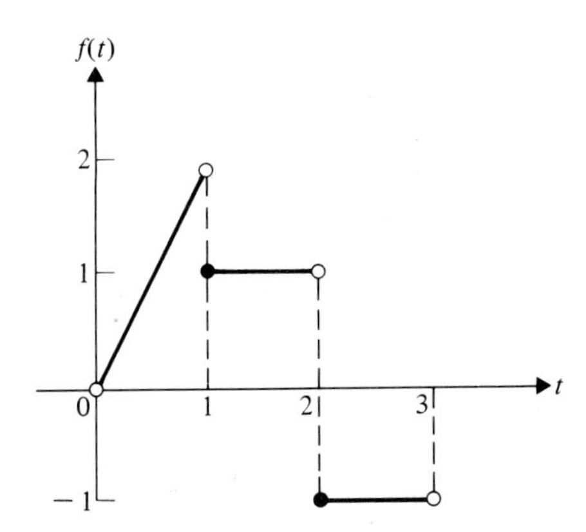

Figure 2.4–1
The function of Example 2.4–1.

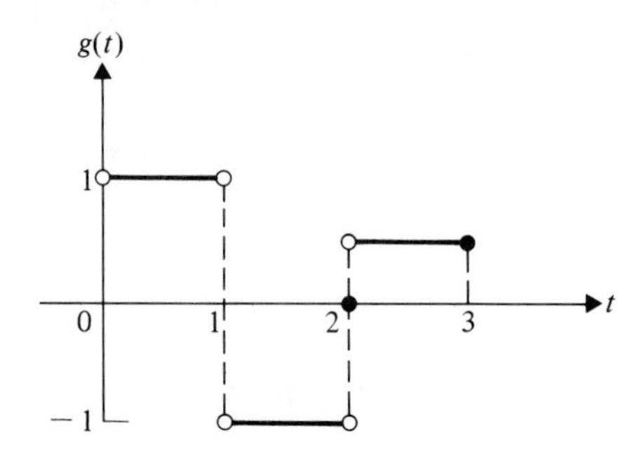

Figure 2.4-2
Piecewise continuous function.

We define the **norm** of f as

$$(f, f) = \|f\| = \left(\int_a^b [f(x)]^2 w(x)\, dx\right)^{1/2}. \tag{2.4-7}$$

From this last equation it follows that $\|f\| = 0$ only when $f = 0$ except perhaps at a finite number of points. Following usual custom*, we will consider two functions to be equal on an interval if they differ at only a finite number of points on the interval. With this understanding we say that Eq. (2.4-7) implies that $\|f\| = 0$ if and only if $f \equiv 0$ on (a, b). We also have

$$\|cf\| = |c|\ \|f\| \tag{2.4-8}$$

for any constant c.

We are now ready to draw an analogy to Eq. (2.4-4) by expressing a function f in terms of an **infinite-dimensional basis** of orthonormal functions. Given that the set of functions

$$\{\phi_i(x),\ i = 1, 2, \ldots\}$$

is *orthonormal* on (a, b), $f(x)$ is piecewise continuous on (a, b), and $w(x) > 0$ on (a, b) except at a finite number of points where it may be zero. Then we can write

$$f(x) = \sum_{i=1}^{\infty} c_i \phi_i(x) \tag{2.4-9}$$

for appropriate constants c_i.

It is generally acknowledged that analogies are useful in helping us to bridge the gap between a well-known concept and an unfamiliar one. In this

---

*This assumption is widely used in finding inverse Laplace transforms where we assume that two functions are equal if they differ by a null function $n(x)$ that has the property:

$$\int_0^x n(t)\, dt = 0$$

for every $x > 0$.

particular case, however, we must pause and ask the question, "What is the meaning of the equality in Eq. (2.4–9)?" Obviously, we must first give some meaning to the constant coefficients c_i; second, we must determine the type of convergence (if any) of the infinite series so that we will have some idea of *how* the series represents $f(x)$. Until we clear up these points, we will change Eq. (2.4–9) to*

$$f(x) \sim \sum_{i=1} c_i \phi_i(x), \tag{2.4–10}$$

where the symbol $\sim$ is to be read, "is represented by." In this section we define the c_i and leave the matter of convergence to Section 2.5.

We define the coefficients c_i in (2.4–10) as follows:

$$c_i = (f, \phi_i) = \int_a^b f(x)\phi_i(x)w(x)\,dx. \tag{2.4–11}$$

Hence (2.4–10) becomes

$$f(x) \sim \sum_{i=1} (f, \phi_i)\phi_i(x), \tag{2.4–12}$$

which is analogous to (2.4–4).

In most of the applications in later chapters of this text the weight function $w(x)$ will be equal to one. Accordingly, we will be using simpler equations than the ones shown as Eqs. (2.4–5), (2.4–7), and (2.4–11).

Key Words and Phrases

natural basis
inner product of functions
Euclidean inner product
piecewise continuous function
jump discontinuity
norm of a function
infinite-dimensional basis

2.4 Exercises

• 1. Use Eq. (2.4–4) to represent the vector $\mathbf{v} = \mathbf{e}_k$ $(1 \le k \le n)$.

2. Prove each of the following properties of the inner product in $\mathbb{R}^n$. If

$$\mathbf{u} = \sum_{i=1}^{n} b_i\mathbf{e}_i, \qquad \mathbf{v} = \sum_{i=1}^{n} c_i\mathbf{e}_i, \qquad \text{and} \qquad \mathbf{w} = \sum_{i=1}^{n} a_i\mathbf{e}_i,$$

*Henceforth we will omit the upper limit on *infinite* series.

then

(a) $(\mathbf{u}, \mathbf{v}) = (\mathbf{v}, \mathbf{u})$;

(b) $(\mathbf{u}, \mathbf{v} + \mathbf{w}) = (\mathbf{u}, \mathbf{v}) + (\mathbf{u}, \mathbf{w})$;

(c) $(\mathbf{u}, c\mathbf{v}) = (c\mathbf{u}, \mathbf{v}) = c(\mathbf{u}, \mathbf{v})$ for an arbitrary constant c;

(d) $(\mathbf{u}, \mathbf{u}) \geq 0$ with the equality holding if and only if $\mathbf{u} = \mathbf{0}$.

3. Compute each of the following limits for the piecewise continuous function $g(t)$ in Fig. 2.4–2.

(a) $\lim_{t \to 0^-} g(t)$ **(b)** $\lim_{t \to 0^+} g(t)$ **(c)** $\lim_{t \to 1^-} g(t)$

(d) $\lim_{t \to 1^+} g(t)$ **(e)** $\lim_{t \to 2^-} g(t)$ **(f)** $\lim_{t \to 2^+} g(t)$

(g) $\lim_{t \to 3^-} g(t)$ **(h)** $\lim_{t \to 3^+} g(t)$

4. If f, g, and h are piecewise continuous functions on (a, b), c is an arbitrary constant, and $w(x) > 0$ on (a, b), prove each of the following properties of the inner product defined in Eq. (2.4–5).

(a) $(f, g) = (g, f)$

(b) $(f, g + h) = (f, g) + (f, h)$

(c) $(f, cg) = (cf, g) = c(f, g)$

5. Prove that

$$\|cf\| = |c|\,\|f\|,$$

using the definition of the norm of a function given in Eq. (2.4–7).

•• **6.** Graph each of the given functions.

(a) $f(t) = \begin{cases} 1, & \text{for } 0 \leq t < 1, \\ -1, & \text{for } 1 \leq t < 2 \end{cases}$

(b) $g(t) = \begin{cases} t, & \text{for } 0 < t \leq 2, \\ 4 - t, & \text{for } 2 \leq t < 4 \end{cases}$

(c) $h(t) = \pi - t$ for $0 < t < \pi$

7. Referring to Exercise 6, evaluate each of the following limits.

(a) $\lim_{t \to 0^+} f(t)$; $\lim_{t \to 1^+} f(t)$; $\lim_{t \to 1^-} f(t)$; $\lim_{t \to 2^-} f(t)$;

(b) $\lim_{t \to 0^+} g(t)$; $\lim_{t \to 2^-} g(t)$; $\lim_{t \to 2^+} g(t)$; $\lim_{t \to 4^-} g(t)$;

(c) $\lim_{t \to 0^+} h(t)$; $\lim_{t \to \pi^-} h(t)$; $\lim_{t \to \pi^+} h(t)$; $\lim_{t \to 2\pi^-} h(t)$;

8. With $w(x) = 1$, compute the inner product of each of the following pairs of functions on the interval $(0, 1)$.

(a) $f = 1, \quad g = x$

(b) $f = 1, \quad g = \frac{5}{2}x^3 - \frac{3}{2}x$

(c) $f = x, \quad g = \frac{5}{2}x^3 - \frac{3}{2}x$

9. What conclusions can you draw from the numerical results of Exercise 8?

10. Verify that the set of functions

$$\{1, x, 1 - 3x^2\}$$

forms an orthogonal set on $(-1, 1)$ with weight function $w(x) = 1$. (*Note*: It is necessary to show that *three* integrals are zero and that three other integrals are different from zero.)

11. Transform the orthogonal set in Exercise 10 into an orthonormal set.

12. Verify that the set of functions

$$\{1, x, 5x^2 - 1\}$$

forms an orthogonal set on $(-1, 1)$ with weight function $w(x) = 1 - x^2$.

13. Transform the orthogonal set in Exercise 12 into an orthonormal set.

14. Express the function $1 + x^2$ in terms of the orthogonal functions in Exercise 10, that is,

$$1 + x^2 = c_1 + c_2x + c_3(1 - 3x^2).$$

Then use Eq. (2.4–11) to compute c_i. Note that the equality *is* appropriate for the representation of $1 + x^2$. Why?

15. Repeat Exercise 14 for the function x^3. The result shows that not all functions can be represented in terms of the orthogonal functions in Exercise 10. (See also Exercise 17.)

••• **16.** Give an example of a function f that fails to be piecewise continuous on (a, b) for each of the following reasons.

(a) (a, b) cannot be partitioned into a finite number of subintervals such that f is continuous on each one;

(b) f does not have a finite left-hand limit on (a, b);

(c) f does not have a finite right-hand limit on (a, b).

17. Show that all quadratic polynomials can be represented by a linear combination of the orthogonal functions in Exercise 10. (*Hint*: Show that $ax^2 + bx + c$ can be so represented.)

18. **(a)** Show that the set of functions

$$\{1, x, x^2\}$$

is not an orthogonal set on $(-1, 1)$ with respect to the weight function $w(x) = 1$.

(b) Find constants a_i so that the set of *three* functions

$$\{a_1, a_2x, a_3x^2 + a_4x + a_5\}$$

forms an orthogonal set on $(-1, 1)$ with weight function $w(x) = 1$.

19. Referring to Exercise 18, construct a set that is orthogonal on $(-1, 1)$ with weight function $w(x) = 1$ and that is different from the set in Exercise 10.

2.5 CONVERGENCE AND COMPLETENESS

In Section 2.4 we considered the representation of a piecewise continuous function $f(x)$ by a series of orthonormal functions. In (2.4–12) we used the notation

$$f(x) \sim \sum_{i=1} (f, \phi_i)\phi_i(x) \qquad \textbf{(2.5–1)}$$

where the $\phi_i(x)$ belonged to a set of orthonormal functions on (a, b) with weight function $w(x)$. We simplify the notation in the remainder of this chapter by assuming that $w(x) \equiv 1$. As a result we can write the inner product in Eq. (2.5-1) as

$$(f, \phi_i) = \int_a^b f(x)\phi_i(x)dx. \tag{2.5-2}$$

Next we examine the representation of a function f by the infinite series in (2.5-1). To this end we define the **partial sum** of the first N terms by

$$S_N(x) = \sum_{i=1}^{N} (f, \phi_i)\phi_i(x). \tag{2.5-3}$$

This allows us to consider the difference

$$|S_N(x) - f(x)| \tag{2.5-4}$$

for various values of N and x. We pointed out in Section 1.6 that if, for an arbitrary $\epsilon > 0$, there is an $N(\epsilon)$ such that

$$|S_N(x) - f(x)| < \epsilon,$$

then the series in (2.5-1) **converges uniformly** to $f(x)$ for all x on (a, b). If, on the other hand, N depends on both x and ϵ, then the series in (2.5-1) **converges pointwise** to $f(x)$. For our present purposes, both of the above types of convergence are too demanding, and we will have to settle for something less, owing to the fact that we are dealing with piecewise continuous functions f.

The criterion that we will use in place of (2.5-4) is the **norm**

$$\|S_N(x) - f(x)\|.$$

We first prove that this norm is a *minimum* on (a, b) for the particular choice of the coefficients shown in Eq. (2.5-3), namely, (f, ϕ_i). Consider the sum

$$\sum_{j=1}^{N} c_j\phi_j(x)$$

where the c_j are to be determined in such a way that

$$\left\| \sum_{j=1}^{N} c_j\phi_j(x) - f(x) \right\|$$

will be minimized. Using the definition of the norm, we have

$$\left\| \sum_{j=1}^{N} c_j\phi_j - f \right\|^2 = \left(\sum_{j=1}^{N} c_j\phi_j - f, \sum_{j=1}^{N} c_j\phi_j - f \right) \tag{2.5-5}$$

$$= \sum_{j=1}^{N} c_j^2 - 2\sum_{j=1}^{N} c_j(f, \phi_j) + (f, f)$$

$$= \sum_{j=1}^{N} [c_j - (f, \phi_j)]^2 - \sum_{j=1}^{N} (f, \phi_j)^2 + \|f\|^2,$$

which is obtained by adding and subtracting the same terms. The last expression is clearly a minimum when

$$c_j = (f, \phi_j), \tag{2. 5-6}$$

and this proves the assertion. The coefficients $c_j, j = 1, 2, \ldots$, defined in Eq. (2.5-6) are called the **generalized Fourier* coefficients** of f, or the Fourier coefficients of f with respect to the orthonormal set $\{\phi_j\}$.

When the coefficients in Eq. (2.5-6) are substituted into Eq. (2.5-5), the result is

$$\left\| \sum_{j=1}^{N} (f, \phi_j)\phi_j - f \right\|^2 = \|f\|^2 - \sum_{j=1}^{N} (f, \phi_j)^2, \tag{2.5-7}$$

which is known as **Bessel's identity.**† Since the left-hand member of Eq. (2.5-7) is nonnegative, it follows that

$$\sum_{j=1}^{N} (f, \phi_j)^2 \le \|f\|^2, \tag{2.5-8}$$

a relation known as **Bessel's inequality**. This last, in turn, implies that the series

$$\sum_{j=1}^{\infty} c_j^2 = \sum_{j=1}^{\infty} (f, \phi_j)^2$$

is *convergent* (Exercise 2); hence

$$\lim_{j\to\infty} c_j = 0.$$

This important result, which we will use in Section 4.5, is stated in the form of the following theorem.

*After Jean B. J. Fourier (1768–1830), a French physicist and mathematician, better known today as Joseph Fourier.

†After Friedrich Wilhelm Bessel (1784–1846), a German mathematician and the founder of modern practical astronomy.

THEOREM 2.5-1 (Riemann-Lebesgue* Theorem.)
The sequence of generalized Fourier coefficients is a null sequence, that is,

$$\lim_{j\to\infty} \mathbf{c}_j = \lim_{j\to\infty} (f, \phi_j) = 0.$$

The additional property

$$\left\| \sum_{j=1} (f, \phi_j)\phi_j - f \right\|^2 = 0 \qquad \textbf{(2.5–9)}$$

and, consequently,

$$f = \sum_{j=1} (f, \phi_j)\phi_j,$$

would be most useful. This conclusion, however, is not always true unless we put further restrictions on f and the orthonormal set $\{\phi_j\}$.

First, we note that to draw conclusions based on the inequality in (2.5–8) requires that meaning be given to

$$\|f\|^2 = \int_a^b [f(x)]^2\,dx.$$

This, in turn, means that the function f must be restricted to the class of **square-integrable functions,**† that is, functions satisfying

$$\int_a^b [f(x)]^2\,dx < \infty.$$

Second, since

$$\int_a^b [\phi_j(x)]^2\,dx = 1,$$

we see from the definition of an orthonormal system that the functions ϕ_j must also be square integrable. Finally, the orthonormal system must be **complete** with respect to a given class of functions‡ in accordance with the following definition.

---

*The result was first proved by Georg F. B. Riemann (1826–1866), a German mathematician, for continuous functions and was later extended by Henri L. Lebesgue (1875–1941), a French mathematician, to more general functions.

†The word "integrable" here refers to integrable in the Riemann sense, that is, the type of integration studied in usual calculus courses. Actually, it is possible to substitute "square integrable in the Lebesgue sense," which results in somewhat more general functions. We will, however, remain with the more familiar Riemann-integrable functions.

‡Examples of a "class of functions" are each of the following: continuous functions, piecewise continuous functions, integrable functions, square-integrable functions, piecewise smooth functions.

DEFINITION 2.5-1

The orthonormal sequence of functions $\{\phi_n\}$ is complete with respect to a given class of functions if

$$\lim_{N \to \infty} \int_a^b [S_N(x) - f(x)]^2 \, dx = 0 \qquad \textbf{(2.5-10)}$$

for every function f in the class. Eq. (2.5-10) is also written

$$\underset{N \to \infty}{l.i.m.} \; S_N(x) = f(x), \qquad \textbf{(2.5-11)}$$

*where l.i.m. is read, "**limit in the mean.**" This type of convergence is called **convergence in the mean.****

We observe that the **completeness property** of an orthonormal system $\{\phi_j\}$ given in Definition 2.5-1 is tied to a certain class of functions. A consequence of this is that no function in the given class, with the exception of one, may be orthogonal to *every* ϕ_j. This exception is the zero function (which is orthogonal to *every* ϕ_j) or, more generally, a **null function.**† This property is called **closure**, which is defined in the next definition.

DEFINITION 2.5-2

*The orthonormal system $\{\phi_j\}$ is **closed** on (a, b), if, for any function f that is piecewise continuous on (a, b), the equation*

$$(f, \phi_j) = \int_a^b f(x)\phi_j(x) \, dx = 0 \qquad \text{for all } j$$

implies that f is a null function on (a, b).

It can be shown (Exercise 8) that an orthonormal system is closed if it is complete. The converse is also true, as is shown in Exercise 10.

Bessel's inequality (2.5-8) can also be written (see Exercise 11)

$$\sum_{j=1}^{\infty} (f, \phi_j)^2 \le \|f\|^2, \qquad \textbf{(2.5-12)}$$

and it can be shown that the equality holds for orthonormal systems that are complete with respect to square-integrable functions at points of continuity of

---

*The name comes from statistics, where the integral in Eq. (2.5-10) is called the **mean-square deviation** of f from S_N. Other names in use are **root mean square** (RMS) **convergence** and **mean square convergence.**

†The function $n(x)$ is a null function on (a, b) if

$$\int_a^b [n(x)]^2 \, dx = 0.$$

Equivalently, a null function is zero at every point where it is continuous.

f. The resulting equation

$$\sum_{j=1}^{\infty}(f, \phi_j)^2 = \|f\|^2, \tag{2.5-13}$$

is called **Parseval's equation.***

We point out that convergence in the mean as defined in Eq. (2.5–10) does not imply pointwise convergence, and neither does pointwise convergence imply convergence in the mean. In other words, the latter is a special type of convergence that, as we shall see, is intimately connected with boundary-value problems.

Key Words and Phrases

partial sum
uniform convergence
pointwise convergence
norm
generalized Fourier coefficient
Bessel's identity
Bessel's inequality
square-integrable function
complete orthonormal system
limit in the mean
convergence in the mean
completeness property
null function
closure
closed orthonormal system
Parseval's equation

2.5 Exercises

• **1.** Explain why

$$\left\| \sum_{j=1}^{N} c_j\phi_j(x) - f(x) \right\|$$

is a minimum when $c_j = (f, \phi_j)$. (*Hint*: See Eq. (2.5–5) and its subsequent development.)

2. Explain why the series following Eq. (2.5–8),

$$\sum_{j=1}^{\infty} c_j^2 = \sum_{j=1}^{\infty} (f, \phi_j)^2,$$

is convergent.

3. Discuss briefly why a piecewise continuous function is necessarily square integrable.

*After Marc-Antoine Parseval des Chênes (1755–1836), a French mathematician.

•• **4.** Show that the set of functions

$$\{\phi_j(x) = \frac{1}{\sqrt{\pi}} \sin jx, \; j = 1, 2, \ldots\}$$

is an orthonormal set on $[0, 2\pi]$ with weight function $w(x) = 1$.

5. Show that the orthonormal system $\{\phi_j(x)\}$ of Exercise 4 is not complete on $[0, 2\pi]$ with respect to the class of continuous functions. (*Hint*: Show that the *continuous* function $\cos x$ cannot be represented by a series of the orthonormal functions. Note that the case $j = 1$ must be examined separately.)

6. Starting with

$$\left\| \sum_{j=1}^{N} c_j\phi_j - f \right\|^2$$

as in Eq. (2.5–5), differentiate partially with respect to c_j, and set the result to zero. Thus find a necessary condition on the c_j for the given square of the norm to be minimum.

••• **7.** Prove that a necessary and sufficient condition that

$$\left\| \sum_{j=1}^{N} c_j\phi_j - f \right\|^2$$

be a minimum is that $c_j = (f, \phi_j)$, where $\{\phi_j\}$ is an orthonormal set on (a, b) and f is piecewise continuous there. (*Hint*: See Exercises 1 and 6.)

8. Prove that if a set of functions is complete, then it is closed. (*Hint*: Use contradiction; assume that there is a *normalized* function $\psi(x)$ such that

$$c_j = \int_a^b \psi(x)\phi_j(x)\, dx = 0$$

for all j. Then obtain

$$\lim_{N\to\infty} \int_a^b \left| \sum_{j=1}^{N} c_j\phi_j - \psi \right|^2 dx = 1,$$

thus contradicting the assumption of completeness.)

9. Prove that

$$\left\| f - g \right\| \geq \left| \|f\| - \|g\| \right|.$$

10. Prove that if a set of functions is closed, then it is complete. (*Hint*: Assume that there is a function f satisfying

$$\int_a^b |f|^2\, dx - \sum_{j=1}^{\infty} |c_j|^2 > 0,$$

where $c_j = (f, \phi_j)$. Then consider

$$g_N(x) = \sum_{j=1}^{N} c_j\phi_j,$$

which converges in the mean to g with $c_j = (g, \phi_j) = (f, \phi_j)$. Hence $\psi = g - f$ is orthogonal to each ϕ_j, and

$$\|\psi\| \geq \left| \, \|g - g_N\| - \|f - g_N\| \, \right| > 0,$$

which shows that ψ can be normalized, and hence the set cannot be closed.)

11. Prove that Bessel's inequality for a *finite* sum as given in (2.5–8) can be extended to *infinite* series as shown in (2.5–12).
12. Read the article, "A Complete Set Which is not a Basis" by J. S. Byrnes in *Amer. Math. Monthly*, May 1972, pp. 510–512.

2.6 SELF-ADJOINT EQUATIONS

Boundary-value problems that involve differential equations having a special form are of particular importance in applied mathematics. It is these differential equations and their properties that we will discuss in this section.

We begin with the linear second-order homogeneous differential equation

$$y'' + a_1(x)y' + a_2(x)y = 0. \tag{2.6–1}$$

The substitutions $u_1 = y$ and $u_2 = y'$ transform Eq. (2.6–1) into a system of two first-order equations

$$\begin{aligned} u_1' &= u_2, \\ u_2' &= -a_2(x)u_1 - a_1(x)u_2. \end{aligned} \tag{2.6–2}$$

This system can be written in matrix form as follows:

$$\mathbf{u}' = A(x)\mathbf{u}, \tag{2.6–3}$$

where

$$\mathbf{u} = \begin{pmatrix} u_1 \\ u_2 \end{pmatrix} \quad \text{and} \quad A(x) = \begin{pmatrix} 0 & 1 \\ -a_2(x) & -a_1(x) \end{pmatrix}.$$

The equation

$$\mathbf{v}' = -A^T(x)\mathbf{v}, \tag{2.6–4}$$

where A^T denotes the transpose of A, is called the **adjoint equation** of Eq. (2.6–3).

There is a tenuous connection between adjoint equations and the adjoint matrix in linear algebra. Let $\Phi(x)$ be a fundamental matrix of solutions of the system

(2.6–2). Such a matrix is necessarily nonsingular.* It can be shown (Exercise 12) that $(\Phi^T)^{-1}$ is the fundamental matrix of solutions of the system (2.6–4).

Since $(\Phi^T)^{-1}$ is the *inverse* of a *transposed* matrix, it was (perhaps inadvisedly) called the *adjoint* of the matrix Φ. The use of the word "adjoint" is unfortunate, since the name was already in use for the matrix formed from the cofactors of a given matrix B that, when multiplied by the reciprocal of det (B), produces B^{-1}. Some authors† in an effort to rectify this confusion define B^{-1} as $(1/\det B)$ adj B, where "adj" denotes "adjunct." This leaves the word "adjoint" for the purpose used in this section.

If we retrace our steps from Eq. (2.6–4), we obtain the system

$$\begin{aligned} v_1' &= \phantom{-v_1 +{}} a_2 v_2 \\ v_2' &= -v_1 + a_1 v_2 \end{aligned}$$

and the second-order equation

$$w'' - (a_1(x)w)' + a_2(x)w = 0. \qquad \textbf{(2.6–5)}$$

We call Eq. (2.6–5) the **adjoint** of Eq. (2.6–1). The relationship is symmetric (Exercise 1), so Eq. (2.6–1) is the adjoint of Eq. (2.6–5).

Of particular importance are differential equations that are **self-adjoint,** that is, equations that are equal to their adjoints. We can see from Eq. (2.6–5) that any equation in which $a_1(x) \equiv 0$ is necessarily self-adjoint. More generally, adjointness is defined in the following way. If a linear homogeneous differential equation is given by

$$a_2(x)y'' + a_1(x)y' + a_0(x)y = 0, \qquad \textbf{(2.6–6)}$$

then its adjoint is

$$[a_2(x)y]'' - [a_1(x)y]' + a_0(x)y = 0. \qquad \textbf{(2.6–7)}$$

Thus (Exercise 2) Eq. (2.6–6) is self-adjoint if and only if

$$a_1(x) = a_2'(x). \qquad \textbf{(2.6–8)}$$

The self-adjointness condition in Eq. (2.6–8) shows that any second-order linear differential equation of the form

$$\frac{d}{dx}\left[a_2(x)\,\frac{dy}{dx}\right] + a_0(x)y = 0$$

is necessarily self-adjoint. In particular, the Sturm–Liouville problem of Section 2.2 is a boundary-value problem involving a self-adjoint differential equation. Moreover, in Section 2.2 we showed that *every* second-order linear

*See E. A. Coddington and N. Levinson, *Theory of Ordinary Differential Equations* (New York: McGraw-Hill, 1955), pp. 67 ff.

†See, for example, B. W. Jones, *Linear Algebra* (San Francisco: Holden-Day, 1973), p. 165.

homogeneous differential equation (2.6–6) can be put into self-adjoint form by multiplying it by the term $\mu(x)/a_2(x)$, where

$$\mu(x) = \exp\left(\int^x \frac{a_1(t)}{a_2(t)}\,dt\right). \tag{2.6–9}$$

In Definition 2.3–1 we defined orthogonality of **real-valued** functions. That definition can be generalized to **complex-valued** functions in the following manner. The set of complex-valued functions

$$\{\phi_i(x),\ i = 1, 2, 3, \ldots\},$$

each of which is continuous on $[a, b]$, is **orthogonal in the hermitian sense*** with real weight function $w(x)$ if

$$\int_a^b \phi_n(x)\,\overline{\phi_m(x)}\,w(x)\,dx = 0, \qquad n \neq m,$$

and **(2.6–10)**

$$\int_a^b \phi_n(x)\,\overline{\phi_n(x)}\,w(x)\,dx \neq 0,$$

where the bar indicates the *complex conjugate.*

Observe that the above definition of hermitian orthogonality reduces to the orthogonality defined in Definition 2.3–1 when the $\phi_i(x)$ are real-valued functions.

EXAMPLE 2.6–1 Show that the functions

$$\{\exp(inx),\ n = 0, \pm 1, \pm 2, \ldots\}$$

are orthogonal in the hermitian sense on the interval $[-\pi, \pi]$ with weight function $w(x) = 1$.

Solution We have

$$\begin{aligned}\int_{-\pi}^{\pi} e^{inx}e^{-imx}\,dx &= \int_{-\pi}^{\pi} e^{i(n-m)x}\,dx \\ &= \left.\frac{e^{i(n-m)x}}{i(n-m)}\right|_{-\pi}^{\pi} \\ &= 0, \qquad \text{if } n \neq m,\end{aligned}$$

and the result is 2π if $n = m$ (Exercise 3). ■

Recall that in Example 2.1–5 we saw a problem that had complex eigenvalues and complex eigenfunctions. In general, a regular Sturm–Liouville

---

*After Charles Hermite (1822–1901), a French mathematician.

problem

$$\frac{d}{dx}\left[r(x)\,\frac{dy}{dx}\right] + [q(x) + \lambda w(x)]y = 0, \qquad a \le x \le b,$$
$$a_1 y(a) + a_2 y'(a) = 0, \tag{2.6-11}$$
$$b_1 y(b) + b_2 y'(b) = 0,$$

is such that $r(x)$, $q(x)$, and $w(x)$ are real-valued functions and the a_i and b_i are real constants.

If $y(x)$ is a complex-valued function and λ is a complex number, then we can take the complex conjugates of the terms in Eq. (2.6–11) to obtain

$$\frac{d}{dx}\left[r(x)\,\frac{d\bar{y}}{dx}\right] + [q(x) + \bar{\lambda} w(x)]\bar{y} = 0, \qquad a \le x \le b,$$
$$a_1 \bar{y}(a) + a_2 \bar{y}'(a) = 0, \tag{2.6-12}$$
$$b_1 \bar{y}(b) + b_2 \bar{y}'(b) = 0.$$

If we multiply the differential equation in (2.6–11) by $\bar{y}(x)$ and the differential equation in (2.6–12) by $y(x)$ and subtract, we have (Exercise 5)

$$\frac{d}{dx}\left[r(x)\,(\bar{y}y' - \bar{y}'y)\right] = w(x)\bar{y}y(\bar{\lambda} - \lambda). \tag{2.6-13}$$

Now integrating this last equation from a to b gives us (Exercise 6)

$$(\bar{\lambda} - \lambda)\int_a^b w(x)\,|y(x)|^2\,dx = 0, \tag{2.6-14}$$

from which we conclude that $\lambda = \bar{\lambda}$. We have proved the following theorem.

THEOREM 2.6-1

The eigenvalues of a regular Sturm–Liouville system are real.

Another important property of the eigenvalues of a regular Sturm–Liouville system is, unfortunately, more difficult to prove. We state it in the following theorem and indicate where the proof can be found.

THEOREM 2.6-2*

The eigenvalues λ_k of a regular Sturm–Liouville system can be ordered,

$$\lambda_1 < \lambda_2 < \lambda_3 < \ldots < \lambda_k < \ldots$$

and

$$\lim_{k\to\infty} \lambda_k = \infty.$$

*See Hans Sagan, *Boundary and Eigenvalue Problems in Mathematical Physics* (New York: John Wiley, 1961), Chapter 5.

Implied in Theorem 2.6–2 is the fact that the set of eigenvalues $\{\lambda_k\}$ is bounded below. It is also true that to each eigenvalue there corresponds an eigenfunction that is unique to within a constant factor. For proofs of these last assertions, see Broman.*

Key Words and Phrases

adjoint equation	self-adjoint equation
adjoint	orthogonal in the hermitian sense

2.6 Exercises

• **1.** Verify that the adjoint of Eq. (2.6–5) is Eq. (2.6–1).

2. Show that Eq. (2.6–6) is self-adjoint if and only if $a_1(x) = a_2'(x)$.

3. Carry out the details in Example 2.6–1.

4. Transform the orthogonal set of Example 2.6–1,

$$\{\exp(inx),\ n = 0, \pm 1, \pm 2, \ldots\},$$

into an orthonormal set.

5. Carry out the details needed to arrive at Eq. (2.6–13).

6. Verify the result in Eq. (2.6–14). (*Hint*: Use the boundary conditions in Eq. 2.6–12.)

•• **7.** Obtain the matrix $A(x)$ shown in Eq. (2.6–3) for each of the following differential equations in which n is a nonnegative integer and λ is a constant.

(a) $y'' + \lambda^2 y = 0$

(b) $x^2y'' + xy' + (x^2 - n^2)y = 0$

(c) $y'' + y' - \lambda y = 0$

(d) $\dfrac{d}{dx}\left[(1 - x^2)\dfrac{dy}{dx}\right] + n(n + 1)y = 0$

8. Using the results of Exercise 7, obtain the transpose $A^T(x)$ and write the adjoint equation of each equation in the form of Eq. (2.6–4).

9. Combine each pair of equations obtained in Exercise 8 into a single second-order equation.

10. Which of the equations of Exercise 7 are self-adjoint?

11. Transform each of the following equations into the form of the Sturm–Liouville equation

$$\frac{d}{dx}\left[r(x)\frac{dy}{dx}\right] + [q(x) + \lambda w(x)]y = 0.$$

---

*Arne Broman, *Introduction to Partial Differential Equations from Fourier Series to Boundary-value Problems* (Reading, Mass.: Addison-Wesley, 1970), Chapter 2.

(a) $y'' + 2xy' + (x + \lambda)y = 0$
(b) $x^2y'' + xy' + \lambda xy = 0$
(c) $(x + 2)y'' + 4y' + (x + \lambda e^x)y = 0$

••• **12.** Suppose that **u** and **ũ** are linearly independent solutions of Eq. (2.6–3) so that the fundamental matrix of solutions of the system (2.6–2) is given by

$$\Phi(x) = \begin{pmatrix} u_1 & \tilde{u}_1 \\ u_2 & \tilde{u}_2 \end{pmatrix}.$$

Show that $(\phi^T)^{-1}$ is the fundamental matrix of solutions of the system (2.6–4).

13. Generalize Example 2.6–1 to show a set orthonormal on $[0, 2c]$.

14. Put the following equation into self-adjoint form.

$$xy'' - 2x^2y' + y = 0.$$

15. Show that Lagrange's identity (Section 2.3) can be written

$$\int_a^b (y_1Ly_2 - y_2My_1)\,dx = a_2y_2'y_1 - y_2(a_2y_1)' + y_2a_1y_1\Big|_a^b,$$

where L and M are adjoint operators and a_2, a_1 are as in Eq. (2.6–6).

16. Write Lagrange's identity for a self-adjoint operator in the form

$$(y_2, Ly_1) = (y_1, Ly_2),$$

which holds for real-valued twice continuously differentiable functions on [a, b] that satisfy the boundary conditions (2.2–7).

17. Let L be the **integral operator** defined by

$$Lx(t) = \int_a^b K(t,s)\,x(t)\,dt,$$

where $x(t)$ belongs to the class of functions that are integrable on $[a, b]$. Define the inner product of such functions in the usual manner, namely as

$$(x,y) = \int_a^b x(t)\,y(t)\,dt.$$

(a) Prove that L is a *linear* integral operator.
(b) Show that if the **kernel** $K(t,s)$ is symmetric, that is, $K(t,s) = K(s,t)$, then $(Lx,y) = (x,Ly)$, that is, L is self-adjoint.
(c) Verify that the **Laplace transform**

$$\mathscr{L}\{x(t)\} = \int_0^\infty x(t)\exp(-st)\,dt$$

is a self-adjoint integral operator.
(d) Repeat part (c) for the **Fourier transform**

$$\mathscr{F}\{x(t)\} = \int_{-\infty}^\infty \exp(ikt)\,x(t)\,dt.$$

18. **(a)** Write Legendre's differential equation $(1 - x^2)y'' - 2xy' + \lambda y = 0$ in the form

$$Ly = -[(1 - x^2)y']' - \lambda y.$$

(b) Show that the Legendre differential operator is self-adjoint.

2.7 OTHER STURM–LIOUVILLE SYSTEMS

Although every linear differential equation can be put into self-adjoint form, not every such equation leads to a *regular* Sturm–Liouville problem. It may happen that in the equation

$$\frac{d}{dx}\left[r(x)\frac{dy}{dx}\right] + [q(x) + \lambda w(x)]y = 0, \qquad a \le x \le b, \tag{2.7–1}$$

the functions $r(x)$, $q(x)$, and $w(x)$ do not have the necessary properties. Or it may be that the boundary conditions do not have the form*

$$\begin{aligned} a_1 y(a) + a_2 y'(a) &= 0,\\ b_1 y(b) + b_2 y'(b) &= 0, \end{aligned} \tag{2.7–2}$$

required for a regular problem.

Since some very important problems belong to the exceptions mentioned above, we need to extend the methods of this chapter to include some special cases. For example, Bessel's differential equation of order n defined on $0 < t < \lambda c$ is (see Eq. (1.5–6))

$$\frac{d^2y}{dt^2} + \frac{1}{t}\frac{dy}{dt} + \left(1 - \frac{n^2}{t^2}\right)y = 0, \qquad n = 0, 1, 2, \ldots. \tag{2.7–3}$$

The substitution $t = \lambda x$ transforms Eq. (2.7–3) into

$$x^2\frac{d^2y}{dx^2} + x\frac{dy}{dx} + (\lambda^2x^2 - n^2)y = 0, \tag{2.7–4}$$

which in self-adjoint form becomes

$$\frac{d}{dx}\left(x\frac{dy}{dx}\right) + \left(\lambda^2 x - \frac{n^2}{x}\right)y = 0, \tag{2.7–5}$$

defined on $0 < x < c$. Equation (2.7–5), however, fails to meet a number of requirements that have been specified for the differential equation in a regular Sturm–Liouville problem (Exercise 1). A similar situation exists in the case of Legendre's differential equation (see Eq. (1.5–9))

$$(1 - x^2)y'' - 2xy' + n(n + 1)y = 0,$$

which in self-adjoint form is

$$\frac{d}{dx}\left[(1 - x^2)\frac{dy}{dx}\right] + n(n + 1)y = 0, \tag{2.7–6}$$

where $\lambda = n(n + 1)$, and we will be interested in solutions defined on $-1 \le x \le 1$ (Exercise 2).

---

*Boundary conditions of this type are called **separated boundary conditions**.

In Section 2.2 we assumed that the functions $r(x)$, $q(x)$, and $w(x)$ have certain properties on the interval $-\infty < a \leq x \leq b < \infty$. For a **regular** Sturm–Liouville problem these properties are the following:

1. $r(x)$ is continuously differentiable, and $r(x) > 0$.
2. $q(x)$ is continuous.
3. $w(x) > 0$ and is continuous.

If any one of these conditions does not hold or if the interval $a \leq x \leq b$ is *unbounded*, then the Sturm–Liouville differential equation (2.7–1) is called **singular**. Thus Eqs. (2.7–5) and (2.7–6) are singular on bounded intervals.

It may also happen that although the Sturm–Liouville equation is nonsingular, the boundary conditions may not have the form shown in Eq. (2.7–2). Consider the following example.

EXAMPLE 2.7–1 Find the eigenvalues and corresponding eigenfunctions of the following problem.

$$y'' + \lambda y = 0, \qquad -c \leq x \leq c,$$
$$y(-c) = y(c), \qquad y'(-c) = y'(c).$$

Solution The boundary conditions here are called **periodic boundary conditions**. They will arise later when we consider boundary-value problems over circular regions in Section 7.1.

We leave it as an exercise to show that $\lambda < 0$ leads to the trivial solution (Exercise 3). For $\lambda > 0$ we have

$$y(x) = c_1 \cos \sqrt{\lambda}x + c_2 \sin \sqrt{\lambda}x$$
$$y'(x) = -c_1 \sqrt{\lambda} \sin \sqrt{\lambda}x + c_2 \sqrt{\lambda} \cos \sqrt{\lambda}x.$$

The condition $y(-c) = y(c)$ results in

$$2c_2 \sin c\sqrt{\lambda} = 0,$$

which can be satisfied either by $c_2 = 0$ or by $\sqrt{\lambda} = n\pi/c$, $n = 1, 2, \ldots$. The condition $y'(-c) = y'(c)$ results in

$$2c_1 \sqrt{\lambda} \sin c\sqrt{\lambda} = 0,$$

which can be satisfied either by $c_1 = 0$ or by $\sqrt{\lambda} = n\pi/c$. Hence the eigenvalues are

$$\lambda_n = \left(\frac{n\pi}{c}\right)^2, \qquad n = 1, 2, \ldots,$$

with corresponding eigenfunctions

$$y_n(x) = a_n \cos\left(\frac{n\pi}{c}x\right) + b_n \sin\left(\frac{n\pi}{c}x\right), \qquad \textbf{(2.7–7)}$$

where the a_n and b_n are not *both* zero but are otherwise arbitrary constants. We leave it as an exercise to show that $\lambda = 0$ is an eigenvalue with corresponding eigenfunction

$$y_0(x) = \frac{1}{2} a_0,$$

where a_0 is an arbitrary constant (Exercise 4). ■

The orthogonality property of the eigenfunctions of Example 2.7–1 will be useful in Chapter 4 in our study of Fourier series. We observe that Eq. (2.7–7) shows that to each eigenvalue λ_n there correspond *two* linearly independent eigenfunctions (Exercise 5).

We can easily prove a theorem similar to Theorem 2.3–1 for the case of periodic boundary conditions.

THEOREM 2.7–1

For a regular Sturm–Liouville problem with periodic boundary conditions and such that r(a) = r(b), eigenfunctions belonging to different eigenvalues are orthogonal on [a, b] with weight function w(x).

Proof *The periodic boundary conditions are in general*

$$y(a) = y(b), \qquad y'(a) = y'(b).$$

From Eq. (2.3–5) we have

$$(\lambda_1 - \lambda_2) \int_a^b w(x)y_1(x)y_2(x)\,dx = r(y_1y_2' - y_2y_1')\Big|_a^b, \tag{2.7–8}$$

and the right-hand member becomes

$$\begin{aligned} &r(b)[y_1(b)y_2'(b) - y_2(b)y_1'(b)] - r(a)[y_1(a)y_2'(a) + y_2(a)y_1'(a)] \\ &\quad = r(a)[y_1(a)y_2'(a) - y_2(a)y_1'(a)] - r(a)[y_1(a)y_2'(a) - y_2(a)y_1'(a)] = 0, \end{aligned}$$

if $\lambda_1 \neq \lambda_2$. □

It is important to note that if $r(x)$ is constant on $[a, b]$, then $r(a) = r(b)$ so that periodic boundary conditions also lead to orthogonal eigenfunctions.

From Eq. (2.7–8) we can see that the condition $r(a) = r(b) = 0$ is enough to show that eigenfunctions belonging to different eigenvalues are orthogonal. Accordingly, the functions $y_n(x)$ that are continuous on $[-1, 1]$ and satisfy Eq. (2.7–6) form an orthogonal set on that interval with weight function $w(x) = 1$. We shall study these functions—called **Legendre polynomials**— in Section 7.3.

We also see from Eq. (2.7–8) that either of the two conditions

$$r(a) = 0, \qquad y_1(b)y_2'(b) - y_2(b)y_1'(b) = 0, \tag{2.7–9}$$

or

$$r(b) = 0, \qquad y_1(a)y_2'(a) - y_2(a)y_1'(a) = 0, \tag{2.7–10}$$

is sufficient to ensure the orthogonality of the eigenfunctions. In general the following theorem covers the case of eigenfunctions of a **singular Sturm–Liouville system.**

THEOREM 2.7–2
The eigenfunctions of a singular Sturm–Liouville system on [a, b] are orthogonal on [a, b] with weight function w(x) if they are square integrable there and if

$$r(x)[y_1(x)y_2'(x) - y_2(x)y_1'(x)]_a^b = 0$$

for every distinct pair of eigenfunctions $y_1(x)$ and $y_2(x)$.

We will see in later chapters that a versatile method of solving boundary-value problems leads to both regular and singular Sturm–Liouville problems. The existence of orthogonal sets of eigenfunctions for these problems will be essential to obtaining solutions to our boundary-value problems.

Key Words and Phrases

separated boundary conditions
singular Sturm–Liouville differential equation
periodic boundary conditions
Legendre polynomials
singular Sturm–Liouville system

2.7 Exercises

• **1.** In what ways does Eq. (2.7–5) fail to meet the requirements for a regular Sturm–Liouville differential equation?

2. In what way does Eq. (2.7–6) fail to meet the requirements for a regular Sturm–Liouville differential equation?

3. Show that $\lambda < 0$ leads to the trivial solution of the problem in Example 2.7–1.

4. Show that $\lambda = 0$ is an eigenvalue of the problem in Example 2.7–1, and find the corresponding eigenfunction.

5. Exhibit the two linearly independent sets of eigenfunctions of the problem in Example 2.7–1.

•• **6.** The **Mathieu equation***

$$\ddot{y} + (\lambda + 16d\cos 2t)y = 0, \qquad 0 \le t \le \pi,$$
$$y(0) = y(\pi), \qquad \dot{y}(0) = \dot{y}(\pi),$$

is a special case of Hill's equation†

$$\ddot{y} + a(t)y = 0,$$

*Émile L. Mathieu (1835–1890), a French applied mathematician.

†George W. Hill (1838–1914), an American astronomer, who studied the motion of the moon.

where $a(t)$ is periodic. Show that if $y_1(t)$ and $y_2(t)$ are periodic solutions of Mathieu's equation with period π and belonging to distinct eigenvalues, then $y_1(t)$ and $y_2(t)$ are orthogonal on $[0, \pi]$ with weight function one.

7. Solve the following problem.

$$y'' + \lambda^2 y = 0, \qquad 0 \le x \le \pi,$$
$$y(0) + y(\pi) = 0, \qquad y'(0) + y'(\pi) = 0.$$

8. Obtain the solution to each of the following singular Sturm–Liouville problems.*

(a) $y'' + \lambda y = 0, \quad 0 < x < \infty,$
$y(0) = 0, \quad |y(x)| < M$

(b) $y'' + \lambda y = 0, \quad 0 < x < \infty,$
$y'(0) = 0, \quad |y(x)| < M$

(c) $y'' + \lambda y = 0, \quad -\infty < x < \infty,$
$|y(x)| < M$

9. Show that zero is not an eigenvalue in any of the problems in Exercise 8.

10. Solutions of *Tchebycheff's differential equation*†

$$(1 - x^2)y'' - xy' + n^2 y = 0, \qquad n = 0, 1, 2, \ldots, \tag{2.7-11}$$

are *Tchebycheff polynomials* defined by

$$T_n(x) = \cos(n \arccos x), \qquad -1 \le x \le 1. \tag{2.7-12}$$

(a) Show that $x = \pm 1$ are regular singular points of Eq. (2.7–11).

(b) Solve Eq. (2.7–11) by a series method (see Section 1.5), and note that the restriction on n is necessary so that the solution will converge for $x = \pm 1$.

(c) Write out the first six Tchebycheff polynomials.

(d) Show that the $T_n(x)$ are orthogonal on $(-1, 1)$ with weight function $(1 - x^2)^{-1/2}$. (*Hint*: Make the substitution $\alpha = \arccos x$ in Eq. 2.7–12).

(e) Explain how Eq. (2.7–11) leads to a singular Sturm–Liouville problem.

(f) Show that the square of the norm is given by

$$\|T_0\|^2 = \pi \quad \text{and} \quad \|T_n\|^2 = \frac{\pi}{2} \quad \text{for } n > 0.$$

(g) Obtain from the $T_n(x)$ an orthonormal series on $(-1, 1)$.

(h) Graph the first four polynomials.

(1) Find out how Tchebycheff polynomials are used to represent a function with minimum error. Read Curtis F. Gerald's *Applied Numerical Analysis* (Reading, Mass.: Addison-Wesley, 1978), Sections 10.5–10.7.

---

*The notation $|y(x)| < M$, where M is some positive constant, means that $y(x)$ is *bounded* on the given interval.

†Pafnuti L. Tchebycheff (1821–1894), a Russian mathematician. Various transliterations of his name, such as Chebishev, are in use.

2.8 NUMERICAL METHODS

When solving **initial-value problems** in ordinary differential equations, we are given the values of the dependent variable and the slope at some initial point (in the case of second-order equations). To carry out a numerical solution for such a problem, we need to find successive values of the unknown function in some way. There are a number of methods for doing this, including the following: Euler's method, Heun's method, Adams–Bashforth method, Adams–Moulton method, and Runge–Kutta methods.*

We have seen that solving **boundary-value problems** in ordinary differential equations presents some additional difficulties. For one thing, a boundary-value problem may have a unique solution, many solutions, or no real solution. For another, starting with the value of the unknown function at one boundary, we must somehow assign the initial value of the slope correctly if we expect to match the value of the function or the slope at another boundary. In other words, we must supply some **missing initial condition** in order to attain a specified value at the terminal point. If the correct initial condition is assigned, the terminal or **target condition** will be satisfied, and we are through. If the target condition is not satisfied, however, then we must adjust the initial condition in some systematic manner until the target condition is met within some specified tolerance.

If the problem involves a *linear* differential equation and *linear* boundary conditions, then only one adjustment of the missing initial condition may be necessary. If the problem is nonlinear, however, then some iterative scheme must be used. In the latter case, questions of **convergence** and **stability** must be considered.

We can best illustrate the foregoing by giving an example using a boundary-value problem that has a *unique* solution that can be found analytically.

EXAMPLE 2.8–1 Solve the boundary-value problem

$$y'' - y = 2x, \qquad y(0) = 0, \qquad y(1) = 1.$$

Solution We can readily check (Exercise 1) that the complementary solution is

$$y_c(x) = c_1 \cosh x + c_2 \sinh x$$

and a particular integral is

$$y_p(x) = -2x.$$

*Brief descriptions of these can be found in Section 1.6 of Ladis D. Kovach, *Advanced Engineering Mathematics* (Reading, Mass.: Addison-Wesley, 1982).

Hence the general solution of the given differential equation is

$$y_g(x) = c_1 \cosh x + c_2 \sinh x - 2x.$$

The condition $y(0) = 0$ implies that $c_1 = 0$ so that

$$y(x) = c_2 \sinh x - 2x.$$

Finally, the condition $y(1) = 1$ leads to

$$c_2 \sinh 1 - 2 = 1,$$

that is,

$$c_2 = 3/\sinh 1,$$

and

$$y(x) = \frac{3}{\sinh 1} \sinh x - 2x. \quad \blacksquare \tag{2.8-1}$$

A graph of Eq. (2.8-1) is shown in Fig. 2.8-1 as a solid curve. It shows that the curve obtained by using points that satisfy Eq. (2.8-1) is correct rather than one of the dashed curves in Fig. 2.8-1.

The boundary-value problem of Example 2.8-1 can be easily converted into an initial-value problem. From Eq. (2.8-1) we have

$$y'(x) = \frac{3}{\sinh 1} \cosh x - 2$$

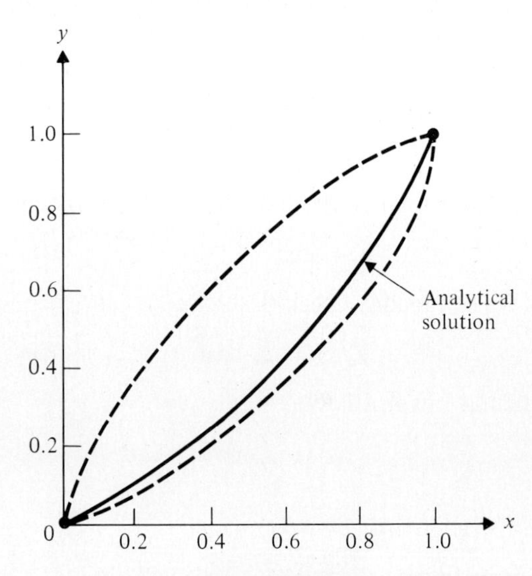

Figure 2.8-1
Analytical solution (Example 2.8-1).

so that

$$y'(0) = \frac{3}{\sinh 1} - 2 \doteq 0.553.$$

Thus the problem of Example 2.8–1 is equivalent to the initial-value problem (Exercise 2)

$$\begin{aligned} &y'' - y = 2x, \qquad 0 < x < 1, \\ &y(0) = 0, \qquad y'(0) = 0.553. \end{aligned}$$

Since we are dealing here with a problem whose solution can be obtained analytically, we do not really have a "missing" initial condition. If, however, we had assumed that $y'(0) = 1$ in the above initial-value problem, then we would have obtained a solution that could not reasonably be expected to pass through the point (1, 1). It can be shown (Exercise 3) that the initial-value problem

$$\begin{aligned} &y'' - y = 2x, \qquad 0 < x < 1, \\ &y(0) = 0, \qquad y'(0) = 1, \end{aligned}$$

has a solution for which $y(1) = 1.526$. On the other hand, the initial-value problem

$$\begin{aligned} &y'' - y = 2x, \qquad 0 < x < 1, \\ &y(0) = 0, \qquad y'(0) = 0.3, \end{aligned}$$

has a solution (Exercise 4) for which $y(1) = 0.703$. Since the problem is a linear one, we can use **linear interpolation** (Exercise 5) to arrive at the correct initial slope, namely, $y'(0) = 0.553$.

The foregoing method of solution is analogous to one that could be used in an artillery problem. Imagine a gun placed at the origin and a target at the point (1, 1). If the gun elevation (initial slope) is too large (1.000), the shell will pass over the target; if the gun elevation is too small (0.3), the shell will fall short of the target. Feedback of the appropriate information gives the artillery crew the necessary data to change the elevation (interpolate) so that the target can be hit. For this reason the method outlined here for solving a boundary-value problem in ordinary differential equations is called the **shooting method**.

We present another example of a linear boundary-value problem, which would be much more difficult to solve analytically. Accordingly, we present a numerical solution. The problem and its solution are from Curtis F. Gerald's *Applied Numerical Analysis*, 2d ed. (Reading, Mass.: Addison-Wesley, 1978), p. 304.

EXAMPLE 2.8–2 Solve the boundary-value problem

$$\frac{d^2x}{dt^2} - \left(1 - \frac{t}{5}\right)x = t, \qquad x(1) = 2, \qquad x(3) = -1. \qquad \textbf{(2.8–2)}$$

Solution Since the differential equation is of second order, the first step in obtaining a numerical solution is to convert the equation to a system of two first-order equations. Introducing a second variable, y we have

$$\frac{dx}{dt} = y, \qquad \frac{dy}{dt} = \left(1 - \frac{t}{5}\right)x + t. \tag{2.8-3}$$

By using primes to denote differentiation with respect to t the system (2.8–3) can be written in recursive form as

$$x_n' = y_n, \qquad y_n' = \left(1 - \frac{t_n}{5}\right)x_n + t_n. \tag{2.8-4}$$

With **Euler's method** the assumption is made that the slope remains essentially constant provided that the change in t is small so that we have

$$x_{n+1} = x_n + hy_n, \qquad y_{n+1} = y_n + h\left[\left(1 - \frac{t_n}{5}\right)x_n + t_n\right], \tag{2.8-5}$$

where h is the constant step size between successive values of t. In Eqs. (2.8–5), $n = 0$ represents the initial values, whereas $n = 1, 2, \ldots$ give successive values of the variables. The disadvantage of Euler's method is that small errors tend to be magnified as the computation progresses. Accordingly, we will use Eqs. (2.8–5) to *predict* each successive value and then *correct* these predictions by using an improved method known as **Heun's method.*** The latter uses the *average value* of the slopes at two adjacent points so that Eqs. (2.8–5) become

$$\tilde{x}_{n+1} = \tilde{x}_n + \frac{h}{2}[\tilde{y}_n + \tilde{y}_{n+1}],$$

$$\tilde{y}_{n+1} = \tilde{y}_n + \frac{h}{2}\left[\left(1 - \frac{t_n}{5}\right)\tilde{x}_n + t_n + \left(1 - \frac{t_{n+1}}{5}\right)\tilde{x}_{n+1} + t_{n+1}\right]. \tag{2.8-6}$$

A numerical method such as the one described above is called the **Euler predictor–corrector method**. We assign $\Delta t = h = 0.2$. Smaller values may give more accurate results, but only up to a certain point, after which round-off errors may deteriorate the accuracy. From the given initial condition $x(1) = 2$ we have $t_0 = 1$, $x_0 = 2$. We *assume* the missing initial slope to be $y_0 = -1.5$. With these values we use Eqs. (2.8–5) with $n = 0$ to obtain

$$x_1 = 2 + (0.2)(-1.5) = 1.700,$$

$$y_1 = -1.5 + 0.2\left[\left(1 - \frac{1}{5}\right)2 + 1\right] = -0.980,$$

*Also called the **modified Euler method**.

Table 2.8–1

Time ($\Delta t = 0.2$)	*Let x′(1) = −1.5*		*Let x′(1) = −3.0*		*Let x′(1) = −3.500*	
	Distance	*Velocity*	*Distance*	*Velocity*	*Distance*	*Velocity*
1.00	2.000	−1.500	2.000	−3.000	2.000	−3.500
1.20	1.751	−0.987	1.449	−2.510	1.348	−3.018
1.40	1.605	−0.478	0.991	−2.068	0.787	−2.599
1.60	1.561	0.043	0.619	−1.665	0.305	−2.221
1.80	1.625	0.594	0.328	−1.252	−0.104	−1.867
2.00	1.803	1.186	0.118	−0.844	−0.443	−1.521
2.20	2.105	1.832	−0.007	−0.417	−0.712	−1.167
2.40	2.542	2.542	−0.045	0.040	−0.908	−0.794
2.60	3.128	3.324	0.013	0.539	−1.026	−0.391
2.80	3.880	4.185	0.175	1.087	−1.060	0.054
3.00	4.811	5.128	0.453	1.693	−1.000	0.547

From Curtis F. Gerald, *Applied Numerical Analysis* (Reading, Mass.: Addison-Wesley, 1978), by permission of the author and publisher.

which is the first pair of predicted values. We then use Eqs. (2.8–6) to obtain

$$\tilde{x}_1 = 2 + \frac{0.2}{2}\,[-1.5 - 0.980] = 1.752,$$

$$\tilde{y}_1 = -1.5 + \frac{0.2}{2}\left[\left(1 - \frac{1}{5}\right)2 + 1 + \left(1 - \frac{1.2}{5}\right)1.752 + 1.2\right]$$
$$= -0.987,$$

the first pair of corrected values. These* and succeeding values are shown in Table 2.8–1. Since the target value is 4.811 and not the desired −1.000, the initial value of the velocity ($y = dx/dt$) is adjusted to −3.0, and the computations are repeated. As is shown in Table 2.8–1, the target distance is now 0.453, which is still incorrect. A graph of these values, shown in Fig. 2.8–2, indicates that linear *extrapolation* can be used to correct the initial velocity to −3.500. The corresponding values of distance and velocity are shown in the last two columns of Table 2.8–1. Graphs of the numerical values of x in Table 2.8–1 are shown in Fig. 2.8–3. ■

We should call attention to a variation of the method illustrated in Example 2.8–2. The point $x(3) = -1$ could have been taken as the initial point with values of h being negative. In this case the target value would have been $x(1) = 2$, and we would assume a missing initial condition, that is, a value of $x'(3)$. This reversal may be advantageous in solving higher-order boundary-value problems.

*Note that slight differences in the third decimal may occur because of the particular computer program being used.

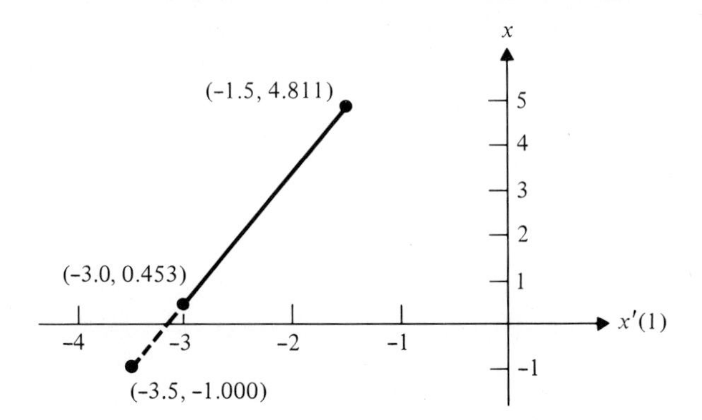

Figure 2.8-2
Linear extrapolation (Example 2.8-2).

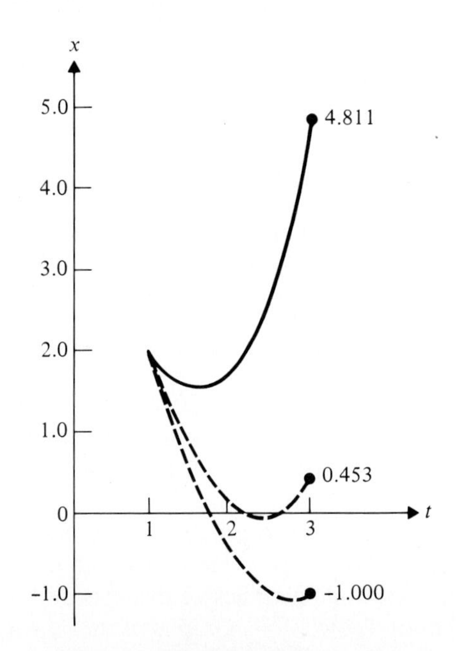

Figure 2.8-3
Numerical values of x (Table 2.8-1).

In the next example we present another method of solving a linear boundary-value problem.

EXAMPLE 2.8-3 The deflections of a beam hinged at both ends, carrying a constant distributed transverse load and equal constant axial compressive loads at the ends, satisfy the two-point boundary-value problem*

$$y'' + (1 + x^2)y = -1,$$
$$y(-1) = y(1) = 0.$$

Solve this problem numerically using a **finite-difference approximation** to y''.

Solution We observe that the derivative

$$y' = \frac{dy}{dx} = \lim_{h \to 0} \frac{y(x + h) - y(x)}{h} = \lim_{h \to 0} \frac{y(x) - y(x - h)}{h}$$

can be approximated at a given point x by the difference quotient

$$\frac{y(x + h) - y(x)}{h} \quad \text{or} \quad \frac{y(x) - y(x - h)}{h},$$

provided that h is sufficiently small. Similarly,

$$\begin{aligned} y'' &= \lim_{h \to 0} \frac{1}{h}[y'(x + h) - y'(x)] \\ &= \lim_{h \to 0} \frac{1}{h}\left[\frac{y(x + h) - y(x)}{h} - \frac{y(x) - y(x - h)}{h}\right] \end{aligned}$$

can be approximated as

$$y'' \doteq \frac{1}{h^2}[y(x + h) - 2y(x) + y(x - h)]$$

or, recursively, as

$$y_i'' = \frac{1}{h^2}[y_{i+1} - 2y_i + y_{i-1}], \tag{2.8-7}$$

where y_i is the value of y at x_i and

$$h = x_{i+1} - x_i,$$

a constant.

By using the finite-difference approximation in Eq. (2.8-7) the differential equation to be solved becomes

$$\frac{1}{h^2}[y_{i+1} - 2y_i + y_{i-1}] + (1 + x_i^2)y_i = -1$$

or

$$y_{i+1} = 2y_i - y_{i-1} - h^2(1 + x_i^2)y_i - h^2. \tag{2.8-8}$$

Since the transverse distributed load is constant and since the axial loads are equal, we expect that the deflections will be **symmetrical** about $x = 0$, that is,

---

*See L. Collatz, *The Numerical Treatment of Differential Equations*, 3d ed., (Berlin: Springer-Verlag, 1960), p. 143.

$y_i = y_{-i}$. Using $h = 0.1$ and assuming $y_0 = 1$, we have for $i = 0$ in Eq. (2.8–8)

$$y_1 = 1 - \frac{0.01}{2}(1 + 0)1 - \frac{0.01}{2} = 0.990,$$

using the fact that $y_1 = y_{-1}$. Putting $i = 1$ into Eq. (2.8–8) produces

$$y_2 = 2(0.990) - 1 - 0.01(1 + 0.01)(0.990) - 0.01 \\ = 0.960.$$

Continuing in this way (Exercise 7), we find that $y_{10} = 0.032$ and not zero as required. Repeating the computations with $y_0 = 0$ (Exercise 8) produces $y_{10} = -0.445$. Linear interpolation now yields the correct result (Exercise 9). ■

There is a shooting method using the adjoint system (see Section 2.6) that finds the missing initial conditions in one pass through the process. For details, see S. M. Roberts and J. S. Shipman, *Two-Point Boundary Value Problems: Shooting Methods* (New York: American Elsevier, 1972), Chapter 3.

Key Words and Phrases

initial-value problem
boundary-value problem
missing initial condition
target condition
linear interpolation
shooting method
Euler's method
Heun's method
modified Euler method
Euler predictor–corrector method
finite-difference approximation

2.8 Exercises

• **1.** Find the complementary solution and a particular integral of the second-order linear nonhomogeneous equation

$$y'' - y = 2x.$$

2. **(a)** Solve the initial-value problem

$$y'' - y = 2x, \quad 0 < x < 1, \\ y(0) = 0, \quad y'(0) = 0.553.$$

(b) Show that $y(1) = 1.000$ using the solution in part (a).

3. **(a)** Solve the initial-value problem

$$y'' - y = 2x, \quad 0 < x < 1, \\ y(0) = 0, \quad y'(0) = 1.$$

(b) Show that $y(1) = 1.526$ using the solution in part (a).

4. **(a)** Solve the initial-value problem

$$y'' - y = 2x, \qquad 0 < x < 1,$$
$$y(0) = 0, \qquad y'(0) = 0.3.$$

(b) Show that $y(1) = 0.703$ using the solution in part (a).

5. Use linear interpolation of the results and conditions in Exercises 3 and 4 to verify that the initial-value problem

$$y'' - y = 2x, \qquad 0 < x < 1,$$
$$y(0) = 0, \qquad y'(0) = 0.553,$$

produces a result for which $y(1) = 1.000$.

6. Continue Example 2.8–2 of the text by computing x_2, y_2, $\tilde{x}_2$, and $\tilde{y}_2$.

7. Continue the computations in Example 2.8–3, and verify that $y_{10} = 0.032$.

8. Repeat the computations in Example 2.8–3 starting with $y_0 = 0$, and verify that $y_{10} = -0.445$.

9. Interpolate linearly between the results and conditions of Exercises 8 and 9 to show that $y_0 = 0.934$ is the correct initial value.

•• 10. Graph the solution of Exercise 3(a), and compare it with Fig. 2.8–1.

11. Solve Example 2.8–3 by using $h = 0.2$.

12. Solve Example 2.8–2 by using the value 0.4 for both Δt and h.

13. Solve the following two-point boundary value problem
(a) analytically
(b) using the shooting method,
(c) using a finite-difference method with $h = 0.25$:

$$y'' - y = 0, \qquad 1 < x < 2,$$
$$y(1) = 1.5431, \qquad y(2) = 3.7622.$$

••• 14. Write the equation of the line shown in Fig. 2.8–2, and explain how the equation can be used to interpolate linearly (or extrapolate) in the problem of Example 2.8–2.

15. Given the *nonlinear* two-point boundary-value problem

$$y'' = 1.5y^2, \qquad y(0) = 4, \qquad y(1) = 1.$$

(a) Using $h = 0.2$, estimate $y'(0)$, and solve the problem using the Euler predictor–corrector method.
(b) Correct the estimate of $y'(0)$ until the target condition is satisfied.
(c) Comment on the interpolation procedure necessary in part (b).

16. With $h = 0.2$, solve the problem of Exercise 15 by using the finite-difference approximation to y''.

17. Consider the linear boundary-value problem in $y(x)$:

$$y'' = y, \qquad y(1) = 1.175, \qquad y(3) = 10.018.$$

(a) By making the change of variable $x = 2t + 1$, show that the interval can be changed from (1, 3) to (0, 1).

(b) Write the differential equation and boundary conditions for the problem in $y(t)$.
(c) Solve the problem in part (b) using $h = 0.2$ and a numerical method of your choice.
(d) Solve the problem analytically, and compare the results with those using the numerical solution in part (c).

18. Consider the following two-point boundary-value problem:

$$\frac{d^2y}{dx^2} - y = x, \qquad a \le x \le b;$$

$$y(a) = c_1, \qquad y(b) = c_2.$$

(a) Divide the interval $[a, b]$ into $(n + 1)$ equally spaced intervals of length h. Show that

$$h = \frac{b - a}{n + 1}.$$

(b) Show that the discrete points of subdivision are given by

$$x_i = x_0 + i\frac{b - a}{n + 1}, \qquad i = 0, 1, \ldots, n + 1.$$

(c) Show that the second derivative at the point x_i may be approximated by

$$\frac{d^2y}{dx^2} \doteq \frac{y_{i+1} - 2y_i + y_{i-1}}{h^2},$$

where $y_i = y(x_i)$.
(d) Show that the finite-difference approximation to the given problem can be written as

$$-y_{i-1} + (2 + h^2)y_i - y_{i+1} = -h^2x_i, \qquad i = 1, 2, \ldots, n.$$

(e) Write the n equations in part (d) in the n unknowns (y_0 and y_{n+1} are given) in matrix form.
(f) Show that the matrix of coefficients in part (e) is a tridiagonal matrix.*

REFERENCES

Coddington, E. A., and N. Levinson, *Theory of Ordinary Differential Equations*. New York: McGraw-Hill, 1955.

Collatz, L., *The Numerical Treatment of Differential Equations*, 3d ed. Berlin: Springer-Verlag, 1960.

Gerald, C. F., *Applied Numerical Analysis*, 2d ed. Reading, Mass.: Addison-Wesley, 1978.

*A tridiagonal matrix has nonzero elements only on the diagonal and in the two subdiagonals adjacent to the diagonal.

Lanczos, C., *Linear Differential Operators*. London: Van Nostrand, 1961.

Roberts, S. M., and J. S. Shipman, *Two-Point Boundary Value Problems: Shooting Methods*. New York: American Elsevier, 1972.

Sagan, H., *Boundary and Eigenvalue Problems in Mathematical Physics*. New York: John Wiley, 1961.

3 Partial Differential Equations

3.1 INTRODUCTION

Ordinary differential equations play a very important role in applied mathematics, since a wide variety of problems from engineering, physics, and other areas can be formulated in terms of ordinary differential equations. But just as it is not always possible to simplify a problem by neglecting friction, air resistance, Coriolis force, etc., we cannot always neglect the presence of other independent variables. Often, time t is a variable and must necessarily be considered in addition to one or more spatial variables.

Whenever more than a *single* independent variable has to be taken into account, it may be possible to formulate a problem in terms of a *partial differential equation.* The remainder of this book is concerned with obtaining solutions to such equations. Much of the terminology used in connection with ordinary differential equations is extended in a natural way to partial differential equations. For example, the *order* of an equation is the same as the highest ordered derivative that appears. The emphasis in this text is on linear partial differential equations of *second* order.

In studying partial differential equations it will be convenient to use a subscript notation. For example, we will write

$$\frac{\partial^2 u}{\partial x^2} = u_{xx}, \qquad \frac{\partial^2 u}{\partial y^2} = u_{yy}, \qquad \frac{\partial^2 u}{\partial x \partial y} = u_{yx} = \frac{\partial}{\partial x}\left(\frac{\partial u}{\partial y}\right), \quad \text{etc.}$$

Since we will be concerned with *linear second-order* partial differential equations, we consider the most general equation of this type,

$$Au_{xx} + Bu_{xy} + Cu_{yy} + Du_x + Eu_y + Fu = G, \qquad \textbf{(3.1-1)}$$

where the coefficient functions $A, B, \ldots, G$ depend on x and y (but *not* on u

or its derivatives). If $G \equiv 0$, the equation is *homogeneous*; otherwise, it is *nonhomogeneous*. It is often convenient to write Eq. (3.1–1) as $Lu = G$, where L indicates the linear differential operator.

By a *solution* of Eq. (3.1–1) we mean a function $u(x, y)$ that satisfies the equation identically. For example, the function

$$u(x, y) = \exp(3x + 4y)$$

satisfies the equation

$$16u_{xx} - 9u_{yy} = 0$$

identically. The **general solution** of Eq. (3.1–1) is the set of all its solutions. A member of this set can be called a **specific solution**. We will be interested mainly in obtaining specific solutions to Eq. (3.1–1), that is, solutions that not only satisfy the equation but satisfy certain other conditions as well.

Linear equations of the form (3.1–1) are categorized as to type in an interesting way. Recall from analytic geometry that the most general *second-degree* equation in two variables is

$$Ax^2 + Bxy + Cy^2 + Dx + Ey + F = 0,$$

where $A, B, \ldots, F$ are constants. Equations of this type represent conic sections (possibly degenerate ones) as follows:

- an ellipse if $B^2 - 4AC < 0$;
- a parabola if $B^2 - 4AC = 0$;
- a hyperbola if $B^2 - 4AC > 0$.

Similarly, *linear* partial differential equations (3.1–1) are called **elliptic, parabolic**, or **hyperbolic** according to whether $B^2 - 4AC$ is negative, zero, or positive, respectively.* Since A, B, and C are functions of x and y in Eq. (3.1–1), it is quite possible for a partial differential equation to be of mixed type. For example,

$$xu_{xx} + yu_{yy} + 2yu_x - xu_y = 0$$

is elliptic if $xy > 0$, hyperbolic if $xy < 0$, and parabolic if $xy = 0$. If x and y are *spatial* coordinates, then the equation is elliptic in the first and third quadrants, hyperbolic in the second and fourth quadrants, and parabolic on the coordinate axes. The theory of equations of **mixed type** was originated by Tricomi† in 1923. In the study of transonic flow the so-called **Tricomi equation**

$$y\psi_{xx} + \psi_{yy} = 0,$$

occurs. This equation is elliptic for $y > 0$ and hyperbolic for $y < 0$. In the study of aerodynamics, where the Tricomi equation occurs, the elliptic region

*See Section 3.5 for a detailed discussion of these types.

†Francesco G. Tricomi (1897–), an Italian mathematician.

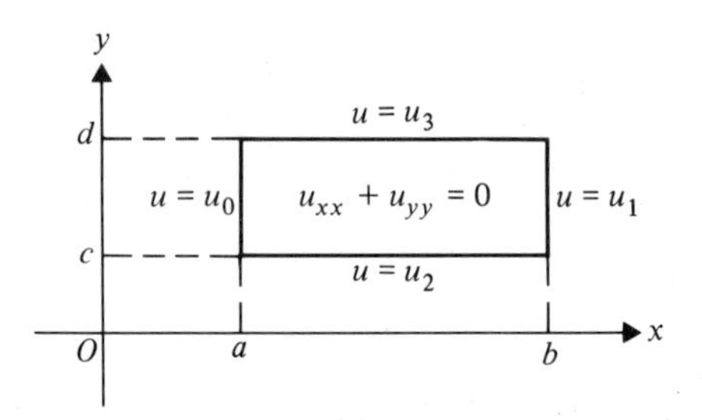

Figure 3.1-1
Laplace's equation (Example 3.1-1).

corresponds to smooth subsonic flow, the parabolic region to a sonic barrier, and the hyperbolic region to supersonic propagation of shock waves.

The classification of a second-order partial differential equation as to type is intimately connected with the type of boundary conditions that must be specified in order to obtain unique, stable solutions.

Although x and y are generally used to designate spatial coordinates, this need not always be the case. One of these could be a time coordinate, for example. The only restriction, in fact, is that x and y be independent variables. If we have spatial independent variables, then by specifying the values of the dependent variable on some boundary—called **boundary conditions**—we can obtain a specific solution. If, on the other hand, one of the independent variables is time, then we must also specify **initial conditions** in order to obtain a specific solution. Problems in which boundary conditions or both boundary and initial conditions are specified are called **boundary-value problems.**

Next we give examples to illustrate the three types of second-order equations. Specific solutions to these will be obtained in ensuing chapters.

EXAMPLE 3.1-1 Elliptic boundary-value problem.

$$\begin{array}{llll} \text{P.D.E.:} & u_{xx} + u_{yy} = 0, & a < x < b, & c < y < d; \\ \text{B.C.:} & u(a, y) = u_0, & u(b, y) = u_1, & c < y < d, \\ & u(x, c) = u_2, & u(x, d) = u_3, & a < x < b, \end{array}$$

where a, b, c, d, u_0, u_1, u_2, u_3 are constants. This elliptic equation is called **Laplace's equation,*** and we shall study it in greater detail in Sections 3.2 and 6.1. The boundary conditions (B.C.) show that $a \le x \le b$ and $c \le y \le d$; hence the values of the unknown function $u(x, y)$ are known on a rectangle in the xy-plane (Fig. 3.1-1). The partial differential equation (P.D.E.) shown indicates that $u(x, y)$ is to satisfy the equation everywhere within the *open* rectangular region. ■

*After Pierre S. de Laplace (1749–1827), a French astronomer and mathematician.

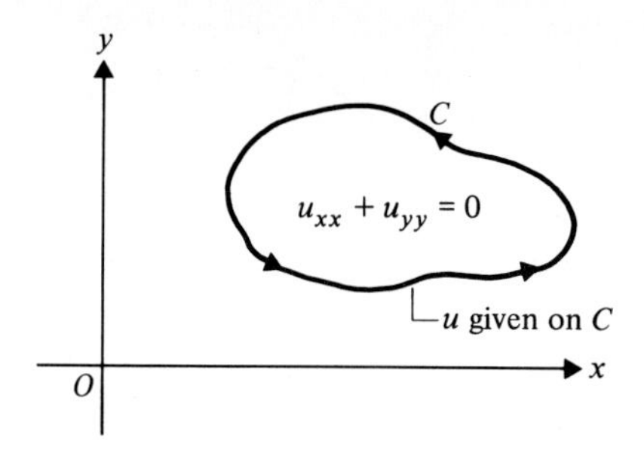

Figure 3.1-2
Dirichlet problem in two dimensions.

We remark that the term "region" may have various meanings in the literature. By a region in the xy-plane we mean a set R of points (x, y) such that each pair (x_1, y_1) and (x_2, y_2) of distinct points of R can be joined by a broken-line path (a polygonal line) all of whose points lie within R. Thus a region is a *connected* set. A region may contain all, none, or part of its boundary. If the region contains *none* of its boundary points, then we call it an open region; if it contains *all* of its boundary points, we call it a *closed region*.

Boundary-value problems of the type in which u satisfies Laplace's equation in an open region and takes on prescribed values on the boundary of the region (Fig. 3.1-2) are called **Dirichlet problems.*** If the above boundary conditions are replaced by

$$\begin{aligned} u_x(a, y) &= u_0, \qquad u_x(b, y) = u_1, \qquad c < y < d, \\ u_y(x, c) &= u_2, \qquad u_y(x, d) = u_3, \qquad a < x < b, \end{aligned}$$

then the normal derivatives u_x and u_y are prescribed, and the problem is called a **Neumann problem.**† In this case the (outward-pointing) **normal derivative,** $\partial u/\partial n$, that is, the directional derivative of u in a direction normal to the boundary, is specified (Fig. 3.1-3). Boundary-value problems may, of course, be of *mixed* type, as we shall see in Section 6.3. Other notations for $u_x(a, y) = u_0$ are in use. Some of these are the following:

$$\left.\frac{\partial u(x, y)}{\partial x}\right|_{x=a} = u_0,$$

$$\lim_{x \to a^+} u_x(x, y) = u_0,$$

$$u_x(a^+, y) = u_0.$$

*After Peter G. L. Dirichlet (1805–1859), a German mathematician.
†After Carl G. Neumann (1832–1925), a German mathematician.

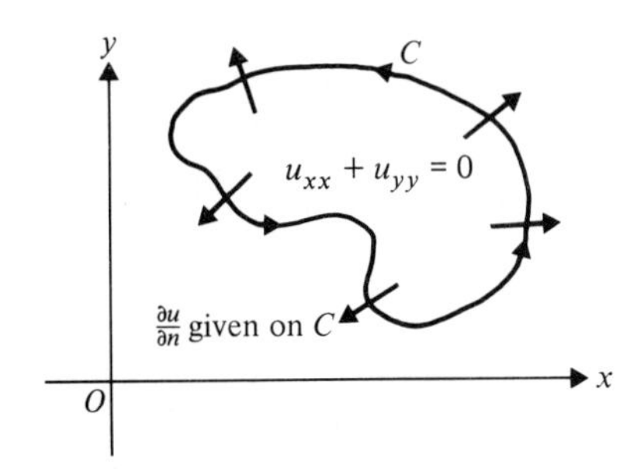

Figure 3.1-3
Neumann problem in two dimensions.

EXAMPLE 3.1-2 Hyperbolic boundary-value problem.

$$
\begin{aligned}
&\text{P.D.E.:} \quad && u_{tt} = a^2 u_{xx}, \quad t > 0, \quad 0 < x < L; \\
&\text{B.C.:} \quad && u(0, t) = u(L, t) = 0, \quad t > 0; \\
&\text{I.C.:} \quad && u(x, 0) = f(x), \quad u_t(x, 0) = g(x), \quad 0 < x < L.
\end{aligned}
$$

Here a is a constant, u is a displacement, and we have both boundary conditions and initial conditions (I.C.) prescribed. This time, u is to satisfy the given equation for positive values of t and for all x in the open interval $0 < x < L$ (Fig. 3.1-4). The boundary conditions state that for positive t, $u = 0$ when $x = 0$ and when $x = L$, whereas the initial conditions give the values of the initial displacement and initial velocity as $f(x)$ and $g(x)$, respectively. The differential equation must be dimensionally correct; hence the constant a has the dimension of velocity (Exercise 1). ■

The partial differential equation in Example 3.1-2 is the **wave equation** in one dimension. We will derive the equation and obtain its general solution in Section 3.3. In Section 5.2 we will consider the infinite $(-\infty < x < \infty)$ case using the Fourier transform.

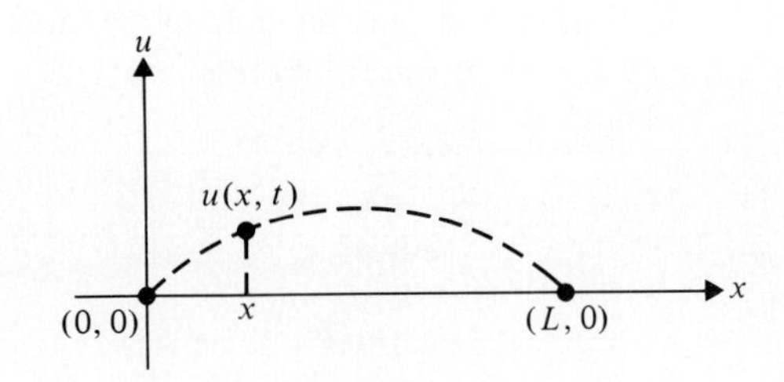

Figure 3.1-4
One-dimensional wave equation (Example 3.1-2).

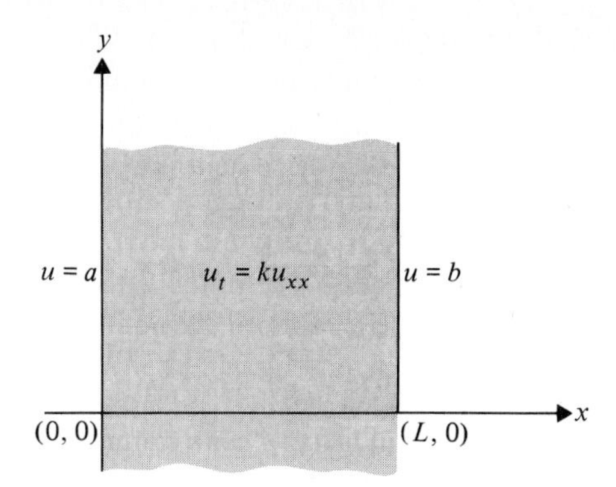

Figure 3.1-5
One-dimensional diffusion equation (Example 3.1-3).

EXAMPLE 3.1-3 Parabolic boundary-value problem.

$$\begin{aligned}
&\text{P.D.E.:} && u_t = ku_{xx}, \quad t > 0, \quad 0 < x < L, \quad k > 0;\\
&\text{B.C.:} && u(0, t) = a, \quad u(L, t) = b, \quad t > 0;\\
&\text{I.C.:} && u(x, 0) = f(x), \quad 0 < x < L.
\end{aligned}$$

Again we have $u(x, t)$ satisfying a partial differential equation on an open interval for all positive t. The boundary conditions prescribe the value of u at the endpoints of the given interval (Fig. 3.1-5), whereas the initial condition specifies the value of u at time $t = 0$. In this example the partial differential equation is the one-dimensional* **diffusion equation**. We will derive it in Section 3.4, solve it in Section 6.3 over finite regions, and solve it in Section 6.4 over semi-infinite regions. ■

The existence of arbitrary *functions*† in the general solution of a linear partial differential equation means that the totality of functions that satisfy such an equation is very large. For example, the functions $u(x, y)$ below

$$\arctan \frac{y}{x}, \qquad e^x \sin y, \qquad \sqrt{x^2 + y^2}, \qquad \sin x \sinh y$$

are quite varied, yet each satisfies Laplace's equation $u_{xx} + u_{yy} = 0$ (Exercise 2). In applications involving partial differential equations, however, we have certain information about a physical system that allows us to find specific solutions. Most of our study of boundary-value problems will be concentrated on this point.

---

*The condition $-\infty < y < \infty$ in Fig. 3.1-5 implies that u varies only in the x-direction.
†In contrast to arbitrary *constants* in the case of ordinary differential equations.

As in the case of ordinary differential equations, the most simple second-order partial differential equations to solve are the homogeneous ones with *constant* coefficients. Consider such an equation in which the first derivatives are absent, namely,

$$au_{xx} + bu_{xy} + cu_{yy} = 0, \tag{3.1-2}$$

and let* $u = f(y + rx)$ where r is a constant. Then $u_x = rf'(y + rx)$ and $u_{xx} = r^2 f''(y + rx)$ where the prime denotes the derivative of the function f with respect to its argument, $y + rx$. Substitution into Eq. (3.1-2) produces

$$(ar^2 + br + c)f''(y + rx) = 0$$

and the **characteristic equation**

$$ar^2 + br + c = 0. \tag{3.1-3}$$

If the roots of Eq. (3.1-3) are real and distinct, say, r_1 and r_2, then the general solution of Eq. (3.1-2) can be written as

$$u(x, y) = f(y + r_1 x) + g(y + r_2 x)$$

where f and g are twice-differentiable but otherwise arbitrary functions. (See Exercises 3 and 4.)

EXAMPLE 3.1-4 Solve the hyperbolic equation

$$a^2 u_{xx} - b^2 u_{yy} = 0$$

where a and b are real constants.

Solution The characteristic equation is

$$a^2 r^2 - b^2 = 0,$$

and the general solution is

$$u(x, y) = f\left(y + \frac{b}{a}x\right) + g\left(y - \frac{b}{a}x\right). \quad \blacksquare \tag{3.1-4}$$

Modifications of the above method can be found in the exercises. It should be mentioned that if the characteristic equation has two equal roots, then the general solution has the form

$$u(x, y) = f(y + rx) + xg(y + rx). \tag{3.1-5}$$

Exercises 18–21 deal with the case where the characteristic equation has complex roots.

---

*This is actually an "educated guess" prompted by the form of Eq. (3.1-2) and the analogous substitution used in Section 1.1.

Key Words and Phrases

general solution
specific solution
elliptic partial differential equation
parabolic partial differential equation
hyperbolic partial differential equation
mixed type partial differential equation
Tricomi equation
boundary condition
initial condition
boundary-value problem
Laplace's equation
Dirichlet problem
Neumann problem
normal derivative
wave equation
diffusion equation
characteristic equation

3.1 Exercises

• **1.** Show that the constant a in the wave equation has the dimension of velocity.

2. Verify that each of the following functions satisfies Laplace's equation.

(a) $\arctan \dfrac{y}{x}$ **(b)** $e^x \sin y$

(c) $\sqrt{x^2 + y^2}$ **(d)** $\sin x \sinh y$

3. Verify that

$$u(x, y) = f(y + r_1x) + g(y + r_2x),$$

where r_1 and r_2 satisfy

$$ar^2 + br + c = 0,$$

is the *general* solution of Eq. (3.1–2).

4. Referring to Exercise 3, verify that

$$u(x, y) = c_1f(y + r_1x) + c_2g(y + r_2x)$$

also satisfies Eq. (3.1–2) for arbitrary constants c_1 and c_2. How does this solution compare with the general solution?

•• **5.** Classify each of the following partial differential equations as elliptic, hyperbolic, or parabolic. Consider the appropriate values of the independent variables in each case.

(a) $u_{xx} + 4u_{xy} + 3u_{yy} + 4u_x - 3u = xy$
(b) $xu_{xx} + u_{yy} - 2x^2u_y = 0$
(c) $u_{xy} - u_x = x \sin y$
(d) $(y^2 - 1)u_{xx} - 2xyu_{xy} + (x^2 - 1)u_{yy} + e^xu_x + u_y = 0$

6. If f and g are twice-differentiable but otherwise *arbitrary* functions, verify that $f(x + at)$, $g(x - at)$, and $f(x + at) + g(x - at)$ are solutions of $u_{tt} = a^2u_{xx}$. (*Hint:* By the chain rule,

$$g_t(x - at) = \frac{dg(x - at)}{d(x - at)} \frac{\partial(x - at)}{\partial t} = -ag'(x - at).)$$

7. Verify that

$$u(x, t) = (c_1 \cos \lambda x + c_2 \sin \lambda x)(c_3 \sin \lambda at + c_4 \cos \lambda at)$$

is a solution of the wave equation $u_{tt} = a^2 u_{xx}$ where c_1, c_2, c_3, c_4, and λ are constants.

8. Show that the function $u(x, t)$ in Exercise 7 becomes $c_5 \cos \lambda at \sin \lambda x$ when the boundary and initial conditions

$$u(0, t) = 0 \quad \text{and} \quad u_t(x, 0) = 0$$

are imposed.

9. Verify that $u = e^{-k\lambda^2 t}(c_1 \cos \lambda x + c_2 \sin \lambda x)$ is a solution of the diffusion equation $u_t = k u_{xx}$ where c_1, c_2, and λ are constants.

10. Verify that

$$u(x, t) = \frac{1}{2a} \int_{x-at}^{x+at} g(s)\, ds$$

is a solution of the wave equation $u_{tt} = a^2 u_{xx}$ satisfying the conditions $u(x, 0) = 0$, $u_t(x, 0) = g(x)$. Use the *Leibniz rule* for differentiating an integral: If

$$u(x, t) = \int_{a(x,t)}^{b(x,t)} f(x, s, t)\, ds,$$

then

$$u_x(x, t) = \int_{a(x,t)}^{b(x,t)} \frac{\partial f(x, s, t)}{\partial x}\, ds + f(x, b, t) \frac{\partial b}{\partial x} - f(x, a, t) \frac{\partial a}{\partial x},$$

with a similar formula for $u_t(x, t)$. (*Note*: The formula for the nth derivative of the product of two functions is also called Leibniz's rule.)

11. For each of the following partial differential equations, (i) give the order, and (ii) state whether or not the equation is linear; if it is not linear, explain why.

(a) $x u_x + y u_y = u$
(b) $u(u_{xx}) + (u_y)^2 = 0$
(c) $u_{xx} - u_{xy} - 2u_{yy} = 1$
(d) $u_{xx} - 2u_y = 2x - e^u$
(e) $(u_x)^2 - x(u_{xy}) = \sin y$

12. **(a)** Show that if $a = 0$ in Eq. (3.1-2), the method given in the text fails. Show, however, that in this case the substitution $u = f(x + ry)$ will produce the general solution.
(b) Obtain the general solution to $u_{xy} - 3u_{yy} = 0$.
(c) Use the substitution $u = f(x + ry)$ to solve

$$u_{xx} + u_{xy} - 6u_{yy} = 0.$$

13. Verify that the general solution of Eq. (3.1-2) is given by Eq. (3.1-5) when the characteristic equation has equal roots.

14. It can be shown* that if $u = F(x, y)$, then the mixed second partial derivatives u_{xy}

*See Robert C. James, *Advanced Calculus* (Belmont, Calif.: Wadsworth, 1966), p. 298.

and u_{yx} are equal whenever they exist and are continuous. Show that the mixed partial derivatives are equal for each of the following functions.

(a) $u = e^x \cos y$

(b) $u = \arctan \dfrac{y}{x}$

(c) $u = e^{xy} \tan xy$

(d) $u = \sqrt{(x + y)/(x - y)}$

15. Obtain a solution of

$$u_{xx} + u_{xy} - 6u_{yy} = 0.$$

16. Obtain the general solution of each of the following equations.

(a) $u_{xx} - 9u_{yy} = 0$
(b) $u_{xx} + 4u_{yy} = 0$
(c) $6u_{xx} + u_{xy} - 2u_{yy} = 0$

17. Classify each of the following equations as elliptic, hyperbolic, or parabolic.

(a) $x^2u_{xx} + 2xyu_{xy} + y^2u_{yy} = 4x^2$
(b) $u_{xx} - (2 \sin x)u_{xy} - (\cos^2 x)u_{yy} - (\cos x)u_y = 0$
(c) $x^2u_{xx} - y^2u_{yy} = xy$
(d) $4u_{xx} - 8u_{xy} + 4u_{yy} = 3$

●●● **18.** Verify that

$$u = f_1(x + iy) + f_2(x - iy)$$

is a solution of $u_{xx} + u_{yy} = 0$.

19. Generalize the result of Exercise 18 to the case where the characteristic equation (3.1-3) has complex roots.

20. Show that

$$u = f_1(y - ix) + xf_2(y - ix) + f_3(y + ix) + xf_4(y + ix)$$

is a solution of

$$u_{xxxx} + 2u_{yyxx} + u_{yyyy} = 0.$$

21. The method given in the text can be extended to some homogeneous linear partial differential equations of order four. Obtain the general solution of each of the following equations.

(a) $\dfrac{\partial^4 u}{\partial x^4} + 2\dfrac{\partial^4 u}{\partial x^2 \partial y^2} + \dfrac{\partial^4 u}{\partial y^4} = 0$*

(b) $\dfrac{\partial^4 u}{\partial x^4} - \dfrac{\partial^4 u}{\partial y^4} = 0$

(c) $\dfrac{\partial^4 u}{\partial x^4} - 2\dfrac{\partial^4 u}{\partial x^2 \partial y^2} + \dfrac{\partial^4 u}{\partial y^4} = 0$

22. Show that if ψ_1 and ψ_2 are any two harmonic functions of x and y, then any func-

---

*This equation, called the *biharmonic equation,* occurs in the study of elasticity and hydrodynamics.

tion ϕ of the form

$$\phi(x, y) = x\psi_1(x, y) + \psi_2(x, y)$$

satisfies the biharmonic equation. (*Note*: A harmonic function is one that satisfies Laplace's equation; the biharmonic equation is given in Exercise 21(a).)

23. Consider Eq. (3.1–1) with $G = 0$. Show that if $u_1(x, y)$ and $u_2(x, y)$ are solutions of this equation, then

$$c_1u_1(x, y) + c_2u_2(x, y)$$

is also a solution for constants c_1 and c_2.

3.2 SEPARATION OF VARIABLES

Laplace's equation,

$$u_{xx} + u_{yy} + u_{zz} = 0, \tag{3.2–1}$$

is one of the classical partial differential equations of mathematical physics. We present it above in the three-dimensional case, and we will examine the two-dimensional case later in this section. The importance of this equation is due to its occurrence in so many branches of science.

In the study of *electrostatics* it is shown that the electric intensity vector **E** due to a collection of stationary charges is given by

$$\mathbf{E} = -\nabla\phi = -(\phi_x\mathbf{i} + \phi_y\mathbf{j} + \phi_z\mathbf{k}),$$

where ϕ is a scalar point function called the *electric potential.* In the above, $\nabla\phi$ is the *gradient* of ϕ, and **i**, **j**, and **k** are unit vectors along the x-, y-, and z-axes, respectively. Further, Gauss' Law states that

$$\nabla\cdot\mathbf{E} = \nabla\cdot(-\nabla\phi) = -(\phi_{xx} + \phi_{yy} + \phi_{zz}) = 4\pi\rho(x, y, z),$$

where $\rho(x, y, z)$ is the charge density. Thus the potential ϕ satisfies the equation

$$\phi_{xx} + \phi_{yy} + \phi_{zz} = -4\pi\rho(x, y, z), \tag{3.2–2}$$

which is known as **Poisson's equation.*** In a region that is free of charges, $\rho(x, y, z) = 0$, and Eq. (3.2–2) reduces to Laplace's equation

$$\phi_{xx} + \phi_{yy} + \phi_{zz} = 0. \tag{3.2–3}$$

In this case it is assumed that the electric potential is due to charges located outside of or on the boundary of the charge-free region. It is significant that *every* potential function that is derived from any electrostatic distribution whatever must satisfy Eq. (3.2–3) in free space.

In an analogous manner, in *magnetostatics* the *magnetic potential* due

*After Siméon D. Poisson (1781–1840), a French mathematical physicist.

to the presence of poles satisfies Eq. (3.2–3) in regions free of poles. Similarly, the *gravitational potential* due to the presence of matter satisfies Eq. (3.2–3) in regions devoid of matter. In *aerodynamics* and *hydrodynamics* the *velocity potential* ϕ has the property $\nabla\phi = \mathbf{v}$, where $\mathbf{v}$ is the velocity vector field. For an idealized fluid—one that is incompressible and irrotational—the velocity potential satisfies Eq. (3.2–3) in those portions of the fluid that contain no sources or sinks. Hence Laplace's equation plays an important role in *potential theory*. For this reason it is often called the **potential equation**, and functions satisfying it are called **potential functions** as well as **harmonic functions.**

Laplace's equation is also prominent in other branches of science. If a membrane of constant density is stretched uniformly over a supporting frame, if no external forces are applied to the membrane except at the frame, and if the frame is given a displacement normal to the plane of the membrane (the xy-plane), then the *static transverse displacement* of the membrane $z(x, y)$ satisfies the two-dimensional Laplace equation. In the derivation it is assumed that z and its derivatives are so small that higher powers of z, z_x, and z_y can be neglected. We shall see in Section 6.1 that the steady-state (independent of time) *temperature* in a substance having constant thermal conductivity and containing no heat sources or sinks also satisfies Laplace's equation.

It should be apparent from the foregoing that the occurrence of Laplace's equation in such diverse fields merits the attention we will pay to its solution. We begin with a simple boundary-value problem involving Laplace's equation in two variables.

EXAMPLE 3.2–1 Solve the following boundary-value problem.

$$\text{P.D.E.:}\quad u_{xx} + u_{yy} = 0, \qquad 0 < x < \pi, \qquad 0 < y < b;$$
$$\text{B.C.:}\quad u(0, y) = g(y), \qquad u(\pi, y) = h(y), \qquad 0 < y < b,$$
$$u(x, 0) = f(x), \qquad u(x, b) = \phi(x), \qquad 0 < x < \pi.$$

Discussion We need to find a function $u(x, y)$ that satisfies the P.D.E. in the open rectangular region $0 < x < \pi$, $0 < y < b$, and that takes on the prescribed values $f(x)$, $\phi(x)$, $g(y)$, $h(y)$ on the boundary of the region (the rectangle) as shown in Fig. 3.2–1. Each vertex of the rectangle belongs to two line

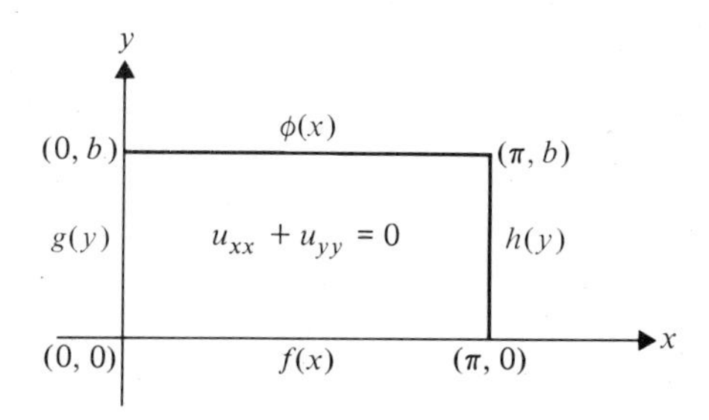

Figure 3.2–1
Boundary-value problem (Example 3.2–1).

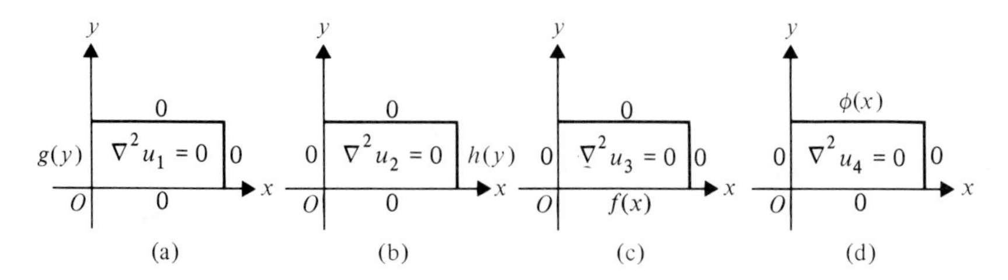

Figure 3.2-2
Applying the principle of superposition (Example 3.2-1).

segments; hence a question arises about continuity at the four vertices. We will leave this question unanswered until Chapter 4. For the present we will not specify boundary conditions at the vertices but only on the *open* intervals $0 < x < \pi$ and $0 < y < b$.

The problem as stated can be considerably simplified because we can apply the *principle of superposition* (see Exercise 22), since Laplace's equation is a linear homogeneous equation. Thus if u_1, u_2, u_3, and u_4 are functions of x and y that satisfy Laplace's equation and the following boundary conditions shown* in Fig. 3.2-2,

$$\begin{aligned}
&\text{(a)} \quad u_1(0, y) = g(y), \quad u_1(\pi, y) = u_1(x, 0) = u_1(x, b) = 0,\\
&\text{(b)} \quad u_2(\pi, y) = h(y), \quad u_2(0, y) = u_2(x, 0) = u_2(x, b) = 0,\\
&\text{(c)} \quad u_3(x, 0) = f(x), \quad u_3(0, y) = u_3(\pi, y) = u_3(x, b) = 0,\\
&\text{(d)} \quad u_4(x, b) = \phi(x), \quad u_4(0, y) = u_4(\pi, y) = u_4(x, 0) = 0,
\end{aligned}$$

then $u = u_1 + u_2 + u_3 + u_4$ will be the solution of the problem in Example 3.2-1. This, in turn, suggests that we need to be concerned only with solving *one* of these problems, since the other three are quite similar. Accordingly, we rephrase the problem of Example 3.2-1. ■

EXAMPLE 3.2-2 Solve the following boundary-value problem.

$$\begin{aligned}
\text{P.D.E.:} \quad & u_{xx} + u_{yy} = 0, \quad 0 < x < \pi, \quad 0 < y < b;\\
\text{B.C.:} \quad & u(0, y) = u(\pi, y) = 0, \quad 0 < y < b,\\
& u(x, b) = 0, \quad u(x, 0) = f(x), \quad 0 < x < \pi.
\end{aligned}$$

Solution Now we have three homogeneous boundary conditions, and u is prescribed on an open interval by $f(x)$, as shown in Fig. 3.2-2(c). We will discuss the nature of $f(x)$ and the values $f(0)$ and $f(\pi)$ in Chapter 4. The significance of the value $x = \pi$ will also become apparent there.

We do not know a priori which of the infinite variety of functions $u(x, y)$ satisfying Laplace's equation will also satisfy the given boundary con-

*The symbol ∇^2, which means $\nabla \cdot \nabla$, is often used for the Laplacian, that is, $\nabla^2 u = u_{xx} + u_{yy}$ in two dimensions and rectangular coordinates. Some writers use the symbol Δu for the Laplacian.

ditions. Because of the simple geometry in this example we will use a method of solution known as **separation of variables.*** We shall see later that this method is particularly useful when most of the boundary conditions are homogeneous. Another advantage is that the method leads to *ordinary, homogeneous* differential equations with constant coefficients, with which we are familiar. (Compare Section 1.1, from which we draw freely in the remainder of this chapter.)

Assume that $u(x, y)$ can be expressed as a *product* of two functions, one a function of x alone and the other a function of y alone. Then

$$u(x, y) = X(x)Y(y) \tag{3.2–4}$$

and

$$u_{xx} = X''Y, \qquad u_{yy} = XY'',$$

where the primes denote *ordinary* derivatives, the differentiation being performed with respect to the arguments of X and Y. Then, substituting into the P.D.E., we have

$$X''Y + XY'' = 0.$$

Next, we recognize that although $u(x, y) = 0$ satisfies the P.D.E., we wish to rule out this **trivial solution.** This, in turn, means that neither $X(x)$ nor $Y(y)$ can be identically zero, so we can divide the last equation by the product XY. Thus

$$\frac{X''}{X} = -\frac{Y''}{Y}, \tag{3.2–5}$$

and the variables have been *separated*, since the left-hand member of Eq. (3.2–5) is a function of x alone and the right-hand member is a function of y alone.

Varying x in Eq. (3.2–5) will change the left-hand member but not the right-hand member; hence the equality can hold, in general, only if both members are constant, that is,

$$\frac{X''}{X} = -\frac{Y''}{Y} = k. \tag{3.2–6}$$

In order to determine the nature of the constant k, we examine the following two-point boundary-value problem:

$$X'' - kX = 0, \qquad X(0) = X(\pi) = 0, \tag{3.2–7}$$

where the conditions $u(0, y) = 0$ and $u(\pi, y) = 0$ have been translated into $X(0) = 0$ and $X(\pi) = 0$, respectively, by using Eq. (3.2–4). (Note that this can be done only with homogeneous conditions.) We now distinguish the three

*Also called the *Fourier method.*

possible cases: $k = 0$, $k > 0$, and $k < 0$. The first two cases lead to the trivial solution (Exercise 2).

The third case can be recognized as a regular Sturm–Liouville problem, which was solved in Example 2.2–1, where we found the eigenvalues $-k = n^2$ and the corresponding eigenfunctions

$$X_n(x) = \sin nx, \qquad n = 1, 2, \ldots. \tag{3.2–8}$$

We can now obtain, for each n, the function $Y_n(y)$ corresponding to $X_n(x)$. From Eq. (3.2–6) we see that Y_n must be a solution to the problem

$$Y_n'' - n^2 Y_n = 0, \qquad Y_n(b) = 0, \qquad n = 1, 2, 3, \ldots. \tag{3.2–9}$$

We have translated the condition $u(x, b) = 0$ to $Y_n(b) = 0$, but the condition $u(x, 0) = f(x)$ cannot be changed to a condition on $Y_n(y)$, since $f(x)$ is not zero. The solutions of the differential equations in (3.2–9) are

$$Y_n(y) = a_n \sinh ny + b_n \cosh ny,$$

and the condition $Y_n(b) = 0$ implies that

$$a_n \sinh nb + b_n \cosh nb = 0.$$

Thus

$$a_n = -\frac{b_n \cosh nb}{\sinh nb} = -b_n \coth nb,$$

and the updated solution becomes

$$\begin{aligned} Y_n(y) &= b_n (\cosh ny - \coth nb \sinh ny) \\ &= \frac{b_n}{\sinh nb} \sinh n(b - y). \end{aligned} \tag{3.2–10}$$

Going back to Eq. (3.2–4), we have

$$u_n(x, y) = \frac{b_n}{\sinh nb} \sin nx \sinh n(b - y), \qquad n = 1, 2, 3, \ldots, \tag{3.2–11}$$

for arbitrary constants b_n. Each of these functions satisfies the given P.D.E. and also the three *homogeneous* boundary conditions (Exercise 3).

It remains to satisfy the *nonhomogeneous* boundary condition, $u(x, 0) = f(x)$. It is clear from the expression

$$u_n(x, 0) = b_n \sin nx = f(x) \tag{3.2–12}$$

that it will not be possible to satisfy this final condition with any *one* of our solutions $u_n(x, y)$ unless $f(x)$ has the form $C_n \sin nx$ for some constant C_n.

We saw in Section 2.3, however, that the eigenfunctions $\sin nx$ of Eq. (3.2–8) form an orthogonal set on $(0, \pi)$ with weight function $w(x) = 1$. Additionally, in Section 2.5 we showed that a square-integrable function $f(x)$ can be represented by the series

$$f(x) \sim \sum_{n=1} b_n \sin nx,$$

where (see Eq. (2.5-2))

$$b_n = (f, \sin nx) = \frac{2}{\pi} \int_0^\pi f(s) \sin ns\, ds, \qquad n = 1, 2, \ldots . \tag{3.2-13}$$

Recall that $\pi/2$ is the square of the norm of the functions $\sin nx$ on $(0, \pi)$.

Accordingly, the final result for the problem of Example 3.2-2 can be written

$$u(x, y) = \frac{2}{\pi} \sum_{n=1} \frac{\sin nx \sinh n(b - y)}{\sinh nb} \int_0^\pi f(s) \sin ns\, ds. \quad \blacksquare \tag{3.2-14}$$

At this point the solution in Eq. (3.2-14) is in the nature of a *formal solution*, since there are still a number of unanswered questions. We will pursue some of these matters in Chapter 4.

We leave it as an exercise (Exercise 5) to show that the solution (3.2-14) satisfies the three homogeneous boundary conditions. In order to give the solution a physical interpretation we present another example.

EXAMPLE 3.2-3 Solve the problem of Example 3.2-2, given that $f(x) = 100$.

Solution One interpretation of this problem is that we seek the potential inside a rectangular region given that the potential of one boundary is 100 volts and the other three boundaries are grounded (the potentials are zero).

We proceed directly to (3.2-13) to obtain

$$\begin{aligned}
b_n &= \frac{200}{\pi} \int_0^\pi \sin ns\, ds \\
&= \frac{200}{\pi} \left[-\frac{\cos ns}{n} \right]_0^\pi \\
&= \frac{200}{\pi n} (1 - \cos n\pi) \\
&= \frac{200}{\pi n} [1 - (-1)^n] \\
&= \begin{cases} 0, & \text{if } n \text{ is even} \\ \dfrac{400}{\pi n}, & \text{if } n \text{ is odd.} \end{cases}
\end{aligned}$$

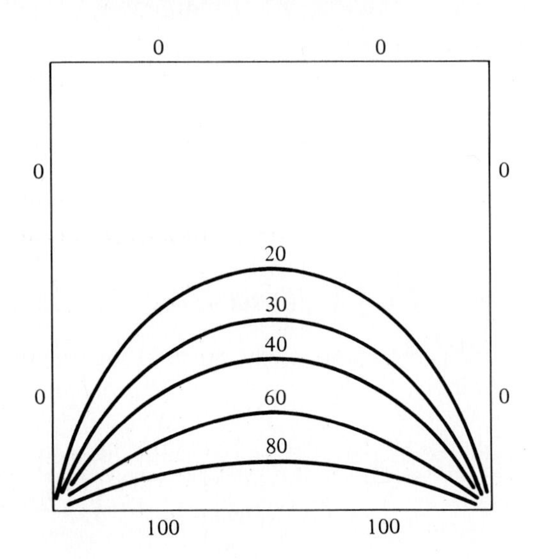

Figure 3.2-3
Equipotential curves (Example 3.2-3).

Hence the solution (3.2-14) becomes

$$u(x, y) = \frac{400}{\pi} \sum_{m=1} \frac{\sin(2m-1)x \sinh(2m-1)(b-y)}{(2m-1)\sinh(2m-1)b}, \quad \textbf{(3.2-15)}$$

where we have replaced n by $2m - 1$. ■

For convenience we take $b = 1$ and use Eq. (3.2-15) to compute some values:

$$\begin{aligned} u\left(\frac{\pi}{2}, 0\right) &= \frac{400}{\pi}\left(1 - \frac{1}{3} + \frac{1}{5} - \frac{1}{7} + - \cdots\right) \\ &= \frac{400}{\pi}\left(\frac{\pi}{4}\right) = 100^*, \\ u\left(\frac{\pi}{2}, \frac{1}{2}\right) &= \frac{400}{\pi}\left(\frac{\sinh 0.5}{\sinh 1} - \frac{\sinh 1.5}{3 \sinh 3} + \frac{\sinh 2.5}{5 \sinh 5} - + \cdots\right) \\ &\doteq \frac{400}{\pi}\left(\frac{0.5211}{1.1752} - \frac{2.1293}{30.054} + \frac{6.0502}{371.015} - + \cdots\right) \\ &\doteq 49.5. \end{aligned}$$

In a similar manner we can compute the values of $u(x, y)$ at other points.

*See Exercise 25 in Section 4.3 for the sum of the alternating series.

Points at which the values of the potential are *equal* can then be joined by smooth curves called **equipotentials**. Some of these are shown in Fig. 3.2–3.

It should be emphasized that the method of separation of variables is not guaranteed to solve *every* linear partial differential equation (Exercise 6).* On the other hand, the method is well suited to obtaining *specific* rather than *general* solutions and can be applied in a great many cases. We will use the method for solving boundary-value problems phrased in various coordinate systems in Chapters 6 and 7.

Key Words and Phrases

Poisson's equation	separation of variables
potential equation	trivial solution
potential function	equipotentials
harmonic function	

3.2 Exercises

• **1.** By differentiating both members of Eq. (3.2–5) partially with respect to y, show that both members must be constant.

2. **(a)** Show that choosing $k = 0$ in the problem of (3.2–7) leads to the trivial solution.

(b) Show that choosing $k > 0$ in the problem of (3.2–7) leads to the trivial solution.

3. Verify that each of the functions $u_n(x, y)$ in Eq. (3.2–11) satisfies the P.D.E. and homogeneous B.C. of Example 3.2–2.

4. Solve Example 3.2–2 *completely* for each of the following cases.

(a) $f(x) = 2 \sin 3x$

(b) $f(x) = 3 \sin 2x + 2 \sin 3x$

(c) $f(x) = \sin 2x \cos x$ (*Hint*: Recall the trigonometric identity

$$2 \sin A \cos B = \sin (A + B) + \sin (A - B).)$$

5. Verify that the expression in Eq. (3.2–14) satisfies the homogeneous boundary conditions of Example 3.2–2.

6. Explain the difficulty encountered in attempting to solve the equation

$$u_{xx} - u_{xy} + 2u_{yy} = 0$$

by the method of separation of variables. (See also Section 3.5.)

•• **7.** Show that each of the following functions are potential functions.

(a) $u = c/r$, where $r = \sqrt{x^2 + y^2 + z^2}$ and c is a constant

(b) $u = c \log r + k$, c and k constants, $r = \sqrt{x^2 + y^2}$

*See also Section 3.5.

(c) $u = \arctan \dfrac{2xy}{x^2 - y^2}$

8. Obtain the solution to the following boundary-value problem.

$$\begin{aligned} \text{P.D.E.:} \quad & u_{xx} + u_{yy} = 0, \quad 0 < x < \pi, \quad 0 < y < b; \\ \text{B.C.:} \quad & u(0, y) = u(\pi, y) = 0, \quad 0 < y < b, \\ & u(x, b) = 0, \quad u(x, 0) = 3 \sin x, \quad 0 < x < \pi. \end{aligned}$$

***9.** In Exercise 8, use $b = 2$, and compute the value of $u(x, y)$ at each of the given points.

(a) $(\pi/2, 0)$ **(b)** $(\pi, 1)$ **(c)** $(\pi/2, 2)$ **(d)** $(\pi/2, 1)$

Use the method of separation of variables of this section to obtain two *ordinary* differential equations in each of Exercises 10–14. Do not attempt to solve the resulting equations.

10. $u_t = ku_{xx}$, where k is a constant.

11. $u_{tt} = c^2 u_{xx}$, where c is a constant.

12. $u_t = ku_{xx} + au$, where a and k are constants.

13. $u_t = ku_{xx} + bu_x$, where b and k are constants.

14. $u_t = ku_{xx} + bu_x + a$, where a, b, and k are constants.

In Exercises 15–17, follow the procedure shown in Example 3.2–2 to obtain solutions, carrying the work as far as possible.

15. Solve the following boundary-value problem.

$$\begin{aligned} \text{P.D.E.:} \quad & u_{xx} + u_{yy} = 0, \quad 0 < x < \pi, \quad 0 < y < b; \\ \text{B.C.:} \quad & u(0, y) = g(y), \quad u(\pi, y) = 0, \quad 0 < y < b, \\ & u(x, 0) = u(x, b) = 0, \quad 0 < x < \pi. \end{aligned}$$

16. Solve the following boundary-value problem.

$$\begin{aligned} \text{P.D.E.:} \quad & u_{xx} + u_{yy} = 0, \quad 0 < x < \pi, \quad 0 < y < b; \\ \text{B.C.:} \quad & u(\pi, y) = h(y), \quad u(0, y) = 0, \quad 0 < y < b, \\ & u(x, 0) = u(x, b) = 0, \quad 0 < x < \pi. \end{aligned}$$

17. Solve the following boundary-value problem.

$$\begin{aligned} \text{P.D.E.:} \quad & u_{xx} + u_{yy} = 0, \quad 0 < x < \pi, \quad 0 < y < b; \\ \text{B.C.:} \quad & u(x, b) = \phi(x), \quad u(x, 0) = 0, \quad 0 < x < \pi, \\ & u(0, y) = u(\pi, y) = 0, \quad 0 < y < b. \end{aligned}$$

18. **(a)** If L is a linear differential operator as in Eq. (3.1–1) and if $u_1, u_2, \ldots, u_n$ are solutions of $Lu = 0$, prove that $c_1u_1 + c_2u_2 + \cdots + c_nu_n$ is also a solution for arbitrary constants c_i.

(b) If v is *any* solution of $Lu = G$, prove that $c_1u_1 + c_2u_2 + \cdots + c_nu_n + v$ is also a solution.

(c) If v is *any* solution of $Lu = G$, u_1 and u_2 are linearly independent solutions of $Lu = 0$, prove that $c_1u_1 + c_2u_2 + v$ is the general solution of $Lu = G$.

••• **19.** In Exercise 12, make the substitution

$$u(x, t) = e^{bt}v(x, t),$$

where b is a constant. Then show that by choosing b properly the variables can be separated in the resulting equation in $v(x, t)$.

20. Explain the difficulty encountered in attempting to solve

$$u_t = ku_{xx} + g(t),$$

where k is a constant, by the method of separation of variables.

21. Obtain the two ordinary differential equations by separating the variables in the Tricomi equation

$$yu_{xx} + u_{yy} = 0.$$

22. Let R be a bounded region in the xy-plane with boundary C, and let u_1 and u_2 be solutions of the Dirichlet problems

$$\nabla^2 u_1 = 0 \quad \text{in } R, \qquad u_1 = g_1 \quad \text{on } C,$$
$$\nabla^2 u_2 = 0 \quad \text{in } R, \qquad u_2 = g_2 \quad \text{on } C.$$

Prove that $u = c_1 u_1 + c_2 u_2$ is a solution of the Dirichlet problem

$$\nabla^2 u = 0 \quad \text{in } R, \qquad u = c_1 g_1 + c_2 g_2 \quad \text{on } C,$$

where c_1 and c_2 are arbitrary constants.

23. Indicate the regions in which each of the functions in Exercise 7 is a potential function.

3.3 THE WAVE EQUATION

Although our main objective is to present various methods of solving boundary-value problems, it may be instructive to see the genesis of one such problem. Accordingly, this section is devoted to a discussion of a physical problem, the simplifying assumptions that are necessary in order to derive a simple partial differential equation, and, finally, obtaining its solution.

Consider a string (a more accurate word would be "wire," since what we are about to describe can be likened to the vibrations of a guitar "string") of length L fastened at two points. It will be convenient to take the x-axis as the position of the string when no external forces are acting. Let $x = 0$ and $x = L$ represent the points at which the string is fastened. When the string is caused to vibrate, a point on the string will, at time t, assume a position with coordinates (x, y). We will be interested in obtaining the equation satisfied by y, as a function of x and t. In other words, if $y(x, t)$ is the vertical displacement of the string at a distance x from the fixed left-hand end at time t, then what is the partial differential equation satisfied by $y(x, t)$?

We first make some assumptions about the **vibrating string** that will simplify the derivation. These are listed and discussed on the next page.

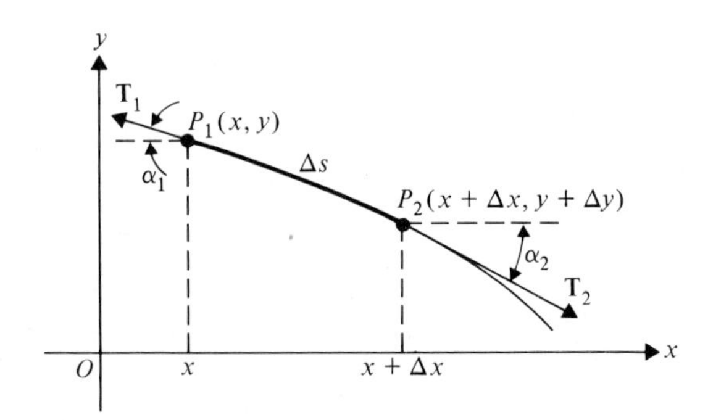

Figure 3.3-1
A portion of the vibrating string.

1. The string is homogeneous, that is, the cross-section and density are constant throughout its length.
2. Each point of the string moves along a line perpendicular to the x-axis.
3. The maximum deflection is "small" in comparison to the length L. This loosely stated assumption can be more clearly understood by saying that y should be on the order of a few millimeters for a string one meter in length.
4. The string is perfectly flexible and under uniform (constant) tension throughout its length.
5. External *forces*, such as air resistance and weight of the string, are ignored.

Next, consider a portion of the string shown *greatly magnified* in Fig. 3.3-1. The coordinates of two neighboring points P_1 and P_2 are (x, y) and $(x + \Delta x, y + \Delta y)$, respectively. Denote the tension at the two points by $\mathbf{T}_1$ and $\mathbf{T}_2$, respectively. These two tension forces necessarily act along the tangents to the curve at the two points, the tangents making angles α_1 and α_2 with the horizontal as shown. Let the length of the portion of string considered be Δs and denote the mass density of the string per unit length by δ. The horizontal components of $\mathbf{T}_1$ and $\mathbf{T}_2$ must be equal; otherwise assumption (2) would be violated. Thus*

$$-T_1 \cos \alpha_1 + T_2 \cos \alpha_2 = 0$$

or

$$T_1 \cos \alpha_1 = T_2 \cos \alpha_2 = T_0, \quad \text{a constant.}$$

The *net* vertical component of tension on the element Δs is

$$T_1 \sin \alpha_1 - T_2 \sin \alpha_2$$

*Using the notation $T_1 = |\mathbf{T}_1|$ and $T_2 = |\mathbf{T}_2|$.

or*

$$T_0(\tan \alpha_1 - \tan \alpha_2) = T_0 \left[-\frac{\partial y(x, t)}{\partial x} + \frac{\partial y(x + \Delta x, t)}{\partial x} \right].$$

The weight of the string is ignored here, since it is small in comparison to the tension T_0. By Newton's second law the sum of the forces on the element Δs must be zero for equilibrium. Hence.

$$T_0 \left[-\frac{\partial y(x, t)}{\partial x} + \frac{\partial y(x + \Delta x, t)}{\partial x} \right] = \delta \Delta s \frac{\partial^2 y(\bar{x}, t)}{\partial t^2}, \qquad \textbf{(3.3–1)}$$

where $\bar{x}$ represents the coordinate of the center of mass of the element Δs. Because of assumption (3), $\Delta s \doteq \Delta x$ so that, dividing both members of Eq. (3.3–1) by Δs and taking the limit as $\Delta x \to 0$, we obtain

$$T_0 \frac{\partial^2 y}{\partial x^2} = \delta \frac{\partial^2 y}{\partial t^2}$$

or

$$\frac{\partial^2 y}{\partial t^2} = a^2 \frac{\partial^2 y}{\partial x^2}, \qquad a^2 = \frac{T_0}{\delta}. \qquad \textbf{(3.3–2)}$$

Equation (3.3–2) is the vibrating-string equation or the wave equation in one (spatial) dimension (compare Example 3.1–2). Since the equation is hyperbolic, its general solution is obtained as shown in Example 3.1–4 and is given by

$$y(x, t) = \phi(x + at) + \psi(x - at), \qquad \textbf{(3.3–3)}$$

where ϕ and ψ are twice-differentiable but otherwise arbitrary functions.

If we now assume that the string is infinite in length† and impose the initial conditions

$$y(x, 0) = f(x), \qquad y_t(x, 0) = 0, \qquad -\infty < x < \infty, \qquad \textbf{(3.3–4)}$$

then we will gain some further insight into the functions ϕ and ψ. From Eq. (3.3–3) we have

$$y_t(x, t) = a\phi'(x + at) - a\psi'(x - at),$$

where primes indicate differentiation with respect to the arguments. For example,

$$\phi'(x + at) = d\phi(x + at)/d(x + at)$$

*Recall from calculus that the derivative at a point is defined as the tangent of the slope angle, which, in turn, is the angle measured from the positive x-axis to the tangent line in a counterclockwise sense.

†This means that we no longer consider the string to be fastened at two points.

so that

$$\frac{\partial\phi(x + at)}{\partial t} = \frac{d\phi(x + at)}{d(x + at)}\frac{\partial(x + at)}{\partial t} = a\phi'(x + at).$$

Then

$$a\phi'(x) - a\psi'(x) = 0,$$

which shows that $\phi'(x) = \psi'(x)$, so that ϕ and ψ differ by at most a constant, $\phi(x) = \psi(x) + C$. Hence

$$y(x, 0) = \phi(x) + \psi(x) = 2\psi(x) + C = f(x)$$

or

$$\psi(x) = \tfrac{1}{2}[f(x) - C]$$

and

$$\phi(x) = \tfrac{1}{2}[f(x) + C].$$

Thus the solution to Eq. (3.3–2) with the initial conditions shown in Eqs. (3.3–4) is

$$y(x, t) = \tfrac{1}{2}[f(x + at) + f(x - at)]. \qquad \textbf{(3.3–5)}$$

We summarize the foregoing in the next example.

EXAMPLE 3.3–1

$$\begin{aligned} \text{P.D.E.:}\quad & y_{tt} = a^2 y_{xx}, && -\infty < x < \infty, \quad t > 0;\\ \text{I.C.:}\quad & y(x, 0) = f(x), && -\infty < x < \infty,\\ & y_t(x, 0) = 0, && -\infty < x < \infty, \end{aligned}$$

has solutions

$$y(x, t) = \tfrac{1}{2}[f(x + at) + f(x - at)]. \quad \blacksquare \qquad \textbf{(3.3–6)}$$

The specific solution in Eq. (3.3–6) was obtained by assuming an initial displacement and *zero* initial velocity. We next consider the case in which the initial displacement is zero and the initial velocity is prescribed.

EXAMPLE 3.3–2

$$\begin{aligned} \text{P.D.E.:}\quad & y_{tt} = a^2 y_{xx}, && -\infty < x < \infty, \quad t > 0;\\ \text{I.C.:}\quad & y(x, 0) = 0, && -\infty < x < \infty,\\ & y_t(x, 0) = g(x), && -\infty < x < \infty. \end{aligned}$$

Solution Our point of departure is again the general solution given in Eq. (3.3–3):

$$y(x, t) = \phi(x + at) + \psi(x - at).$$

This time the condition $y(x, 0) = 0$ implies $\phi(x) = -\psi(x)$, so that

$$y(x, t) = \phi(x + at) - \phi(x - at),$$

and

$$y_t(x, t) = a\phi'(x + at) + a\phi'(x - at).$$

Hence $y_t(x, 0) = g(x)$ produces

$$\phi'(x) = \frac{1}{2a} g(x)$$

and

$$\phi(x) - \phi(\alpha) = \frac{1}{2a} \int_\alpha^x g(s)\, ds,$$

using the fundamental theorem of the integral calculus. The general solution is thus transformed into

$$y(x, t) = \frac{1}{2a} \left[\int_\alpha^{x+at} g(s)\, ds - \int_\alpha^{x-at} g(s)\, ds \right]$$

$$= \frac{1}{2a} \int_{x-at}^{x+at} g(s)\, ds. \quad \blacksquare \qquad (3.3\text{–}7)$$

Using superposition (Exercise 3), we may obtain the solution to the following example.

EXAMPLE 3.3–3

P.D.E.: $y_{tt} = a^2 y_{xx}, \quad -\infty < x < \infty, \quad t > 0;$
I.C.: $y(x, 0) = f(x), \quad -\infty < x < \infty,$
$y_t(x, 0) = g(x), \quad -\infty < x < \infty,$

has solution

$$y(x, t) = \frac{1}{2}[f(x + at) + f(x - at)] + \frac{1}{2a} \int_{x-at}^{x+at} g(s)\, ds. \quad \blacksquare \qquad (3.3\text{–}8)$$

This is called **D'Alembert's solution.**

In order to gain further insight into the solution given in Eq. (3.3–8), consider an initial displacement $f(x)$ shown in Fig. 3.3–2. If the string has zero initial velocity and is given an initial displacement as shown by the (idealized) rectangular pulse, then the figure shows the string at time $t = 0$. The position of the pulse at a later time can be calculated. For example, at time $t = 1/2a$ we would graph

$$y = \tfrac{1}{2}[f(x + \tfrac{1}{2}) + f(x - \tfrac{1}{2})]$$

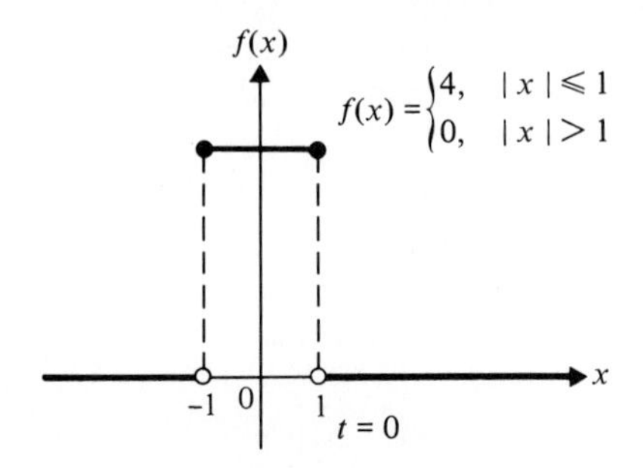

Figure 3.3-2
Initial displacement.

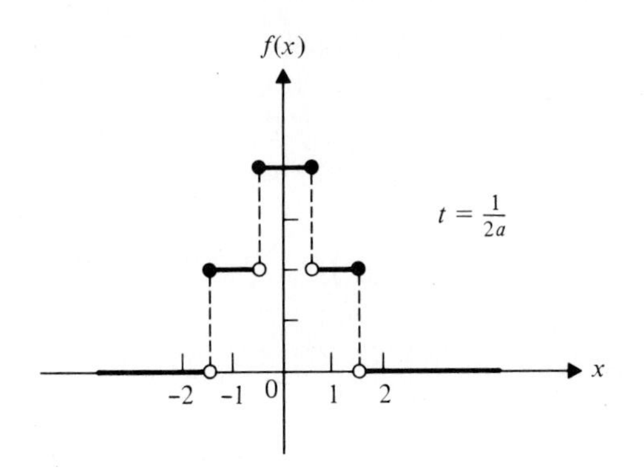

Figure 3.3-3
Displacement at $t = 1/2a$.

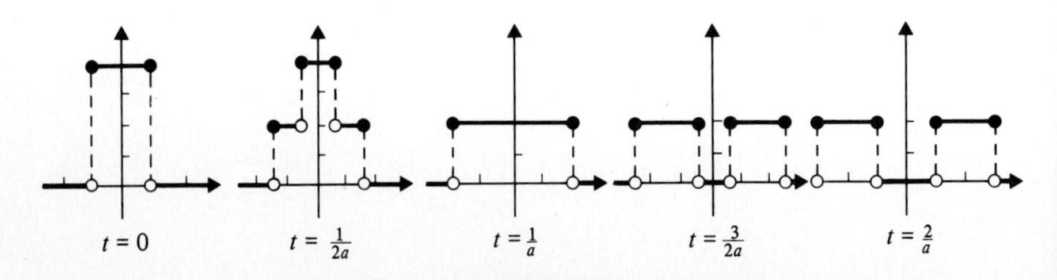

Figure 3.3-4
The time history of the vibrating string.

by using the following values to obtain the result shown in Fig. 3.3-3:

x	0	$\pm 1/2$	± 1	$\pm 3/2$	± 2
y	4	4	2	2	0

Using $t = 1/a$, $3/2a$, and $2/a$, we obtain graphs that show the propagation of the initial displacement of the string in two directions. This is shown in Fig. 3.3–4.

When the string has finite length, say, L, the solution becomes somewhat more complicated because of reflections at the boundaries. We illustrate this case in the next example.

EXAMPLE 3.3–4 Solve the following problem and graph the result.

$$\begin{aligned} \text{P.D.E.:} \quad & y_{tt} = a^2 y_{xx}, \quad 0 < x < L, \quad t > 0; \\ \text{B.C.:} \quad & y(0, t) = y(L, t) = 0, \quad t > 0; \\ \text{I.C.:} \quad & y(x, 0) = f(x), \quad 0 < x < L, \\ & y_t(x, 0) = 0, \quad 0 < x < L. \end{aligned}$$

Solution This is the case of a string of finite length L, fixed at both ends, and set into motion from rest by giving the string an initial displacement $f(x)$. In order to ensure compatibility of the boundary and initial conditions we assume that $f(0) = f(L) = 0$.

We begin with the general solution

$$y(x, t) = \phi(x + at) + \psi(x - at). \tag{3.3–3}$$

When we apply the initial conditions, we have

$$\begin{aligned} y_t(x, t) &= a\phi(x + at) - a\psi(x - at), \\ y_t(x, 0) &= a\phi(x) - a\psi(x) = 0, \end{aligned}$$

that is,

$$\phi(x) = \psi(x)$$

and

$$y(x, 0) = 2\phi(x) = f(x).$$

Thus

$$\phi(x) = \psi(x) = \frac{1}{2} f(x), \qquad 0 \leq x \leq L.$$

We note that $f(x)$ is defined only on the interval $[0, L]$, while t ranges from zero to infinity. Hence the solution of Eq. (3.3–3) can be meaningful only if we define $\phi(x)$ for $0 \leq x < \infty$ and $\psi(x)$ for $-\infty < x < \infty$. We can use the boundary conditions to obtain these extended definitions. Applying the boundary conditions to Eq. (3.3–3) gives us

$$\phi(at) + \psi(-at) = 0 \tag{3.3–9}$$

and

$$\phi(L + at) + \psi(L - at) = 0. \tag{3.3–10}$$

Putting $x = -at$ in Eq. (3.3–9) produces

$$\psi(x) = -\phi(-x) = -\frac{1}{2}f(-x), \qquad (3.3\text{–}11)$$

which shows how the definition of $\psi(x)$ can be extended to the interval $-L \leq x \leq 0$. Thus we have

$$\psi(x) = -\frac{1}{2}f(-x), \qquad -L \leq x \leq 0.$$

In a similar manner we put $x = L + at$ in Eq. (3.3–10) to obtain

$$\phi(x) = -\psi(2L - x).$$

But $\psi(x)$ has already been defined on the interval $-L \leq x \leq L$, so that the last equation now extends $\phi(x)$ to the interval $-L \leq 2L - x \leq L$, that is (Exercise 5),

$$\phi(x) = -\psi(2L - x), \qquad L \leq x \leq 3L.$$

Next, we return to Eq. (3.3–11) to obtain

$$\psi(x) = -\phi(x), \qquad -3L \leq x \leq -L.$$

Continuing in this way, we find that

$$\begin{aligned} \phi(x) &= -\psi(2L - x), \qquad 3L \leq x \leq 5L, \\ \psi(x) &= -\phi(x), \qquad -5L \leq x \leq -3L, \end{aligned}$$

and so on.

If $f(x)$ is as shown in Fig. 3.3–5, which is the case of a string of length L plucked at its center, then the dashed lines show the extensions of $\phi(x)$ and $\psi(x)$. A time history of the plucked string is shown in Fig. 3.3–6. For each value of t we show the corresponding values of $\phi(x)$ and $\psi(x)$ and their sum $y(x)$. ■

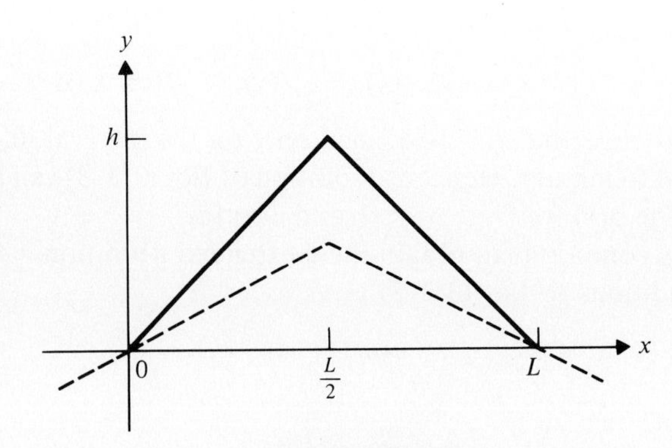

Figure 3.3–5
A string plucked at its center.

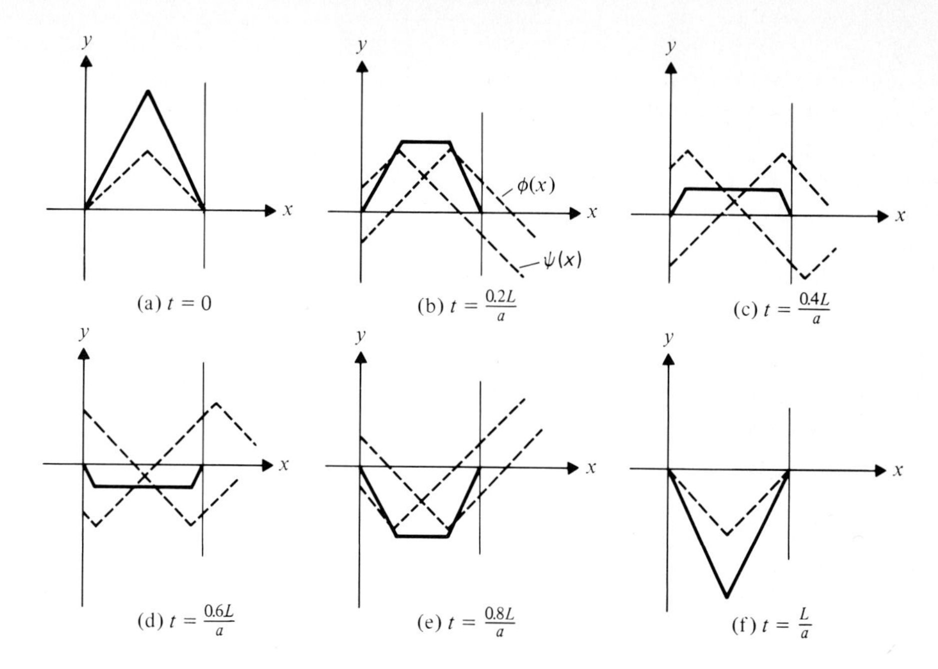

Figure 3.3-6
The time history of the plucked string.

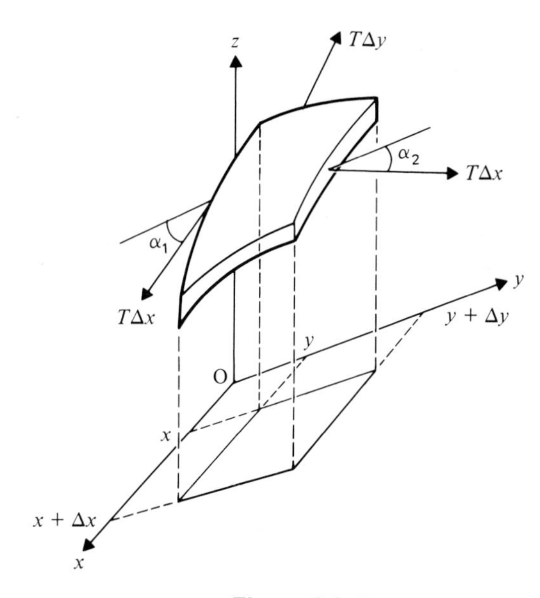

Figure 3.3-7
A portion of the vibrating membrane.

We conclude this section with a brief discussion of the two-dimensional wave equation. A physical model of this generalization of the vibrating string is a **vibrating membrane** such as, for example, a drumhead. The simplifying assumptions made earlier for the case of the vibrating string can be easily extended to the vibrating membrane (see Exercise 6).

In Fig. 3.3–7 we show a portion of the membrane. Since the deflection z is "small," so are the angles α_1 and α_2. If T is the force per unit length, then the total forces on the two edges are approximately $T\Delta x$ and $T\Delta y$ as shown. The y-components of the forces on the two edges parallel to the x-axis are given by $T\Delta x \cos \alpha_1$ and $T\Delta x \cos \alpha_2$, respectively. Since the angles α_1 and α_2 are small, these y-components are essentially equal, and we can assume that each point of the membrane moves only in the z-direction. The net resultant vertical force parallel to the xz-plane is thus

$$T\Delta x(\sin \alpha_2 - \sin \alpha_1),$$

which is approximately equal to

$$T\Delta x(\tan \alpha_2 - \tan \alpha_1)$$

or

$$T\Delta x[z_y(x_1, y + \Delta y) - z_y(x_2, y)],$$

where x_1 and x_2 are values of x between x and $x + \Delta x$. Similarly, the net resultant vertical force parallel to the yz-plane is

$$T\Delta y[z_x(x + \Delta x, y_1) - z_x(x, y_2)],$$

where y_1 and y_2 are values of y between y and $y + \Delta y$.

By Newton's second law we must have

$$T\Delta x[z_y(x_1, y + \Delta y) - z_y(x_2, y)] + T\Delta y[z_x(x + \Delta x, y_1) - z_x(x, y_2)] = \rho\Delta x\Delta y \frac{\partial^2 \bar{z}}{\partial t^2} \tag{3.3–12}$$

for equilibrium, where ρ is the mass density per unit area, $\Delta x\Delta y$ is the approximate area, and $\bar{z}$ is the deflection at some point within. Dividing both sides of Eq. (3.3–12) by $\rho\Delta x\Delta y$, and taking the limit as Δx and Δy simultaneously approach zero, produces

$$\frac{\partial^2 z}{\partial t^2} = \frac{T}{\rho}\left(\frac{\partial^2 z}{\partial x^2} + \frac{\partial^2 z}{\partial y^2}\right)$$

or

$$z_{tt} = c^2(z_{xx} + z_{yy}), \tag{3.3–13}$$

where $c^2 = T/\rho$. Equation (3.3–13) is the **two-dimensional wave equation** or the *vibrating membrane equation*. Note that Eq. (3.3–13) can be written (see also Exercise 7) as

$$z_{tt} = c^2\nabla^2 z.$$

We will find solutions to this equation in Section 6.2.

Key Words and Phrases

vibrating string	vibrating membrane
D'Alembert's solution	two-dimensional wave equation

3.3 Exercises

• **1.** **(a)** Referring to Fig. 3.3–1, show that

$$\mathbf{T}_1 = -T_1 \cos \alpha_1 \mathbf{i} + T_1 \sin \alpha_1 \mathbf{j},$$

which resolves the tension at P_1 into horizontal and vertical components.

(b) Resolve $\mathbf{T}_2$ in a similar fashion.

(c) By equating the horizontal components of $\mathbf{T}_1$ and $\mathbf{T}_2$, show that

$$T_1 \cos \alpha_1 = T_2 \cos \alpha_2 = T_0, \text{ a constant.}$$

(d) Show that the net vertical component of tension on the element Δs is

$$T_1 \sin \alpha_1 - T_2 \sin \alpha_2.$$

2. Carry out the details needed to arrive at Eq. (3.3–2). (*Hint*: Recall from calculus the definition

$$\frac{\partial y}{\partial x} = \lim_{\Delta x \to 0} \frac{y(x + \Delta x, t) - y(x, t)}{\Delta x};$$

then replace the y's by y_x.)

3. Obtain D'Alembert's solution to the wave equation in Example 3.3–3 without using the principle of superposition; that is, apply the initial conditions to the general solution.

4. Verify the graphs shown in Fig. 3.3–4 for the values of t indicated.

5. In Example 3.3–4, verify that

$$\phi(x) = -\psi(2L - x), \qquad L \le x \le 3L.$$

6. List simplifying assumptions for the vibrating membrane analogous to those given for the vibrating string in the text.

7. Write the three-dimensional wave equation (the vibrating bell equation).

8. **(a)** Show that the constant c in Eq. (3.3–13) must have the dimension of velocity in order that the equation be correct dimensionally.

(b) Show that $c^2 = T/\rho$ is correct dimensionally.

•• **9.** Solve Eq. (3.3–2) with the initial conditions

$$y(x, 0) = \sin x, \qquad y_t(x, 0) = 0.$$

10. Solve Eq. (3.3–2) with the initial conditions

$$y(x, 0) = 0, \qquad y_t(x, 0) = \cos x.$$

11. Solve Eq. (3.3–2) with the initial conditions

$$y(x, 0) = \sin x, \qquad y_t(x, 0) = \cos x.$$

12. If an infinite string is given an initial displacement defined by

$$f(x) = \begin{cases} a(ax + 1), & -\frac{1}{a} \le x \le 0, \\ a(1 - ax), & 0 \le x \le \frac{1}{a}, \\ 0, & \text{otherwise}, \end{cases}$$

and initial velocity zero, show the graph of the solution $y(x, t)$ of Eq. (3.3–2) for

(a) $t = 0$ **(b)** $t = \frac{1}{2a}$ **(c)** $t = \frac{1}{a}$ **(d)** $t = \frac{3}{2a}$

(*Hint*: Take $a = 1$.)

13. Verify that the initial displacement $f(x)$ of Exercise 12 travels with a speed a both to the right and to the left.

14. Solve Eq. (3.3–2) with the initial conditions $y(x, 0) = 0$, $y_t(x, 0) = f(x)$ defined in Exercise 12.

15. Solve Eq. (3.3–2) for a finite string with the initial conditions

$$y(x, 0) = 0, \qquad 0 < x < L,$$
$$y_t(x, 0) = 2 \cos \frac{3\pi x}{L}, \qquad 0 < x < L.$$

16. Verify that if $f(x)$ is a twice-differentiable function and $g(x)$ is a differentiable function, then the solution (3.3–8) satisfies Eq. (3.3–2).

17. Explain why the result obtained in Eq. (3.3–7) is independent of the constant used as the lower limit in the previous step.

18. A variation of the method of separation of variables consists of making the substitution

$$y(x, t) = X(x)e^{i\omega t}.$$

Solve the following problem by this method.

$$\begin{aligned} &\text{P.D.E.:} && y_{tt} = a^2 y_{xx}, \quad 0 < x < L, \quad t > 0; \\ &\text{B.C.:} && \left.\begin{aligned} y(0, t) &= 0, \\ y(L, t) &= 0, \end{aligned}\right\} \quad t > 0; \\ &\text{I.C.:} && y(x, 0) = 3 \sin \frac{2\pi x}{L}, \quad 0 < x < L, \\ & && y_t(x, 0) = 0, \quad 0 < x < L. \end{aligned}$$

Interpret the problem physically, and explain why the separation constant appears to be missing.

19. If an external force per unit length of magnitude F is applied to the string, show that the resulting partial differential equation becomes

$$y_{tt}(x, t) = a^2 y_{xx}(x, t) + F/\delta.$$

20. If the external force in Exercise 19 consists of the weight of the string, show that the equation there becomes

$$y_{tt}(x, t) = a^2 y_{xx}(x, t) - g,$$

where g is the acceleration due to gravity.

21. If a string is vibrating in a medium that provides a damping coefficient b ($b > 0$), show that the partial differential equation becomes

$$y_{tt}(x, t) = a^2 y_{xx}(x, t) - \frac{b}{\delta} y_t(x, t).$$

22. Obtain the static (independent of time) displacements $y(x)$ of points of a string of length L which hangs at rest under its own weight. Show that the string hangs in a parabolic arc and find the maximum displacement.

••• **23.** In the general solution of the wave equation show that

$$y(x, t) = -y\left(L - x, t + \frac{L}{a}\right)$$

for some positive constant L. Interpret this result physically.

24. In the one-dimensional wave equation, make the substitution $\tau = at$, and obtain

$$y_{\tau\tau} = y_{xx}.$$

25. In Example 3.3–4 of the plucked string, show that $y(x, t) = 0$ when $t = 0.5/a$.

26. Read the paper, "The Equations for Large Vibrations of Strings" by Stuart S. Antman in *Amer. Math. Monthly* (**87**), 5, May 1980, pp. 359–370.

3.4 THE DIFFUSION EQUATION

The process of **diffusion** is a familiar one in the physical and biological sciences. If a salt crystal is placed in a beaker of water, the salt begins to dissolve, and the high salt content of the water near the crystal gradually spreads out (or diffuses) until eventually the salt concentration is the same throughout the container. When two inert gases are brought into contact with each other, their molecules intermingle (or diffuse) until the mixture is homogeneous. When food is digested, it is changed chemically to a form that will permit the nutrients to pass through the intestinal wall (or diffuse) into the blood stream.

Diffusion of *matter* is a relatively slow process, since it depends on the random motion of molecules. Diffusion of *energy*, on the other hand, may be quite different. The propagation of long electromagnetic waves in a good conductor is an example of the diffusion of energy at a speed of 3×10^{10} cm/sec.

In this section we derive the **diffusion equation**, that is, the equation that governs the process of diffusion under certain simplifying assumptions. Our task will be made much simpler if we consider the diffusion of energy in the form of heat. It is known that if one end of a homogeneous isotropic metal bar is heated, energy will be transported to the other end. This is yet another example of diffusion, called **thermal diffusion** or, more commonly, **heat conduction**.

The first significant study of the mathematical theory of heat conduction was done by the French physicist and mathematician, Joseph Fourier. His book *La théorie analytique de la chaleur* (The analytical theory of heat), published in Paris in 1822, not only provided an explanation of thermal diffusion, but also laid the foundations for what was to be called **Fourier's series**. Further historical details can be found in I. Grattan-Guinness's *Joseph Fourier 1768–1830* (Cambridge, Mass.: The MIT Press, 1972) and in his *The Development of the Foundations of Mathematical Analysis from Euler to Riemann* (Cambridge, Mass.: The MIT Press, 1970).

We consider a material that is *thermally isotropic*, that is, one whose density, specific heat, and thermal conductivity are independent of direction. We assume also the following facts regarding thermal conductivity, all of which are based on experimental evidence.

1. Heat flows from a region of higher temperature to a region of lower temperature.
2. Heat flows through an area at a rate proportional to the temperature gradient normal to the area.
3. The change in heat of a body as its temperature changes is proportional to its mass and to the change in temperature.

Now consider a region V bounded by a closed surface S (Fig. 3.4–1). If we denote the temperature at the point $P(x, y, z)$ by $u(x, y, z, t)$, then the rate at which heat flows outward from the region through a surface element dS is given by

$$\frac{dQ}{dt} = -K \frac{du}{dn} \, dS = -K(\nabla u \cdot \mathbf{n}) \, dS,$$

where $\mathbf{n}$ is the unit *outward* normal to dS at P', K is a proportionality constant called the **thermal conductivity** of the material, and Q is the amount of heat transferred across the element dS. The derivative du/dn is the **directional derivative** of u in the direction of n. Hence the net rate of heat flow *into* V is given by

$$K \oiint_S \nabla u \cdot \mathbf{n} \, dS,$$

the surface integral being taken over the entire surface S. If there are no heat

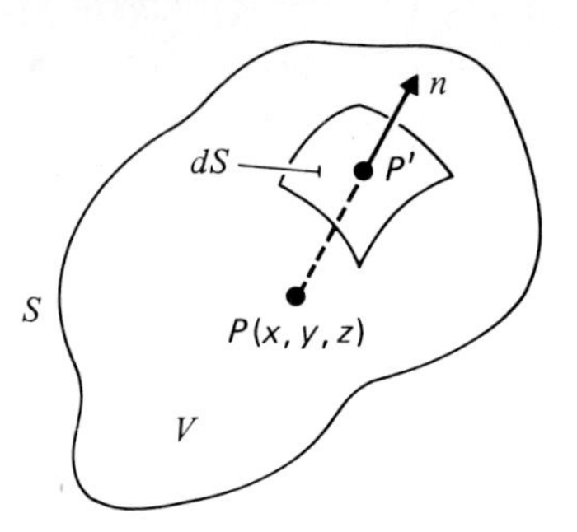

Figure 3.4–1
Heat flow across a surface.

sources or sinks in V, then this inward flow must be balanced by the amount of heat lost by the body, namely,

$$\iiint_V c\rho \frac{\partial u}{\partial t}\, dV,$$

where the *constants* c and ρ are the specific heat and the density of the material, respectively.

Equating the two integrals, we have

$$K \oiint_S \nabla u \cdot \mathbf{n}\, dS = c\rho \iiint_V \frac{\partial u}{\partial t}\, dV, \tag{3.4–1}$$

and applying the divergence theorem* gives us

$$K \iiint_V \nabla^2 u\, dV = c\rho \iiint_V \frac{\partial u}{\partial t}\, dV,$$

This last can be written

$$\iiint_V \left(K\nabla^2 u - c\rho \frac{\partial u}{\partial t}\right) dV = 0,$$

and since the above holds for an *arbitrary* region V, the integrand must be identically zero, and we have

$$K\nabla^2 u = c\rho \frac{\partial u}{\partial t}$$

*See Section 5.6 of the author's *Advanced Engineering Mathematics* (Reading, Mass.: Addison-Wesley, 1982).

or

$$u_t = k\nabla^2 u, \qquad k = \frac{K}{c\rho}, \tag{3.4–2}$$

where k is the **thermal diffusivity*** of the material. Equation (3.4–2) is the **heat equation** or the *diffusion equation.*

We close this section with an example illustrating the one-dimensional version of Eq. (3.4–2).

EXAMPLE 3.4–1 The ends of a bar of length L are kept at temperature zero, and the initial temperature distribution is given by $f(x)$. Find the temperature at any point of the bar at any time t.

Solution We translate the given problem into mathematical language as follows:

$$\begin{aligned} \text{P.D.E.:} &\quad u_t = ku_{xx}, \quad 0 < x < L, \quad t > 0; \\ \text{B.C.:} &\quad u(0, t) = u(L, t) = 0, \quad t > 0; \\ \text{I.C.:} &\quad u(x, 0) = f(x), \quad 0 < x < L. \end{aligned}$$

The problem is shown diagrammatically in Fig. 3.4–2. Since the partial differential equation and the boundary conditions are homogeneous, we use the method of separation of variables (see Section 3.2). The substitution $u(x, t) = X(x)T(t)$ transforms the P.D.E. into

$$\frac{T'}{kT} = \frac{X''}{X} = -\lambda^2,$$

where the separation constant has been chosen to be *negative*. There are two reasons for this choice. First, the problem in $X(x)$ is a Sturm–Liouville problem that has only the trivial solution if the separation constant is positive or zero; second, the problem in $T(t)$ must have a solution that approaches zero as $t \to \infty$ from physical considerations.

The Sturm–Liouville problem

$$X'' + \lambda^2 X = 0, \qquad X(0) = 0, \qquad X(L) = 0,$$

has eigenvalues n^2, $n = 1, 2, \ldots$, with corresponding eigenfunctions

$$X_n(x) = \sin \frac{n\pi}{L}x, \qquad n = 1, 2, \ldots. \tag{3.4–3}$$

The differential equation

$$T' + \frac{kn^2\pi^2}{L^2}T = 0$$

*In other than thermal applications, k is called the *diffusivity coefficient* or the *transport coefficient*.

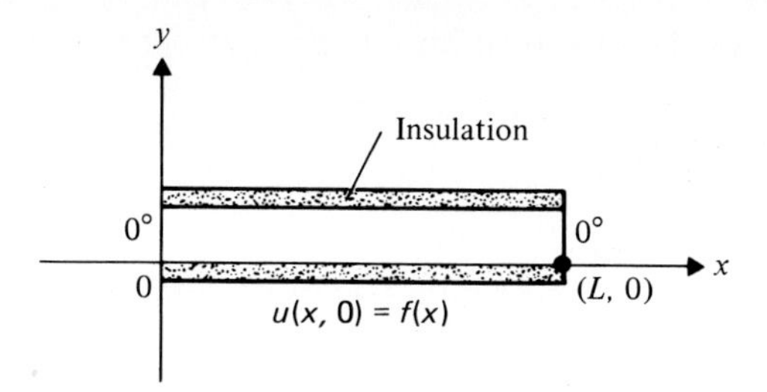

Figure 3.4-2
Heat flow (Example 3.4-1).

has solutions

$$T_n(t) = \exp\left(-\frac{kn^2\pi^2}{L^2}t\right), \qquad n = 1, 2, \ldots. \tag{3.4-4}$$

In Eqs. (3.4-3) and (3.4-4) we have set the arbitrary constants equal to unity, since the functions $X_n(x)$ and $T_n(t)$ are known only to within an arbitrary constant.

Although products of the form

$$u_n(x, t) = b_n \sin\frac{n\pi}{L}x \exp\left(-\frac{kn^2\pi^2}{L^2}t\right)$$

satisfy the diffusion equation, we cannot hope (in general) to satisfy the given initial condition (I.C.) with such a product. Hence we consider

$$u(x, t) = \sum_{n=1} b_n \sin\frac{n\pi}{L}x \exp\left(-\frac{kn^2\pi^2}{L^2}t\right), \tag{3.4-5}$$

a linear combination of products, and try to find the b_n. We have

$$u(x, 0) = \sum_{n=1} b_n \sin\frac{n\pi}{L}x = f(x), \tag{3.4-6}$$

and it can be seen that by placing suitable restrictions on $f(x)$ we can obtain the b_n from

$$b_n = \frac{2}{L}\int_0^L f(s) \sin\frac{n\pi}{L}s\,ds. \tag{3.4-7}$$

This is possible because the set of functions

$$\left\{\sin\frac{n\pi}{L}x, \qquad n = 1, 2, \ldots\right\}$$

is orthogonal on the interval $(0, L)$ with weight function $w(x) = 1$. (See Section 2.3 and Exercise 17.)

Hence the final solution of the problem can be written as

$$u(x, t) = \frac{2}{L} \sum_{n=1} \sin \frac{n\pi}{L} x \exp\left(-\frac{kn^2\pi^2}{L^2} t\right) \int_0^L f(s) \sin \frac{n\pi}{L} s \, ds. \tag{3.4-8}$$

We point out that representing $f(x)$ by an infinite series of sine functions may require that some restrictions be placed on $f(x)$. It will also be necessary to examine conditions under which the series in Eq. (3.4–8) converges and can be differentiated. These questions will be considered in Chapter 4. ■

Note that from Eq. (3.4–8) we have

$$\lim_{t \to \infty} u(x, t) = 0$$

provided that the b_n are bounded.* We call $u(x, t) = 0$ the **steady-state solution** of the problem in Example 3.4–1. This is the solution after the effect of the initial temperature distribution has been dissipated. Since the steady-state solution is independent of time, it can be obtained easily by setting u_t equal to zero in the diffusion equation, that is, by solving

$$u''(x) = 0, \qquad u(0) = 0, \qquad u(L) = 0.$$

The solution shown in Eq. (3.4–8), on the other hand, is called the **transient solution**.

Key Words and Phrases

diffusion
diffusion equation
thermal diffusion
heat conduction
thermal conductivity
directional derivative
thermal diffusivity
heat equation
steady-state solution
transient solution

3.4 Exercises

• **1.** Referring to Eq. (3.4–2), show that if there is a source of heat continuously distributed throughout V given by $f(x, y, z, t)$, then the diffusion equation becomes

$$k\nabla^2 u + \frac{f}{c\rho} = \frac{\partial u}{\partial t}.$$

2. Show that the equation obtained in Exercise 1 becomes Poisson's equation if u is independent of t and Laplace's equation if, additionally, there are no heat sources present.

*See Section 2.5.

3. In the diffusion equation (3.4–2), u is given in °C and t in seconds. Determine the units of k, and K, if c is given in calorie-sec/(g °C).

4. In obtaining Eq. (3.4–2) it was assumed that K was constant. If K is a function of position, show that the diffusion equation becomes

$$\nabla \cdot K \nabla u = c\rho \frac{\partial u}{\partial t}.$$

5. Show that the Sturm–Liouville problem

$$X''(x) - \lambda^2 X(x) = 0, \qquad X(0) = 0, \qquad X(L) = 0,$$

has only the trivial solution when λ is real (including zero).

6. Explain why the solution of the problem of Example 3.4–1 must have the property

$$\lim_{t \to \infty} u(x, t) = 0.$$

7. Obtain the eigenfunctions $X_n(x)$ in Eq. (3.4–3).

8. Verify that the functions

$$u_n(x, t) = \sin \frac{n\pi}{L} x \exp \left(-\frac{kn^2\pi^2}{L^2} t \right), \qquad n = 1, 2, \ldots,$$

satisfy the partial differential equation and the boundary conditions of Example 3.4–1.

•• **9.** Find the specific solution for the problem of Example 3.4–1 for each of the following cases.

(a) $f(x) = 3 \sin \frac{2\pi}{L} x$

(b) $f(x) = 2 \sin \frac{3\pi}{L} x$

(c) $f(x) = 2 \sin \frac{3\pi}{L} x + 3 \sin \frac{2\pi}{L} x$

10. **(a)** Solve the problem of Example 3.4–1 for the case

$$u(x, 0) = 0°, \qquad 0 < x < L.$$

(b) Explain the result on the basis of physical principles.

11. Obtain the *steady-state* solution to the problem of Example 3.4–1 for each of the following cases.

(a) $u(0, t) = 0, \quad u(L, t) = 100$
(b) $u(0, t) = 100, \quad u(L, t) = 0$
(c) $u(0, t) = u(L, t) = 100$
(d) $u(0, t) = 50, \quad u(L, t) = 100$

••• **12.** Show that the one-dimensional diffusion equation is parabolic everywhere.

13. Let $u(x, t) = f(ax + bt)$ and show that f must be of the form

$$\exp (ax + ka^2 t),$$

where $a \neq 0$ in order to satisfy $u_t = ku_{xx}$.

14. In Exercise 13, put $a = i\omega$, where $i^2 = -1$ and ω is a real positive constant. Thus show that

$$u(x, t) = (A \cos \omega x + B \sin \omega x) \exp(-k\omega^2 t)$$

satisfies the one-dimensional heat equation and also has the required property

$$\lim_{t \to \infty} u(x, t) = 0.$$

15. Make the substitution $t = c\tau$ in Eq. (3.4–2), and then show that for a suitable choice of c the equation becomes

$$u_\tau = \nabla^2 u.$$

16. Explain how the rate of diffusion of respiratory gases in humans is rendered more rapid. (*Hint*: What does the rate of diffusion depend on?)

17. We have seen that the set of functions

$$\{\sin nx, \qquad n = 1, 2, \ldots\}$$

is orthogonal on the interval $(0, \pi)$ with weight function $w(x) = 1$. By making the change of variable

$$x = \frac{\pi s}{L},$$

show that the set of functions

$$\left\{\sin \frac{n\pi}{L} s, \qquad n = 1, 2, \ldots\right\}$$

is orthogonal on the interval $(0, L)$ with weight function $w(s) = 1$.

3.5 CANONICAL FORMS

In Section 3.1 we discussed the classification of second-order linear partial differential equations, that is, equations of the type

$$A(x, y)u_{xx} + B(x, y)u_{xy} + C(x, y)u_{yy} + D(x, y)u_x + E(x, y)u_y + F(x, y)u = G(x, y). \tag{3.5–1}$$

The **principal part** of Eq. (3.5–1) consists of the portion

$$A(x, y)u_{xx} + B(x, y)u_{xy} + C(x, y)u_{yy},$$

and the **discriminant** is the quantity $B^2 - 4AC$. In regions of the xy-plane where $B^2 - 4AC < 0$ the equation (3.5–1) is said to be of **elliptic type**; where $B^2 - 4AC > 0$ it is said to be of **hyperbolic type**; and where $B^2 - 4AC = 0$ it is said to be of **parabolic type**.

We consider first the effect of a coordinate transformation given by

$$\begin{aligned} \xi &= \xi(x, y) \\ \eta &= \eta(x, y). \end{aligned} \tag{3.5-2}$$

We have (Exercise 1)

$$\begin{aligned}
u_x &= u_\xi \xi_x + u_\eta \eta_x, \\
u_y &= u_\xi \xi_y + u_\eta \eta_y, \\
u_{xx} &= u_{\xi\xi}\xi_x^2 + 2u_{\xi\eta}\xi_x\eta_x + u_{\eta\eta}\eta_x^2 + u_\xi \xi_{xx} + u_\eta \eta_{xx}, \\
u_{xy} &= u_{\xi\xi}\xi_x\xi_y + u_{\xi\eta}(\xi_x\eta_y + \xi_y\eta_x) + u_{\eta\eta}\eta_x\eta_y + u_\xi\xi_{xy} + u_\eta\eta_{xy}, \\
u_{yy} &= u_{\xi\xi}\xi_y^2 + 2u_{\xi\eta}\xi_y\eta_y + u_{\eta\eta}\eta_y^2 + u_\xi\xi_{yy} + u_\eta\eta_{yy}.
\end{aligned}$$

With these substitutions, Eq. (3.5–1) becomes

$$\hat{A}(x, y)u_{\xi\xi} + \hat{B}(x, y)u_{\xi\eta} + \hat{C}(x, y)u_{\eta\eta} + \hat{D}(x, y)u_\xi + \hat{E}(x, y)u_\eta + F(x, y)u = G(x, y), \tag{3.5-3}$$

where (Exercise 2)

$$\hat{A}(x, y) = A\xi_x^2 + B\xi_x\xi_y + C\xi_y^2, \tag{3.5-4}$$

$$\hat{B}(x, y) = 2A\xi_x\eta_x + B(\xi_x\eta_y + \xi_y\eta_x) + 2C\xi_y\eta_y, \tag{3.5-5}$$

$$\hat{C}(x, y) = A\eta_x^2 + B\eta_x\eta_y + C\eta_y^2, \tag{3.5-6}$$

$$\hat{D}(x, y) = A\xi_{xx} + B\xi_{xy} + C\xi_{yy} + D\xi_x + E\xi_y, \tag{3.5-7}$$

$$\hat{E}(x, y) = A\eta_{xx} + B\eta_{xy} + C\eta_{yy} + D\eta_x + E\eta_y. \tag{3.5-8}$$

If $B \neq 0$, then Eq. (3.5–1) can be changed into Eq. (3.5–3) with $\hat{B} = 0$ by a coordinate transformation involving rotation of axes. Let

$$\begin{aligned} x &= \xi \cos\theta - \eta \sin\theta, \\ y &= \xi \sin\theta + \eta \cos\theta. \end{aligned} \tag{3.5-9}$$

Then Eq. (3.5–5) becomes (Exercise 3)

$$-A \sin 2\theta + B \cos 2\theta + C \sin 2\theta = 0$$

or

$$\cot 2\theta = \frac{A - C}{B}. \tag{3.5-10}$$

EXAMPLE 3.5–1 Use rotation of axes to eliminate the mixed partial derivative in

$$u_{xx} + u_{xy} + u_{yy} + u_x = 0.$$

Solution We readily see (Exercise 4) that the given equation is everywhere elliptic. Then, using Eq. (3.5–10), we have $\cot 2\theta = 0$, that is, $\theta = \pi/4$. Applying this value to Eqs. (3.5–9) produces

$$\begin{bmatrix} x \\ y \end{bmatrix} = \begin{bmatrix} \frac{1}{\sqrt{2}} & -\frac{1}{\sqrt{2}} \\ \frac{1}{\sqrt{2}} & \frac{1}{\sqrt{2}} \end{bmatrix} \begin{bmatrix} \xi \\ \eta \end{bmatrix},$$

in matrix form. Since the 2×2 matrix is an **orthogonal matrix**, that is, one whose inverse is equal to its transpose, we have

$$\begin{bmatrix} \xi \\ \eta \end{bmatrix} = \begin{bmatrix} \frac{1}{\sqrt{2}} & \frac{1}{\sqrt{2}} \\ -\frac{1}{\sqrt{2}} & \frac{1}{\sqrt{2}} \end{bmatrix} \begin{bmatrix} x \\ y \end{bmatrix}. \tag{3.5–11}$$

We can use Eq. (3.5–11) to compute $\hat{A}$, $\hat{B}$, $\hat{C}$, and $\hat{D}$ as given in Eqs. (3.5–4) through (3.5–7) to obtain (Exercise 5)

$$\frac{3}{2} u_{\xi\xi} + \frac{1}{2} u_{\eta\eta} = -\frac{\sqrt{2}}{2} u_{\xi}. \quad ■ \tag{3.5–12}$$

In Example 3.5–1 the principal part of the given partial differential equation, that is,

$$u_{xx} + u_{xy} + u_{yy},$$

can be compared to a **quadratic form** with matrix

$$M = \begin{pmatrix} 1 & 1/2 \\ 1/2 & 1 \end{pmatrix}.$$

Eliminating the u_{xy}-term is equivalent to diagonalizing M (see below). Since M is a symmetric matrix, it can be diagonalized by means of an orthogonal matrix. The resulting diagonal matrix displays the eigenvalues of M on its diagonal. We leave it as an exercise to show that the eigenvalues of M are 3/2 and 1/2 (Exercise 17).

Any nonzero vector $\mathbf{x}$ in the plane that satisfies the equation $M\mathbf{x} = \lambda\mathbf{x}$ is called an eigenvector of M, and the corresponding scalar λ is called the **eigenvalue** belonging to $\mathbf{x}$. A square matrix P having the property that its inverse is equal to its transpose P^T (rows and columns interchanged) is called an *orthogonal matrix*. A symmetric (rows equal to corresponding columns) matrix M can be *diagonalized*, meaning that an orthogonal matrix P can be found such that P^TMP is a diagonal matrix (one having zeros everywhere except possibly along the main diagonal).

We remark that the coordinate transformation (3.5–2) is assumed to be such that ξ and η are twice continuously differentiable and that the determinant

$$J = \begin{vmatrix} \xi_x & \xi_y \\ \eta_x & \eta_y \end{vmatrix},$$

called the **Jacobian*** of the transformation, is different from zero in the region of interest. Under these conditions it can be shown that the sign of the discriminant is invariant, since (Exercise 16)

$$\hat{B}^2 - 4\hat{A}\hat{C} = J^2(B^2 - 4AC). \tag{3.5–13}$$

If an equation is of *hyperbolic* type, then it can be transformed into a particularly simple form by using a coordinate transformation that will make $\hat{A} = \hat{C} = 0$. We note from Eqs. (3.5–4) and (3.5–6) that both equations will then have the form

$$A\phi_x^2 + B\phi_x\phi_y + C\phi_y^2 = 0.$$

Assuming $\phi_y \neq 0$, this last can be written as

$$A\left(\frac{\phi_x}{\phi_y}\right)^2 + B\left(\frac{\phi_x}{\phi_y}\right) + C = 0. \tag{3.5–14}$$

Along any curve where $\phi(x, y)$ is constant, we have

$$d\phi = \phi_x dx + \phi_y dy = 0,$$

that is,

$$\frac{\phi_x}{\phi_y} = -\frac{dy}{dx}.$$

Thus Eq. (3.5–14) becomes

$$A\left(\frac{dy}{dx}\right)^2 - B\left(\frac{dy}{dx}\right) + C = 0,$$

with solutions

$$\frac{dy}{dx} = \frac{B \pm \sqrt{B^2 - 4AC}}{2A}, \tag{3.5–15}$$

called **characteristic equations** of Eq. (3.5–1). Solutions of these equations are called **characteristic curves** (or **characteristics**) of Eq. (3.5–1). Along the characteristics the partial differential equation takes on its simplest form. It is these so-called **canonical forms** (or normal forms) that we will investigate in more detail.

---

*After Carl G. J. Jacobi (1804–1851), a German mathematician. The notation $\partial(\xi, \eta)/\partial(x, y)$ is also used for the above Jacobian.

Parabolic Equations

In this case, $B^2 - 4AC = 0$, and Eqs. (3.5–15) reduce to

$$\frac{dy}{dx} = \frac{B}{2A}. \tag{3.5–16}$$

Now, A and C cannot both be zero (Why?), so assume that $A \neq 0$. Then the solution of Eq. (3.5–16), namely,

$$\psi(x, y) = \text{constant}$$

provides the value of $\xi = \psi(x, y)$, and $\xi =$ constant defines a *single* family of characteristics. The value of η is arbitrary provided that it is twice continuously differentiable and the aforementioned Jacobian is nonzero.

EXAMPLE 3.5–2 Transform the differential equation

$$x^2 u_{xx} + 2xyu_{xy} + y^2 u_{yy} = 4x^2$$

to canonical form, and then solve the resulting equation.

Solution The given equation is parabolic everywhere, and Eq. (3.5–16) becomes

$$\frac{dy}{dx} = \frac{y}{x}.$$

This last differential equation has solutions given by $y/x =$ constant; hence we make the coordinate transformation

$$\xi = \frac{y}{x},$$

$$\eta = y.$$

Thus $\hat{A} = \hat{B} = 0$ and $\hat{C} = \eta^2$ (Exercise 6), so the given partial differential equation becomes

$$\eta^2 u_{\eta\eta} = \frac{4\eta^2}{\xi^2} \quad \text{or} \quad u_{\eta\eta} = \frac{4}{\xi^2}.$$

Integrating with respect to η, we get

$$u_\eta = \frac{4\eta}{\xi^2} + f(\xi)$$

and

$$u(\xi, \eta) = \frac{2\eta^2}{\xi^2} + \eta f(\xi) + g(\xi).$$

Hence in terms of the original variables we have the *general solution*

$$u(x, y) = 2x^2 + yf\left(\frac{y}{x}\right) + g\left(\frac{y}{x}\right). \quad \blacksquare$$

Note that Eq. (3.5–14) can be rewritten as

$$A + B\left(\frac{\phi_y}{\phi_x}\right) + C\left(\frac{\phi_y}{\phi_x}\right)^2 = 0,$$

which leads to

$$\frac{dx}{dy} = \frac{B \pm \sqrt{B^2 - 4AC}}{2C} \tag{3.5–17}$$

instead of Eqs. (3.5–15). If $A = 0$ and $C \neq 0$, then the characteristic equations (3.5–17) can be used. Thus for parabolic equations the canonical form is given by either

$$u_{\xi\xi} = H(u_\xi, u_\eta, u, \xi, \eta) \tag{3.5–18}$$

or

$$u_{\eta\eta} = H(u_\xi, u_\eta, u, \xi, \eta), \tag{3.5–19}$$

where H represents a linear function of terms in the indicated variables.

Hyperbolic Equations

In the case of hyperbolic equations the characteristic equations

$$\frac{dy}{dx} = \frac{B \pm \sqrt{B^2 - 4AC}}{2C}$$

have two real and distinct solutions whenever $A \neq 0$. Denote these solutions by

$$\psi_1(x, y) = c_1$$

and

$$\psi_2(x, y) = c_2.$$

Then the coordinate transformation

$$\xi = \psi_1(x, y)$$
$$\eta = \psi_2(x, y)$$

results in $\hat{A} = \hat{C} = 0$ and the canonical form

$$u_{\xi\eta} = H(u_\xi, u_\eta, u, \xi, \eta). \tag{3.5–20}$$

A similar result can be obtained if $A = 0$, but $C \neq 0$.

EXAMPLE 3.5–3 Transform the differential equation

$$u_{xx} - (2\sin x)u_{xy} - (\cos^2 x)u_{yy} - (\cos x)u_y = 0$$

to canonical form, and solve the resulting equation.

Solution Here $B^2 - 4AC = 4$, so that the equation is hyperbolic everywhere. The solutions of the characteristic equations

$$\frac{dy}{dx} = -\sin x \pm 1$$

are

$$\psi_1(x, y) = \cos x + x - y = c_1$$

and

$$\psi_2(x, y) = \cos x - x - y = c_2.$$

Hence we make the coordinate transformation

$$\xi = \cos x + x - y,$$
$$\eta = \cos x - x - y.$$

Then $\hat{A} = \hat{C} = \hat{D} = \hat{E} = 0$ and $\hat{B} = -4$, so that the canonical form becomes $u_{\xi\eta} = 0$. From this we have $u_\xi = h(\xi)$ and $u = g(\xi) + f(\eta)$, that is,

$$u(x, y) = f(\cos x - x - y) + g(\cos x + x - y),$$

where f and g are twice-differentiable but otherwise arbitrary functions of the indicated variables. ■

The next example illustrates the case of a hyperbolic equation in which $B = 0$.

EXAMPLE 3.5–4 Transform the equation

$$x^2u_{xx} - y^2u_{yy} = 0$$

to canonical form, and solve the resulting equation.

Solution Here we have $B^2 - 4AC = 4x^2y^2 > 0$, showing that the equation is hyperbolic everywhere except on the x- and y-axes. We consider a region that does not include any part of either axis, for example, a region in the first quadrant. The characteristic equations

$$\frac{dy}{dx} = \pm\frac{y}{x}$$

have solutions (Exercise 8) $y = c_1x$ and $y = c_2/x$; hence we make the coordi-

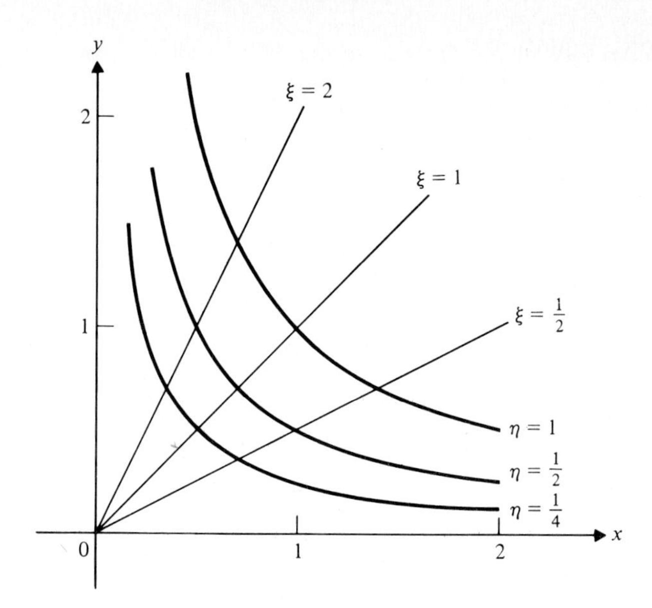

Figure 3.5-1
Characteristic curves (Example 3.5-4).

nate transformation

$$\xi = \frac{y}{x}$$

$$\eta = xy.$$

Then $\hat{A} = \hat{C} = \hat{E} = 0$, $\hat{B} = -4y^2 = -4\xi\eta$, and $\hat{D} = 2y/x = 2\xi$. Thus the given partial differential equation becomes (Exercise 9)

$$u_{\xi\eta} - \frac{1}{2\eta} u_\xi = 0, \tag{3.5-21}$$

which is equivalent to the canonical form shown in Eq. (3.5-20). We can solve Eq. (3.5-21) by observing that it is a first-order linear differential equation in u_ξ. Then we obtain the solution (Exercise 10)

$$u(\xi, \eta) = \eta^{1/2} f(\xi) + g(\eta),$$

that is,

$$u(x, y) = \sqrt{xy}\, f\left(\frac{y}{x}\right) + g(xy). \quad \blacksquare$$

The characteristic curves are shown in Fig. 3.5-1.

Elliptic Equations

In the case of elliptic equations the characteristic equations (3.5–17) have *complex* conjugate solutions. Hence in order to obtain *real* characteristics it is necessary to make two coordinate transformations. We illustrate with an example.

EXAMPLE 3.5–5 In the region where the equation

$$xu_{xx} + u_{yy} = x^2$$

is elliptic, transform the equation to canonical form.

Solution Since $B^2 - 4AC = -4x$, the equation is elliptic in the half-plane $x > 0$. In this region the characteristic equations are

$$\frac{dy}{dx} = \pm ix^{-1/2}$$

with solutions

$$y = 2ix^{1/2} + c_1 \quad \text{and} \quad y = -2ix^{1/2} + c_2.$$

We first make the coordinate transformation

$$\sigma = y + 2x^{1/2}i,$$
$$\tau = y - 2x^{1/2}i,$$

but since these are complex-valued, we make a second coordinate transformation defined by

$$\sigma = \xi + i\eta,$$
$$\tau = \xi - i\eta.$$

As a result of these two transformations we have (Exercise 11)

$$\xi = y \quad \text{and} \quad \eta = 2x^{1/2}.$$

Then $\hat{A} = \hat{C} = 1$, $\hat{B} = \hat{D} = 0$, and $\hat{E} = -\frac{1}{2}x^{-1/2}$ so that the canonical form

$$u_{\xi\xi} + u_{\eta\eta} = \frac{1}{\eta}u_\eta + \frac{\eta^4}{16}$$

is obtained (Exercise 12). ■

We observe that, in general, the canonical form for an elliptic equation can be written as

$$u_{\xi\xi} + u_{\eta\eta} = H(u_\xi, u_\eta, u, \xi, \eta). \tag{3.5–22}$$

The simplest example of an elliptic equation is Laplace's equation.

In summary, parabolic equations have a single family of characteristic curves given by $\xi =$ constant; hyperbolic equations have characteristic curves given by ξ = constant and η = constant, which form a curvilinear coordinate system; elliptic equations have no real characteristic curves.

Key Words and Phrases

principal part	quadratic form
discriminant	eigenvalue
elliptic type	Jacobian
hyperbolic type	characteristic equations
parabolic type	characteristic curves
orthogonal matrix	canonical form

3.5 Exercises

• **1.** Use the coordinate transformation defined in Eqs. (3.5–2) to compute each of the following partial derivatives.

(a) u_x and u_y

(b) u_{xx}, u_{xy}, and u_{yy} (*Hint*: Note that

$$\frac{\partial u}{\partial x} = \frac{\partial u}{\partial \xi}\frac{\partial \xi}{\partial x} + \frac{\partial u}{\partial \eta}\frac{\partial \eta}{\partial x}$$

provides a formula

$$\frac{\partial(\ )}{\partial x} = \frac{\partial(\ )}{\partial \xi}\frac{\partial \xi}{\partial x} + \frac{\partial(\ )}{\partial \eta}\frac{\partial \eta}{\partial x}$$

in which any function of x and y, such as u_x or u_y, may be inserted. There is a similar formula from part (a) that can be used to obtain derivatives with respect to y.)

2. Verify that the coefficients $\hat{A}$, $\hat{B}$, etc. are given by Eqs. (3.5–4) through (3.5–8).

3. **(a)** Solve Eqs. (3.5–9) for ξ and η in terms of x and y.

(b) Use the results in part (a) to compute $\hat{B}(x, y)$ from Eq. (3.5–5).

(c) Show that setting $\hat{B}(x, y) = 0$ in part (b) leads to

$$\cot 2\theta = \frac{A - C}{B},$$

provided that $B \neq 0$.

4. Show that the partial differential equation in Example 3.5–1 is elliptic everywhere.

5. Carry out the details to arrive at Eq. (3.5–12).

6. Use the coordinate transformation

$$\xi = \frac{y}{x} \quad \text{and} \quad \eta = y$$

to compute $\hat{A}$, $\hat{B}$, $\hat{C}$, $\hat{D}$, and $\hat{E}$ in Example 3.5–2.

7. Compute $\hat{A}$, $\hat{B}$, $\hat{C}$, $\hat{D}$, and $\hat{E}$ in Example 3.5–3.

8. Show that the solutions of

$$\frac{dy}{dx} = \pm\frac{y}{x}$$

are $y = c_1x$ and $y = c_2/x$.

9. Obtain Eq. (3.5–21) in Example 3.5–4.

10. Solve Eq. (3.5–21). (*Hint*: Make the substitution $v = u_\xi$.)

11. In Example 3.5–5, fill in the details necessary to arrive at

$$\xi = y \quad \text{and} \quad \eta = 2x^{1/2}.$$

12. In Example 3.5–5, compute $\hat{A}$, $\hat{B}$, $\hat{C}$, $\hat{D}$, and $\hat{E}$, and then obtain the canonical form shown.

•• **13.** By differentiation, verify that the following are general solutions of the partial differential equations given in the examples of this section.

(a) $u(x, y) = 2x^2 + yf\left(\frac{y}{x}\right) + g\left(\frac{y}{x}\right)$. (Example 3.5–2)

(b) $u(x, y) = f(\cos x - x - y) + g(\cos x + x - y)$. (Example 3.5–3)

(c) $u(x, y) = \sqrt{xy}\, f\left(\frac{y}{x}\right) + g(xy)$. (Example 3.5–4)

14. For each of the following equations, determine the regions where the equation is elliptic, hyperbolic, or parabolic.

(a) $u_{xx} + yu_{xy} + xu_{yy} = e^{x+y}$
(b) $y^2u_{xx} - u_{yy} + u = 0$
(c) $xu_{xx} + 2u_{xy} + yu_{yy} + xu_x + u_y = 0$
(d) $(1 - x^2)u_{xx} - 2xyu_{xy} - (1 - y^2)u_{yy} = 0$
(e) $16u_{xx} + x^2u_{yy} = 0$
(f) $4u_{xx} - 8u_{xy} + 4u_{yy} = 1$
(g) $x^2u_{xx} - y^2u_{yy} = xy$

15. Compute the Jacobian

$$J\left(\frac{\xi, \eta}{x, y}\right)$$

for each of the following transformations.

(a) $\xi = \frac{1}{\sqrt{2}}(x + y), \quad \eta = \frac{1}{\sqrt{2}}(-x + y)$
(b) $\xi = y/x, \quad \eta = y$
(c) $\xi = \cos x + x - y, \quad \eta = \cos x - x - y$
(d) $\xi = y/x, \quad \eta = xy$
(e) $\xi = y, \quad \eta = 2x^{1/2}$

••• **16.** Prove that

$$\hat{B}^2 - 4\hat{A}\hat{C} = J^2(B^2 - 4AC),$$

where J is the Jacobian of the transformation.

17. Find the eigenvalues of the matrix

$$M = \begin{pmatrix} 1 & 1/2 \\ 1/2 & 1 \end{pmatrix}.$$

(*Hint*: See Section 4.4 of the author's *Advanced Engineering Mathematics* (Reading, Mass.: Addison-Wesley, 1982).)

18. By rotating the x- and y-axes through an appropriate angle, transform each of the following equations to one that does not have a u_{xy} term.

(a) $2u_{xx} + 4u_{xy} + 5u_{yy} = 0$

(b) $u_{xx} + 4u_{xy} + 4u_{yy} = 0$ (Note that the equation need not be elliptic to use this technique.)

4 Fourier Series

4.1 INTRODUCTION

This chapter and the next are transitional ones in our study of boundary-value problems. Up to this point we have reviewed some of the basic concepts in ordinary differential equations (Chapter 1). We will continue to make frequent use of this material, with Sections 1.3 and 1.5 playing a prominent role in the remaining chapters. We will also refer to Sturm–Liouville problems and the representation of functions by a series of orthonormal functions (Chapter 2) in future chapters. We will consider more varied boundary-value problems than the ones discussed in Chapter 3. Examples 3.2–2 and 3.4–1 dealt with the potential and diffusion equations, respectively. These equations together with the wave equation will be treated in greater detail (in higher dimensions, in various coordinate systems, etc.) in later chapters.

As a preliminary to more complicated types of boundary-value problems, however, we present a branch of mathematical analysis that had its origins in the eighteenth century. As early as 1713, Taylor suggested that the function

$$\sin \frac{\pi}{L} x$$

could be used to explain the steady-state motion of a vibrating string of length L fixed at both ends. In 1749 and 1754, D'Alembert and Euler published papers in which the expansion of a function in a series of cosines was discussed. A problem from astronomy at that time involved the expansion of the reciprocal of the distance between two planets in a series of cosines of multiples of the angle between the radius vectors. In 1811,* Fourier developed the idea of expanding a function in a series of sines and cosines to the point at which it became generally useful.

4.2 FOURIER COEFFICIENTS

In Section 2.4 we represented a piecewise continuous function $f(x)$ by a series of orthonormal functions. Given that the set

$$\{\phi_i(x), \qquad i = 1, 2, \ldots\}$$

is orthonormal on $[a, b]$ with weight function $w(x)$, that is,

$$\int_a^b \phi_i(x)\phi_j(x)w(x)\,dx = \delta_{ij}.$$

Then

$$f(x) \sim \sum_{i=1} c_i\phi_i(x), \tag{4.2-1}$$

where

$$c_i = (f, \phi_i) = \int_a^b f(x)\phi_i(x)w(x)\,dx \tag{4.2-2}$$

and the symbol $\sim$ means "is represented by, in the sense of convergence in the mean" (see Eq. (2.5-10)).

It was shown in Section 2.3 that the sets

$$\{\sin nx, \qquad n = 1, 2, \ldots\}$$

and

$$\{\cos nx, \qquad n = 0, 1, 2, \ldots\}$$

are orthogonal on $[0, \pi]$ with weight function one. By dividing each function in the above sets by its norm we obtain the orthonormal sets

$$\{\sqrt{2/\pi}\sin nx, \qquad n = 1, 2, \ldots\} \tag{4.2-3}$$

and

$$\left\{\frac{1}{\sqrt{\pi}}, \quad \sqrt{2/\pi}\cos nx, \qquad n = 1, 2, \ldots\right\}. \tag{4.2-4}$$

*Fourier's paper was submitted in 1807 but was rejected by Lagrange, one of the examiners, presumably because he failed to understand how the motion of a vibrating string, for example, could be periodic in a spatial coordinate.

We leave it as an exercise to show that the set

$$\left\{\frac{1}{\sqrt{2\pi}}, \frac{1}{\sqrt{\pi}} \cos nx, \frac{1}{\sqrt{\pi}} \sin nx, \qquad n = 1, 2, \ldots\right\} \tag{4.2-5}$$

is an orthonormal set on $[-\pi, \pi]$ with weight function one (Exercise 2).

By using the orthonormal set (4.2-5) we can obtain the **Fourier series representation** of $f(x)$:

$$f(x) \sim \frac{1}{2} a_0 + \sum_{n=1} a_n \cos nx + b_n \sin nx. \tag{4.2-6}$$

In this representation the **Fourier coefficients** are given by

$$a_n = \frac{1}{\pi} \int_{-\pi}^{\pi} f(x) \cos nx \, dx, \qquad n = 0, 1, 2, \ldots \tag{4.2-7}$$

and

$$b_n = \frac{1}{\pi} \int_{-\pi}^{\pi} f(x) \sin nx \, dx, \qquad n = 1, 2, \ldots. \tag{4.2-8}$$

We observe that if a function $f(x)$ is to have a valid Fourier series representation as given by (4.2-6) on an interval extending beyond $[-\pi, \pi]$, then it must be **periodic**. This is due to the fact that the individual terms of the series are, themselves, periodic. Recall that a function is *periodic of period* p if*

$$f(x + p) = f(x)$$

for all values of x. Each term in the representation of (4.2-6) is periodic of period 2π. (Note that a constant function $f(x) = k$ satisfies the above definition of periodicity for any value of p.) Thus a function must be periodic of period 2π in order for the Fourier representation (4.2-6) to hold outside of $[-\pi, \pi]$. This requirement is not a restriction in case we are interested only in representing the function on $(0, \pi)$, since then it is unimportant to what values the series might converge outside this interval.

A second property that the function $f(x)$ must have is that it must be **piecewise smooth**. A function $f(x)$ is said to be piecewise smooth if $f(x)$ and $f'(x)$ are *both* piecewise continuous. See Fig. 4.2-1 for an example of a piecewise smooth function. (See also Exercise 4.)

Sufficient (but not necessary) conditions for obtaining a Fourier series representation of a function are given in Theorem 4.2-1. These conditions are

*The smallest value of p for which the relation holds is called the **fundamental period** of f.

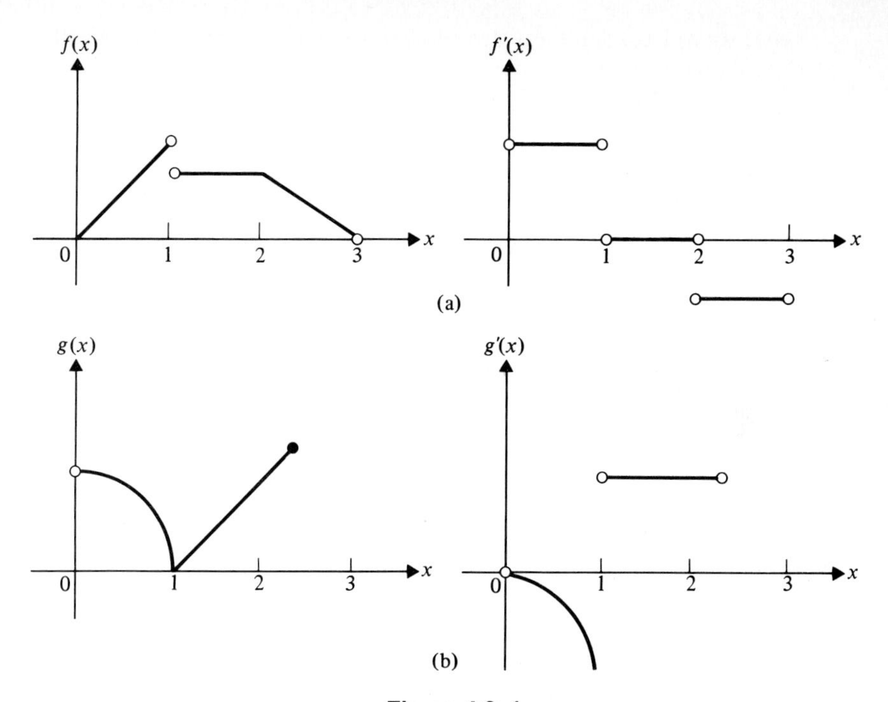

Figure 4.2-1
(a) *f*(*x*) and *f*′(*x*) are both piecewise continuous; (b) *g*(*x*) is piecewise continuous, but *g*′(*x*) is not.

called the **Dirichlet conditions**, since they appeared in papers of Dirichlet published in 1829* and 1837.†

THEOREM 4.2-1
If $f(x)$ is a periodic function of period 2π and piecewise smooth for $-\pi \leq x \leq \pi$, then the Fourier series representation (4.2-6) of $f(x)$ converges to $f(x)$ at all points where $f(x)$ is continuous and converges to the average of the left- and right-hand limits of $f(x)$ where $f(x)$ is discontinuous.

At the present time the convergence problem for a Fourier series is still unsolved. The Dirichlet conditions are known not to be necessary, and necessary conditions that have been obtained are known not to be sufficient. Suffice it to say that the conditions stated in Theorem 4.2-1 are satisfied by a large number of the functions that we encounter in applied mathematics;

*"Sur la convergence des séries trigonométriques qui servent à representer une fonction arbitraire entre des limites données," *J. rei. ang. Math.*, *4* (1829), pp. 157-169.

†"Über die Darstellung ganz willkürlicher Funktionen durch Sinus- und Cosinusreihen," *Report. Phys.*, *1* (1837), pp. 152-174.

hence we will not delve any deeper into the theory at this point. We illustrate in the following examples how Fourier series representations are obtained.

EXAMPLE 4.2-1 Find the Fourier series representation of the function

$$f(x) = \begin{cases} 0, & \text{for } -\pi < x < 0, \\ x, & \text{for } 0 < x < \pi, \\ \pi/2, & \text{for } x = \pm\pi, \end{cases}$$

$$f(x + 2\pi) = f(x).$$

Solution A sketch of the function is shown in Fig. 4.2-2. Note that the function satisfies the Dirichlet conditions of Theorem 4.2-1. Using Eq. (4.2-7), we have

$$a_n = \frac{1}{\pi}\int_{-\pi}^{\pi} f(s)\cos ns\,ds = \frac{1}{\pi}\int_0^{\pi} s\cos ns\,ds$$

$$= \frac{1}{\pi}\left(\frac{s}{n}\sin ns + \frac{1}{n^2}\cos ns\right)\Bigg|_0^{\pi}$$

$$= \frac{1}{\pi n^2}(\cos n\pi - 1)$$

$$= \frac{1}{\pi n^2}[(-1)^n - 1] = \begin{cases} 0, & \text{if } n \text{ is even}, \\ -\dfrac{2}{\pi n^2}, & \text{if } n \text{ is odd}. \end{cases}$$

The integration can be done by parts, or tables may be used. We have also used the relation $\cos n\pi = (-1)^n$ in the computation. Note that the above scheme for obtaining a_n is not valid for $n = 0$. (Why?) We can, however, put $n = 0$ *before* the integration so that

$$a_0 = \frac{1}{\pi}\int_0^{\pi} s\,ds = \frac{\pi}{2}.$$

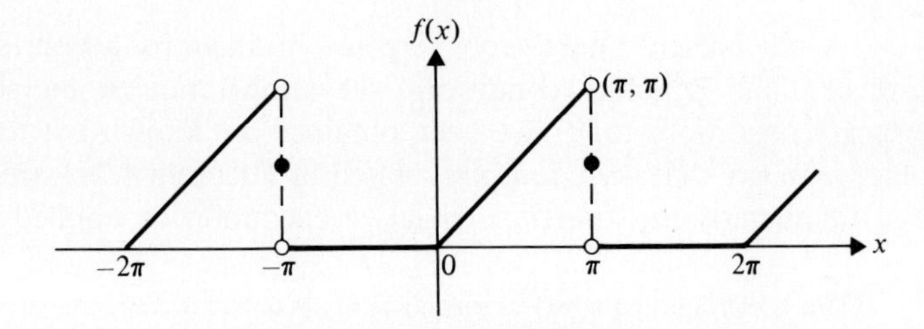

Figure 4.2-2
The function of Example 4.2-1.

The constant term in the Fourier series, namely, $1/2a_0$, is the **average value of the function** being represented over the given interval.* Continuing by using Eq. (4.2–8), we have

$$b_n = \frac{1}{\pi}\int_{-\pi}^{\pi} f(s)\sin ns\, ds = \frac{1}{\pi}\int_{0}^{\pi} s\sin ns\, ds$$

$$= \frac{1}{\pi}\left(-\frac{s}{n}\cos ns + \frac{1}{n^2}\sin ns\right)\Big|_0^{\pi}$$

$$= \frac{1}{n}(-\cos n\pi) = \frac{(-1)^{n+1}}{n}.$$

Thus the function can be represented as

$$f(x) = \frac{\pi}{4} - \frac{2}{\pi}\left(\frac{\cos x}{1^2} + \frac{\cos 3x}{3^2} + \frac{\cos 5x}{5^2} + \cdots\right)$$

$$+ \left(\frac{\sin x}{1} - \frac{\sin 2x}{2} + \frac{\sin 3x}{3} - + \cdots\right),$$

or, if we use summation notation,

$$f(x) = \frac{\pi}{4} + \sum_{n=1}^{\infty}\left[-\frac{2}{\pi}\frac{\cos(2n-1)x}{(2n-1)^2} + \frac{(-1)^{n+1}}{n}\sin nx\right]. \quad \blacksquare \qquad \textbf{(4.2–9)}$$

A remark is necessary now about the use of the equality in Eq. (4.2–9). For those values of x for which $f(x)$ is continuous, the equality is appropriate in the sense that the series converges to the function. In other words, the more terms we add, the closer the sum will be to the functional value at that point. This pointwise convergence does not hold, however, at points of discontinuity, since there the convergence is to the average of the left- and right-hand limits. Hence we will use the symbol $\sim$ rather than the equality symbol for the Fourier series representation of a function. Accordingly, Eq. (4.2–9) becomes

$$f(x) \sim \frac{\pi}{4} + \sum_{n=1}^{\infty}\left[-\frac{2}{\pi}\frac{\cos(2n-1)x}{(2n-1)^2} + \frac{(-1)^{n+1}}{n}\sin nx\right],$$

and we read the symbol $\sim$ as "has the representation" in the sense of Theorem 4.2-1.

We can graph approximations to the infinite series in Eq. (4.2–9) in the

---

*Recall from calculus that the average (or mean) value of a continuous function $f(x)$ on the interval $a \le x \le b$ is given by

$$\frac{1}{b-a}\int_a^b f(x)\,dx.$$

form of *finite* partial sums. Let the sum of the first N terms of the infinite series be denoted by S_N. Then we have

$$S_N = \frac{\pi}{4} + \sum_{n=1}^{N}\left[-\frac{2}{\pi}\frac{\cos(2n-1)x}{(2n-1)^2} + \frac{(-1)^{n+1}}{n}\sin nx\right]. \tag{4.2-10}$$

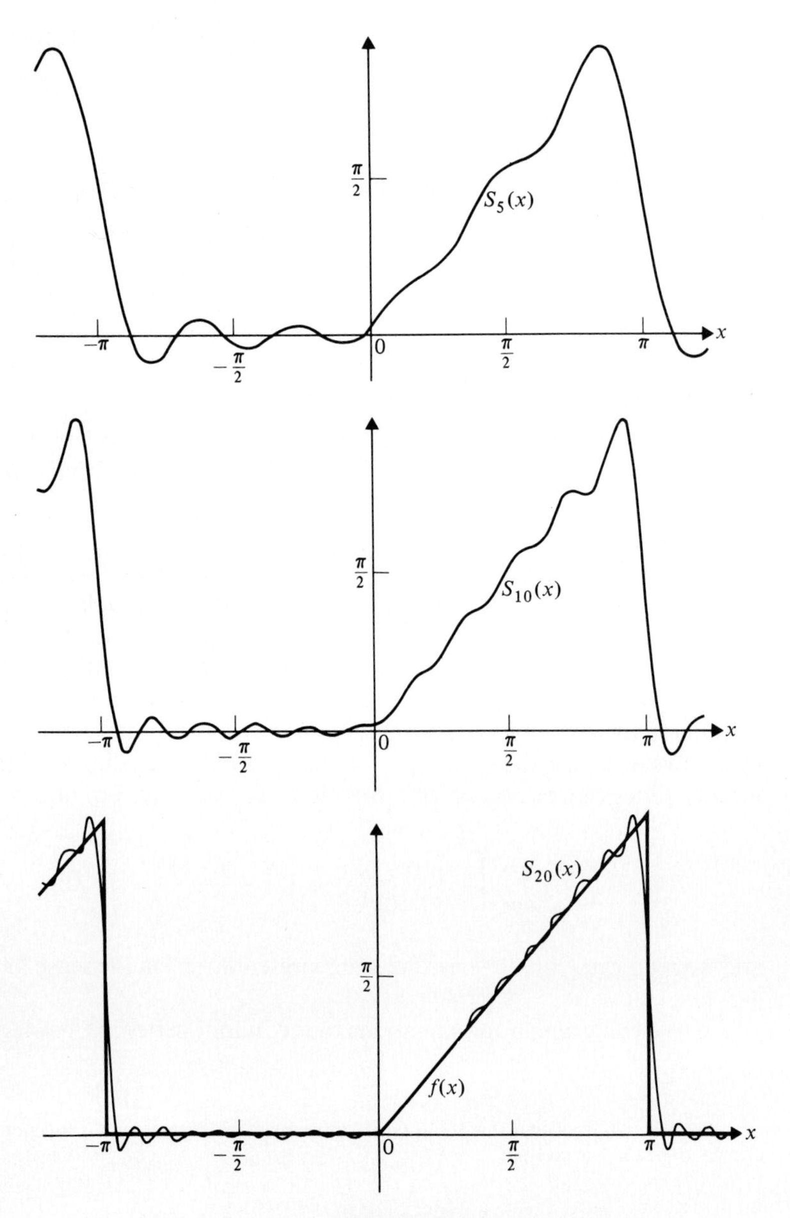

Figure 4.2-3
The graph of Eq. (4.2-10) for N = 5, 10, 20.

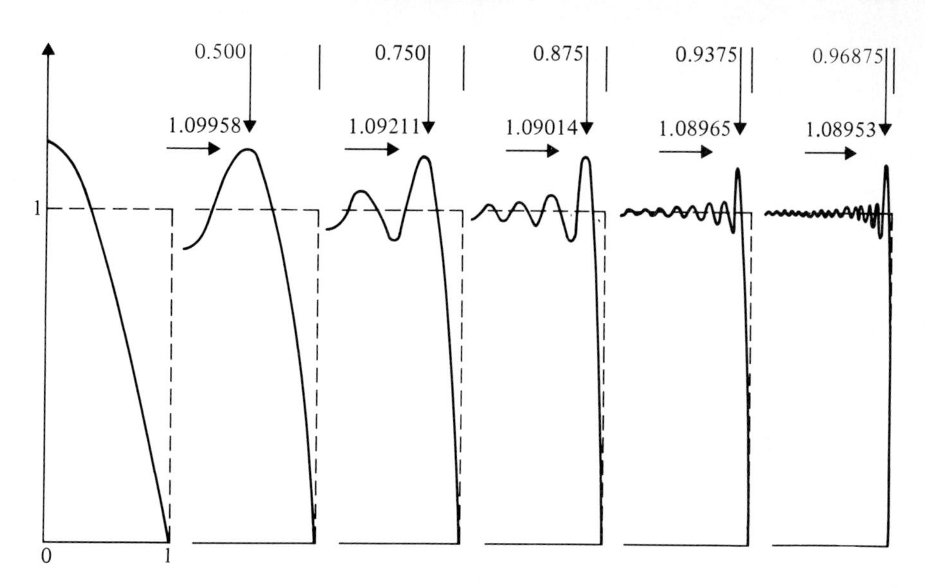

Figure 4.2-4
The Gibbs phenomenon. (From Kurt B. Wolf, *Integral Transforms in Science and Engineering* (New York: Plenum Press, 1979), by permission of the author and publisher.)

Graphs of Eq. (4.2–10) are shown in Fig. 4.2–3 for $N = 5$, 10, and 20. Note the fairly good correspondence of S_5 to the function being represented.

In Fig. 4.2–3 the **overshoot** at $x = \pi^-$ and the **undershoot** at $x = \pi^+$ are characteristics of Fourier (and other) series representations at points of discontinuity and are known as the **Gibbs phenomenon.*** The overshoot and undershoot together generally amount to about 18 percent of the distance between the functional values at a discontinuity. This phenomenon persists even though a large number of terms are summed.† A detailed computer-generated diagram of the Gibbs phenomenon is shown in Fig. 4.2–4. Here the function $f(x) = 1$ is being represented by S_N with $N = 2, 4, 8, 16, 32$, and 64. With increasing N the highest peak moves closer to the discontinuity at $x = 1$, but the overshoot remains approximately 1.09.

EXAMPLE 4.2–2 Obtain the Fourier series representation of the function

$$f(x) = \begin{cases} x + 2, & \text{for } -2 < x < 0, \\ 1, & \text{for } 0 < x < 2, \end{cases}$$

$$f(x + 4) = f(x).$$

---

*Josiah W. Gibbs (1839–1903), an American mathematical physicist, pointed this out in 1899 (*Nature, 59*, p. 606). In 1906, Maxime Bôcher (1867–1918), an American mathematician, gave a mathematical explanation of the phenomenon (*Annals of Math., 2* (7), p. 81).

†See also David Shelupsky, "Derivation of the Gibbs Phenomenon," *Amer. Math. Monthly, 87*, No. 3 (March 1980), pp. 210–212.

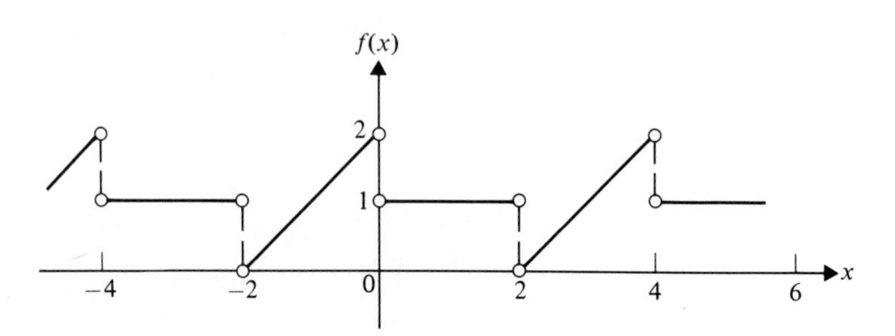

Figure 4.2-5
The function of Example 4.2-2.

Solution This function shown in Fig. 4.2-5 meets all the requirements of Theorem 4.2-1 except that the period is 4 instead of 2π. We can easily remedy this, however, by making a scale change according to the proportion

$$\frac{x}{\pi} = \frac{t}{2}.$$

Then $x = \pi t/2$, $dx = \pi dt/2$ so that Eqs. (4.2-7) and (4.2-8) become

$$a_n = \frac{1}{2}\int_{-2}^{2} f(s) \cos \frac{n\pi s}{2}\, ds, \qquad n = 0, 1, 2, \ldots,$$

and

$$b_n = \frac{1}{2}\int_{-2}^{2} f(s) \sin \frac{n\pi s}{2}\, ds, \qquad n = 1, 2, \ldots,$$

respectively. Observe that the above change affects only the *independent* variable. Hence we have

$$\begin{aligned} a_n &= \frac{1}{2}\int_{-2}^{0} (s + 2) \cos \frac{n\pi s}{2}\, ds + \frac{1}{2}\int_{0}^{2} \cos \frac{n\pi s}{2}\, ds \\ &= \frac{4}{n^2\pi^2} \quad \text{if } n \text{ is odd, zero otherwise;} \\ a_0 &= 2 \\ b_n &= \frac{1}{2}\int_{-2}^{0} (s + 2) \sin \frac{n\pi s}{2}\, ds + \frac{1}{2}\int_{0}^{2} \sin \frac{n\pi s}{2}\, ds \\ &= -\frac{2}{n\pi} \quad \text{if } n \text{ is even, zero otherwise.} \end{aligned}$$

Thus

$$f(x) \sim 1 + \frac{4}{\pi^2} \sum_{m=1}^{\infty} \frac{1}{(2m-1)^2} \cos \frac{(2m-1)\pi x}{2} - \frac{1}{\pi} \sum_{m=1}^{\infty} \frac{\sin m\pi x}{m}. \quad \blacksquare$$

(4.2–11)

Some useful information can be gained from the Fourier series representation in (4.2–11). If we put $x = 0$, we obtain

$$f(0) = 1 + \frac{4}{\pi^2} \sum_{m=1}^{\infty} \frac{1}{(2m-1)^2}.$$

We know, however, from Theorem 4.2–1 that $f(0) = 3/2$. (Why?) Hence

$$\sum_{m=1}^{\infty} \frac{1}{(2m-1)^2} = \frac{\pi^2}{8}, \tag{4.2–12}$$

a result that can be verified by other means (Exercise 20).

We caution the student that the problem of convergence of Fourier series is not a simple one. From Fig. 4.2–3 it might be tempting to conclude that a Fourier series converges (perhaps even uniformly) to the function it represents whenever the function is continuous. This, however, is not the case, as we shall see in Section 4.5.

Key Words and Phrases

Fourier series representation	Dirichlet conditions
Fourier coefficients	average value of a function
periodic function	overshoot
fundamental period	undershoot
piecewise smooth function	Gibbs phenomenon

4.2 Exercises

• **1.** Obtain the orthonormal sets shown in (4.2–3) and (4.2–4) from the orthogonal sets given in the text.

2. Show that the set in (4.2–5) is orthonormal on $[-\pi, \pi]$ with weight function one.

3. Verify that the Fourier coefficients shown in Eqs. (4.2–7) and (4.2–8) apply to the Fourier series representation in (4.2–6). (*Hint*: Use Eq. 4.2–2.)

4. Show that the function

$$g(x) = \begin{cases} \sqrt{1 - x^2}, & \text{for } 0 < x < 1, \\ x, & \text{for } 1 < x < 2, \end{cases}$$

is piecewise continuous but not piecewise smooth on (0, 2).

5. Obtain the Fourier coefficients shown for a function $f(x)$ that is piecewise smooth on $[-L, L]$ and periodic of period $2L$:

$$a_n = \frac{1}{L}\int_{-L}^{L} f(x)\cos\frac{n\pi}{L}x\,dx, \qquad n = 0, 1, 2, \ldots, \tag{4.2–13}$$

$$b_n = \frac{1}{L}\int_{-L}^{L} f(x)\sin\frac{n\pi}{L}x\,dx, \qquad n = 1, 2, \ldots. \tag{4.2–14}$$

•• 6. The smallest value of p for which $f(x + p) = f(x)$ is called the *fundamental period* of the function $f(x)$. Find the fundamental period of each of the following functions.

(a) $\sin \frac{1}{2}x$ **(b)** $\cos 2x$ **(c)** $\cos 3\pi x$ **(d)** $\sin \pi x$

(e) $\cos \frac{\pi}{2}x$

7. Prove that if a function is periodic of period p, then it is also periodic of period np, $n = \pm 1, \pm 2, \ldots$.

8. Show that the function $f(x) = c$, where c is a constant, is periodic of period p for any value of p.

9. Obtain the Fourier series representations of each of the following functions.

(a) $$f(x) = \begin{cases} x, & \text{for } -\pi < x < 0, \\ 0, & \text{for } 0 < x < \pi, \end{cases}$$
$f(x + 2\pi) = f(x)$

(b) $$f(x) = \begin{cases} 1, & \text{for } -2 < x < 0, \\ x - 2, & \text{for } 0 < x < 2, \end{cases}$$
$f(x + 4) = f(x)$

(c) $$f(x) = \begin{cases} 2, & \text{for } -1 < x < 0, \\ 0, & \text{for } 0 < x < 1, \end{cases}$$
$f(x + 2) = f(x)$

(d) $$f(x) = \begin{cases} 0, & \text{for } -1 < x < 0, \\ \sin \pi x, & \text{for } 0 < x < 1, \end{cases}$$
$f(x + 2) = f(x)$

(e) $$f(x) = \begin{cases} 0, & \text{for } -\frac{1}{2} < x < 0, \\ x^2, & \text{for } 0 < x < \frac{1}{2}, \end{cases}$$
$f(x + 1) = f(x)$

(f) $f(x) = \cos \pi x$, for $-1 < x < 1$,
$f(x + 2) = f(x)$

(g) $$f(x) = \begin{cases} x, & \text{for } 0 \le x < \pi, \\ 0, & \text{for } \pi < x \le 2\pi, \end{cases}$$
$f(x + 2\pi) = f(x)$

(h) $f(x) = x^2$, for $-\pi \le x \le \pi$,
$f(x + 2\pi) = f(x)$

(i) $f(x) = \begin{cases} \dfrac{2}{\pi}x + 1, & \text{for } -\pi \le x \le 0, \\ 1 - \dfrac{2}{\pi}x, & \text{for } 0 \le x \le \pi, \end{cases}$

$f(x + 2\pi) = f(x)$

(j) $f(x) = \begin{cases} 0, & \text{for } -5 < x < 0, \\ 3, & \text{for } 0 < x < 5, \end{cases}$

$f(x + 10) = f(x)$

10. Assume that the Fourier series representation has been obtained for the following function.

$$f(x) = \begin{cases} 2, & \text{for } -1 < x < 0, \\ 1, & \text{for } 0 < x < 1, \\ 3, & \text{for } x = 1, \\ 2(2 - x), & \text{for } 1 < x < 2, \end{cases}$$

$$f(x + 3) = f(x)$$

To what value will the Fourier series converge at each of the following points?

(a) $x = -1$ **(b)** $x = 0$ **(c)** $x = 1$ **(d)** $x = 2$

(e) $x = \frac{3}{2}$ **(f)** $x = 3$ **(g)** $x = -2$

***11.** Use Eq. (4.2–10) to compute $f(\pi/2)$ for $N = 5$, 10, and 20, and compare these values with the true value.

***12.** Use the result (4.2–11) to compute $f(1)$ taking m to 5, 10, and 20, and compare these values with the true value.

13. To what value will the Fourier series Eq. (4.2–9) converge at each of the following values of x?

(a) $x = -\pi$ **(b)** $x = 0$ **(c)** $x = \pi$ **(d)** $x = -\dfrac{3\pi}{2}$

(e) $x = 3\pi$

14. To what value will the Fourier series (4.2–11) converge at each of the following values of x?

(a) $x = -2$ **(b)** $x = -1$ **(c)** $x = 0$ **(d)** $x = 1$

(e) $x = 2$ **(f)** $x = 3$

15. Show that the set

$$\left\{ \frac{1}{\sqrt{2c}}, \frac{1}{\sqrt{c}} \cos \frac{n\pi}{c} x, \frac{1}{\sqrt{c}} \sin \frac{m\pi}{c} x, n, m = 1, 2, \ldots \right\}$$

is an orthonormal set on $[a, a + 2c]$ with weight function one for any value of a.

16. Referring to Exercise 15, show that the coefficients in the Fourier series representation of a function that is piecewise smooth on $[a, a + 2c]$ are given by

$$a_n = \frac{1}{c} \int_a^{a+2c} f(s) \cos \frac{n\pi s}{c} \, ds, \qquad n = 0, 1, 2, \ldots \qquad \textbf{(4.2–15)}$$

---

*Calculator problem.

and

$$b_n = \frac{1}{c}\int_a^{a+2c} f(s)\sin\frac{n\pi s}{c}\,ds, \qquad n = 1, 2, \ldots . \tag{4.2-16}$$

*** 17.** Use 20 terms of the series in Eq. (4.2-12) to compute $\pi^2/8$. What is the relative error?

••• 18. Verify that the expansion

$$(\arcsin x)^2 = \frac{1}{2}\sum_{n=1}^{\infty}\frac{(n-1)^2}{(2n)!}(2x)^{2n}$$

is valid for $-1 \le x \le 1$.

19. Let $x = \frac{1}{2}$ in the series expansion of Exercise 18, and obtain

$$\sum_{n=1}^{\infty}\frac{1}{n^2} = \frac{\pi^2}{6}.$$

20. Use the result in Exercise 19 to show that

$$\sum_{n=1}^{\infty}\frac{1}{(2n-1)^2} = \frac{\pi^2}{8}.$$

21. Demonstrate that the set of trigonometric functions

$$\left\{\frac{1}{\sqrt{2\pi}}, \frac{\cos nx}{\sqrt{\pi}}, n = 1, 2, \ldots\right\}$$

is not **complete** with respect to the class of *continuous* functions on $[-\pi, \pi]$ by showing that $f(x) = \sin nx$, $n = 1, 2, \ldots$, cannot be represented by a series of functions in the given set.

22. Show that the formula for the average value of a continuous function over an interval is applicable to piecewise continuous functions.

23. Generalize Theorem 4.2-1 for a function $f(x)$ that is periodic of period $2L$ and piecewise smooth for $a \le x \le a + 2L$, where a is a constant.

24. Study the solution to Exercise 19 given in A. M. Yaglom and I. M. Yaglom, *Challenging Mathematical Problems with Elementary Solutions*, Vol. 2 (San Francisco: Holden-Day, 1967), problem 145a.

25. Read the paper "Still Another Elementary Proof That $\sum 1/k^2 = \pi^2/6$," by Daniel P. Giesy in *Math. Mag. 45* (1972), pp. 148–149.

26. Obtain the Fourier series expansions of the following functions.

(a) $f(x) = e^x$, for $0 < x < 2\pi$, $f(x + 2\pi) = f(x)$

(b) $f(x) = \cos ax$, for $-\pi < x < \pi$, $f(x + 2\pi) = f(x)$

27. Using the result of Exercise 26(a), find the sum of the series

$$\sum_{n=1}\frac{1}{n^2+1}.$$

4.3 COSINE, SINE, AND EXPONENTIAL SERIES

In many of the applications of Fourier series a function $f(x)$ is defined on an interval $0 < x < L$. We may then represent this function either as a series consisting only of sine terms or as one consisting only of cosine terms.* This is accomplished by making either an odd or an even **periodic extension** of the given function, respectively.

A function $F(x)$ is called an **odd function** if it has the property

$$F(-x) = -F(x).$$

Odd functions have graphs that are symmetrical about the origin as shown in Fig. 4.3-1. Some examples of odd functions are x, x^3, x^5, $\sin x$, $\tan x$, $\sinh x$, and $\csc x$.

A function $F(x)$ is called an **even function** if it has the property

$$F(-x) = F(x).$$

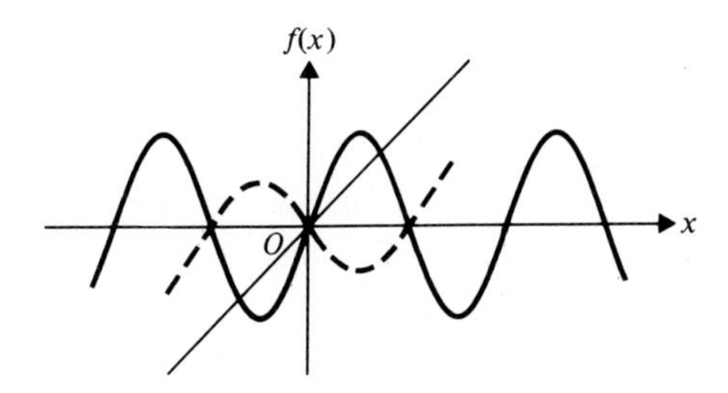

Figure 4.3-1
Odd functions.

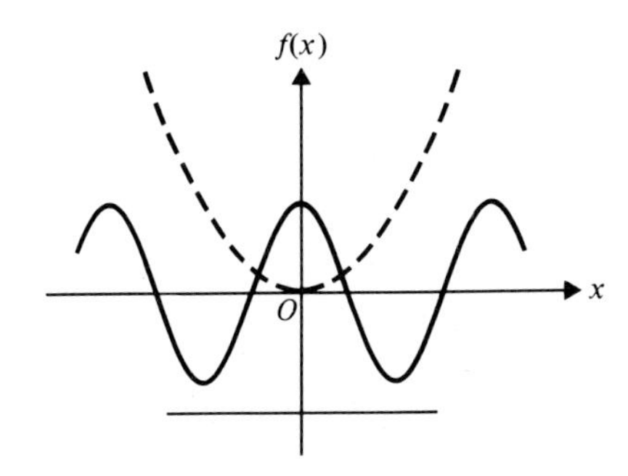

Figure 4.3-2
Even functions.

*We will see later that certain boundary conditions will lead to these specific requirements.

Even functions have graphs that are symmetrical about the line $x = 0$ as shown in Fig. 4.3–2. Some examples of even functions are 1, x^2, x^4, $\cos x$, $\cosh x$, and $\sec x$.

Generalizing the formulas (4.2–6), (4.2–7), and (4.2–8) of Section 4.2, we have (Exercise 1)

$$f(x) \sim \frac{1}{2}a_0 + \sum_{n=1}^{\infty} a_n \cos\frac{n\pi x}{L} + b_n \sin\frac{n\pi x}{L} \tag{4.3–1}$$

where

$$a_n = \frac{1}{L}\int_{-L}^{L} f(s) \cos\frac{n\pi s}{L}\, ds, \qquad n = 0, 1, 2, \ldots, \tag{4.3–2}$$

and

$$b_n = \frac{1}{L}\int_{-L}^{L} f(s) \sin\frac{n\pi s}{L}\, ds, \qquad n = 1, 2, \ldots. \tag{4.3–3}$$

Note that these three reduce to Eqs. (4.2–6), (4.2–7), and (4.2–8) when $L = \pi$ as they should.

Cosine Series

If $f(x)$ is defined on $0 < x < L$ and is to be represented by a series of cosines, we make an **even periodic extension** of $f(x)$, as shown in Fig. 4.3–3. The resulting function, defined by

$$f(x) = \begin{cases} f(x), & 0 < x < L, \\ f(-x), & -L < x < 0, \end{cases}$$

$$f(x + 2L) = f(x), \qquad f(0) = \lim_{\epsilon \to 0^+} f(\epsilon),$$

$$f(L) = \lim_{\epsilon \to 0} f(L - \epsilon),$$

is an even periodic function; hence $b_n \equiv 0$, $n = 1, 2, \ldots$, and Eq. (4.3–2) can

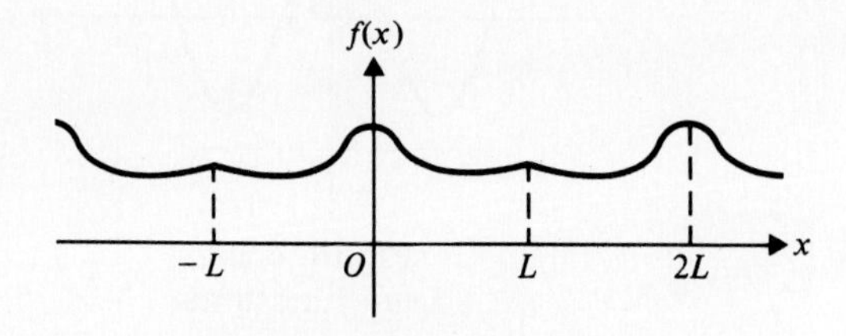

Figure 4.3–3
An even periodic extension.

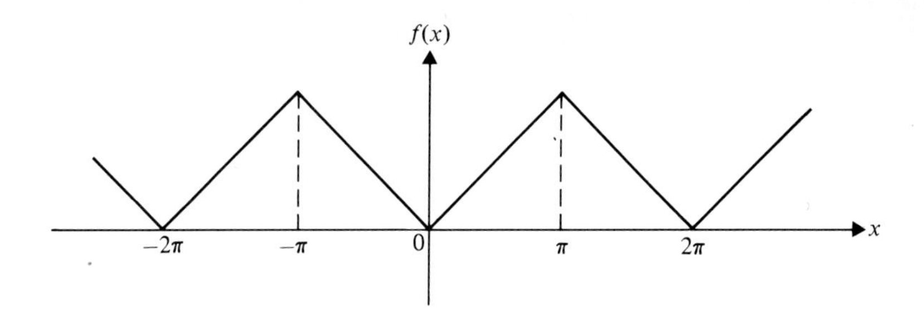

Figure 4.3–4
The even periodic extension in Example 4.3–1.

be simplified to (Exercises 2 and 3)

$$a_n = \frac{2}{L}\int_0^L f(s)\cos\frac{n\pi s}{L}\,ds, \qquad n = 0, 1, 2, \ldots. \tag{4.3–4}$$

EXAMPLE 4.3–1 Obtain the Fourier cosine series representation of the function $f(x) = x$, $0 \le x \le \pi$.

Solution The even periodic extension of the function is shown in Fig. 4.3–4.

Using Eq. (4.3–4), we have (Exercise 4)

$$a_n = \frac{2}{\pi}\int_0^\pi s\cos ns\,ds$$

$$a_n = \begin{cases} 0, & \text{if } n \text{ is even,} \\ -\dfrac{4}{\pi n^2}, & \text{if } n \text{ is odd,} \end{cases}$$

$$a_0 = \frac{2}{\pi}\int_0^\pi s\,ds = \pi.$$

Hence the required series representation can be written as

$$f(x) \sim \frac{\pi}{2} - \frac{4}{\pi}\sum_{m=1}^{\infty}\frac{\cos(2m-1)x}{(2m-1)^2}. \quad \blacksquare \tag{4.3–5}$$

Sine Series

If $f(x)$ is defined on $0 < x < L$ and is to be represented by a series of sines, we make an **odd periodic extension** of $f(x)$, as shown in Fig. 4.3–5. The resulting

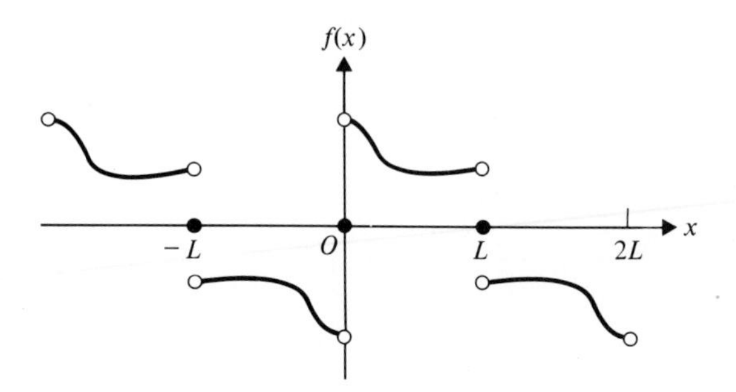

Figure 4.3-5
An odd periodic extension.

function, defined by

$$f(x) = \begin{cases} f(x), & 0 < x < L, \\ -f(-x), & -L < x < 0, \end{cases}$$

$$f(x + 2L) = f(x), \qquad f(0) = f(L) = 0,$$

is an odd periodic function; hence $a_n \equiv 0$, $n = 0, 1, 2, \ldots$, and the formula (4.3-3) can be simplified to (Exercises 5 and 6)

$$b_n = \frac{2}{L}\int_0^L f(s) \sin \frac{n\pi s}{L}\, ds, \qquad n = 1, 2, 3, \ldots. \tag{4.3-6}$$

EXAMPLE 4.3-2 Obtain the Fourier sine series representation of the function $f(x) = x$, $0 \le x < \pi$, $f(\pi) = 0$.

Solution The odd periodic extension of the function is shown in Fig. 4.3-6. Using Eq. (4.3-6), we have

$$b_n = \frac{2}{\pi}\int_0^\pi s \sin ns\, ds = \frac{2}{n}(-1)^{n+1}$$

so that the required series can be written as (Exercise 7)

$$f(x) \sim 2\sum_{n=1}^{\infty} \frac{(-1)^{n+1} \sin nx}{n}. \quad \blacksquare \tag{4.3-7}$$

It is to be noted that although the expressions (4.2-9), (4.3-5), and (4.3-7) are all different, they all represent the function $f(x) = x$ on the interval $0 < x < \pi$. This versatility of Fourier series representation makes the technique a valuable one in applied mathematics. We have already seen that some boundary-value problems require that a given function be represented by a series of sines (compare Eq. (3.2-12)), and we shall see in Chapters 6 and 7 that

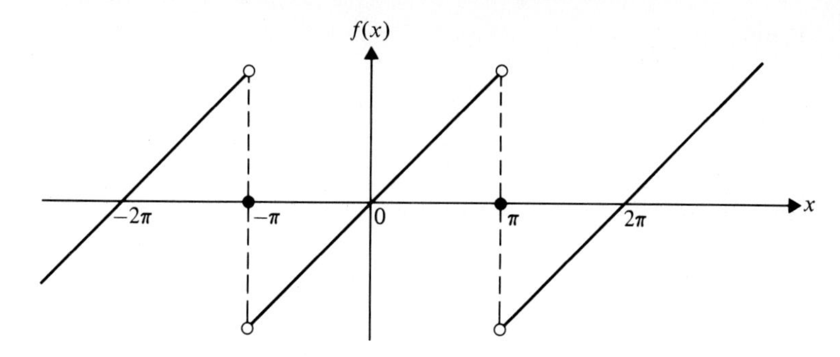

Figure 4.3-6
The odd periodic extension in Example 4.3-2.

other problems may require a series of cosines. The exercises at the end of this section are designed to illustrate both types of representations.

Exponential Series

Another useful form of Fourier series is the exponential form or complex form. This is obtained from the standard trigonometric form (see Eq. (4.2-7))

$$f(x) \sim \frac{1}{2}a_0 + \sum_{n=1}^{\infty} a_n \cos nx + b_n \sin nx$$

by using Euler's formulas

$$e^{inx} = \cos nx + i \sin nx \qquad \text{and} \qquad e^{-inx} = \cos nx - i \sin nx.$$

We have, by adding and subtracting these formulas,

$$\cos nx = \frac{1}{2}(e^{inx} + e^{-inx}) \qquad \text{and} \qquad \sin nx = \frac{1}{2i}(e^{inx} - e^{-inx}),$$

so that we can write

$$f(x) \sim \frac{1}{2}a_0 + \frac{1}{2}\sum_{n=1}^{\infty} a_n(e^{inx} + e^{-inx}) - ib_n(e^{inx} - e^{-inx})$$

$$= \frac{1}{2}a_0 + \frac{1}{2}\sum_{n=1}^{\infty} (a_n - ib_n)e^{inx} + (a_n + ib_n)e^{-inx}.$$

If we now define **complex Fourier coefficients** by

$$c_0 = \tfrac{1}{2}a_0, \qquad c_n = \tfrac{1}{2}(a_n - ib_n), \qquad c_{-n} = \tfrac{1}{2}(a_n + ib_n), \qquad n = 1, 2, \ldots, \tag{4.3-8}$$

then we have

$$f(x) \sim \sum_{n=-\infty}^{\infty} c_n e^{inx}, \tag{4.3-9}$$

which is the **exponential form** (or complex form) of Fourier series. It is commonly used in physics and engineering because of its notational simplicity.

In the above development of the exponential form we have assumed that $f(x)$ had period 2π, but this was done only to simplify the notation. For period $2L$ we would write

$$f(x) \sim \sum_{n=-\infty}^{\infty} c_n e^{in\pi x/L}$$

with a corresponding modification of the c_n. Details of this as well as the computation of the formula

$$c_n = \frac{1}{2\pi}\int_{-\pi}^{\pi} f(s)\, e^{-ins}\, ds, \qquad n = 0, \pm 1, \pm 2, \ldots, \tag{4.3-10}$$

from the a_n and b_n are left to the exercises. It would be advisable at this point to review the subject of **hermitian orthogonality** in Section 2.6 (see especially Example 2.6-1).

Key Words and Phrases

periodic extension
odd function
even function
cosine series
even periodic extension
sine series
odd periodic extension
exponential series
complex Fourier coefficients
exponential form
hermitian orthogonality

4.3 Exercises

• **1.** **(a)** By making the change of variable $x = \pi s/L$, verify that Eqs. (4.2-7) and (4.2-8) become Eqs. (4.3-2) and (4.3-3).
(b) Show that Eq. (4.3-1) is the equivalent of Eq. (4.2-6) when the change of variable in part (a) is made.

2. Show that if $f(x)$ is an even function, then $b_n \equiv 0$, $n = 1, 2, \ldots$, in Eq. (4.3-3).

3. Show that if $f(x)$ is an even function, then the formula for a_n, Eq. (4.3-2), can be simplified as shown in Eq. (4.3-4).

4. In Example 4.3-1, compute a_0 and a_n, $n = 1, 2, \ldots$.

5. Show that if $f(x)$ is an odd function, then $a_n \equiv 0$, $n = 0, 1, 2, \ldots$, in Eq. (4.3–2).

6. Show that if $f(x)$ is an odd function, then the formula for b_n, Eq. (4.3–3), can be simplified as shown in Eq. (4.3–6).

7. In Example 4.3–2, compute b_n, $n = 1, 2, \ldots$.

8. Verify that c_0, c_n, and c_{-n} as defined in Eq. (4.3–8) are given by Eq. (4.3–10).

9. State the formula for c_n when the period is $2L$.

•• 10. Verify that each of the following functions is an odd function by showing that each has the property $F(-x) = -F(x)$.

(a) x^3 (b) $\tan x$ (c) $\csc x$ (d) $\sinh x$

(e) $x \exp(-x^2)$

11. Verify that each of the following functions is an even function by showing that each has the property $F(-x) = F(x)$.

(a) 1 (b) $\cos x$ (c) $\sec x$ (d) $\cosh x$

(e) $x \tan x$ (f) $\cos x \exp(-x^2)$

12. Show that each of the following functions is neither odd nor even.

(a) $ax^2 + bx + c$ (b) $\log x$ (c) e^x (d) $x^2/(1 + x)$

13. Obtain the following multiplication table pertaining to multiplication of odd and even functions:

×	Odd	Even
Odd	Even	Odd
Even	Odd	Even

(*Note*: The entry in the first line and second column of the table shows that the product of an odd function and an even function is an odd function.)

In Exercises 14–19, obtain (a) the Fourier sine and (b) the Fourier cosine representations of the given function.

14. $f(x) = 2 - x, \quad 0 < x \le 2$

15. $f(x) = a, \quad 0 < x \le 3, \quad a > 0$

16. $f(x) = x^2, \quad 0 \le x < 1$

17. $f(x) = e^x, \quad 0 < x < 2$

18. $f(x) = \sin \pi x, \quad 0 \le x < 1$

19. $f(x) = \cos x, \quad 0 < x \le \pi/2$

20. Represent the function of Exercise 16 by a Fourier series that contains both sines and cosines. (*Hint*: Define a periodic extension that is neither odd nor even. There are an infinite number of ways of doing this.)

21. Find a Fourier sine series representation of the function $x - 1$ on the open interval $1 < x < 2$. (*Note*: The solution is not unique.)

22. Find a Fourier cosine series representation of the function $x - 1$ on the open interval $1 < x < 2$. (*Note*: The solution is not unique.)

23. **(a)** Find the Fourier series representation of the function

$$f(x) = \begin{cases} x + 1, & -1 \leq x \leq 0, \\ -x + 1, & 0 \leq x \leq 1, \end{cases}$$

$$f(x + 2) = f(x).$$

(b) Use the result of part (a) to show that

$$1 + \frac{1}{3^2} + \frac{1}{5^2} + \cdots = \frac{\pi^2}{8}.$$

24. Find the Fourier sine series representation of the function $f(x) = 1, 0 < x < \pi$, $f(0) = f(\pi) = 0$.

25. Obtain each of the following series representations.

(a) $\dfrac{\pi}{4} = 1 - \dfrac{1}{3} + \dfrac{1}{5} - \dfrac{1}{7} + - \cdots$

(b) $\dfrac{\pi}{4} = \dfrac{\sqrt{2}}{2}\left(1 + \dfrac{1}{3} - \dfrac{1}{5} - \dfrac{1}{7} + + - - \cdots\right)$

(*Hint*: Use the result of Exercise 24 with appropriate values for x.)

26. Consider the following triangular function:

$$f(x) = \begin{cases} 2 - x, & 1 < x < 2, \\ x - 2, & 2 < x < 3, \end{cases}$$

$$f(x + 2) = f(x).$$

Obtain the Fourier series representation of this function.

27. Given the function

$$f(x) = \begin{cases} 1, & 0 < x < 2, \\ -1, & -2 < x < 0, \end{cases}$$

$f(x + 4) = f(x), f(0) = f(2) = 0$, write the exponential form of the Fourier series representation of this function.

28. Given the function

$$f(x) = 1, \qquad -\infty < x < \infty,$$

write the exponential form of the Fourier series representation of this function.

••• **29.** Show that although most functions are neither odd nor even, every function defined on $(-c, c)$ can be expressed as the sum of an even function and an odd function by using the identity

$$f(x) \equiv \tfrac{1}{2}[f(x) + f(-x)] + \tfrac{1}{2}[f(x) - f(-x)].$$

30. Apply the formula in Exercise 29 to the function $\exp x$ to show that

$$\exp x = \cosh x + \sinh x.$$

31. If $f(x) = x - x^2$ when $0 < x < 1$ and $f(x)$ is periodic of period 1, show that $f(x)$ is an even function.

32. Step functions occur in modeling off–on controls in mechanical systems. Such a function is given by

$$f(x) = (-1)^n h, \qquad n = 0, \pm 1, \pm 2, \ldots, n < x < n + 1.$$

(a) Sketch the function.
(b) Obtain the Fourier series representation of this function.
(c) Sketch the first term of the Fourier series representation.
(d) Sketch the first two terms of the Fourier series representation.

33. Show that the set

$$\left\{ \sqrt{2/L} \sin \frac{n\pi x}{L}, \qquad n = 1, 2, \ldots, \right\}$$

is orthonormal on $[0, L]$. (*Hint*: Use the fact that

$$\sin \frac{n\pi x}{L} = \frac{e^{in\pi x/L} - e^{-in\pi x/L}}{2i}.\Big)$$

34. Prove that a table for division, similar to the one in Exercise 13, holds for odd and even functions if certain restrictions are made.

35. Prove that the sum (and difference) of even (odd) functions is an even (odd) function.

36. Given the function

$$f(x) = 2 - x, \qquad 0 < x < 2,$$

define a function whose Fourier sine series representation will converge to $f(x)$ for *all finite* values of x. (*Note*: The solution is not unique.)

37. Prove that if $f(x)$ and $f'(x)$ are both even functions, then $f(x)$ is a constant.

38. Deduce from the result to Exercise 16 that

$$\sum_{n=1}^{\infty} \frac{(-1)^{n+1}}{n^2} = \frac{\pi^2}{12}.$$

39. Obtain the first five terms of the Fourier sine series representation of the function:

$$f(x) = \begin{cases} 0, & \text{for } 0 \le x < \pi/2, \\ 1, & \text{for } \pi/2 < x \le \pi. \end{cases}$$

40. Obtain the Fourier series representation of the following function:

$$f(x) = \begin{cases} 1 + x, & \text{for } 0 < x \le 1, \\ 3 - x, & \text{for } 1 \le x < 2, \end{cases}$$

$$f(x + 2) = f(x).$$

4.4 APPLICATIONS

In this section we illustrate how Fourier series are used to solve boundary-value problems. Wherever possible we point out the physical interpretation of each problem. We also discuss any assumptions that have to be made, especially with regard to boundary conditions.

EXAMPLE 4.4–1 Find the steady-state temperatures in the interior of a rectangular plate $0 < x < \pi$, $0 < y < b$, if the edge $y = 0$ has a temperature distribution given by $f(x) = 64x(\pi - x)$, $0 < x < \pi$, and the other three edges are kept at zero.

Solution We assume that the faces of the rectangular plate are perfectly insulated so that the problem is one of heat conduction (or diffusion) in the xy-plane. The diffusion equation in two dimensions is

$$u_t = k(u_{xx} + u_{yy}),$$

but since we are seeking the steady-state temperatures, the equation to be solved is the potential equation

$$u_{xx} + u_{yy} = 0.$$

In other words, the temperature u is independent of t and depends only on the spatial variables x and y.

We can phrase the problem in mathematical terms in the following way.

P.D.E.: $u_{xx} + u_{yy} = 0, \quad 0 < x < \pi, \quad 0 < y < b;$

B.C.: $u(0, y) = u(\pi, y) = 0, \quad 0 < y < b,$

$u(x, b) = 0, \quad u(x, 0) = 64x(\pi - x), \quad 0 < x < \pi.$

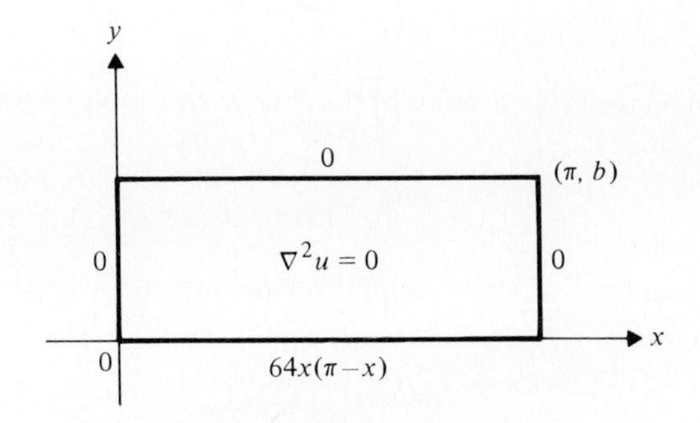

Figure 4.4–1
Boundary-value problem (Example 4.4–1).

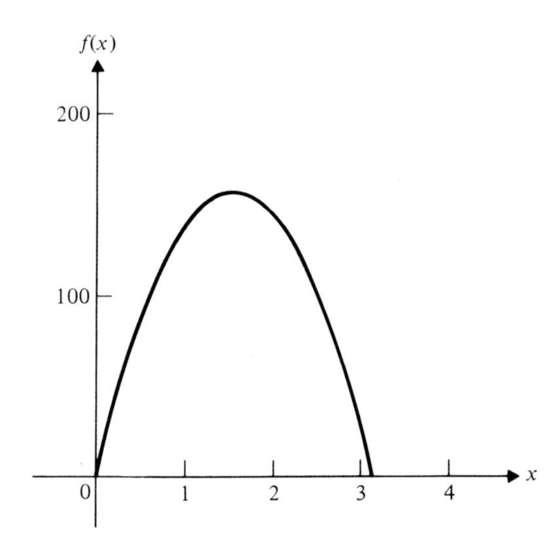

Figure 4.4–2
Temperature distribution (Example 4.4–1).

Figure 4.4–1 shows a sketch of this boundary-value problem; the temperature distribution $f(x)$ is shown in Fig. 4.4–2.

We solve the problem by the method of separation of variables as in Example 3.2–2 to obtain the n solutions given in Eq. (3.2–11)

$$u_n(x, y) = \frac{b_n}{\sinh nb} \sin nx \sinh n(b - y), \qquad n = 1, 2, \ldots, \quad \textbf{(4.4–1)}$$

where the b_n are arbitrary constants. Each of the functions in Eq. (4.4–1) satisfies the given P.D.E. and also the *homogeneous* boundary conditions. Hence we take for our complete solution a linear combination of solutions, that is,

$$u(x, y) = \sum_{n=1}^{\infty} \frac{b_n}{\sinh nb} \sin nx \sinh n(b - y).$$

When we apply the nonhomogeneous boundary condition, we have

$$u(x, 0) = \sum_{n=1}^{\infty} b_n \sin nx = 64x(\pi - x).$$

But this last is a Fourier sine series representation of the function $64x(\pi - x)$ on the interval $0 < x < \pi$. We can use Eq. (4.3–6) to find the b_n as outlined in

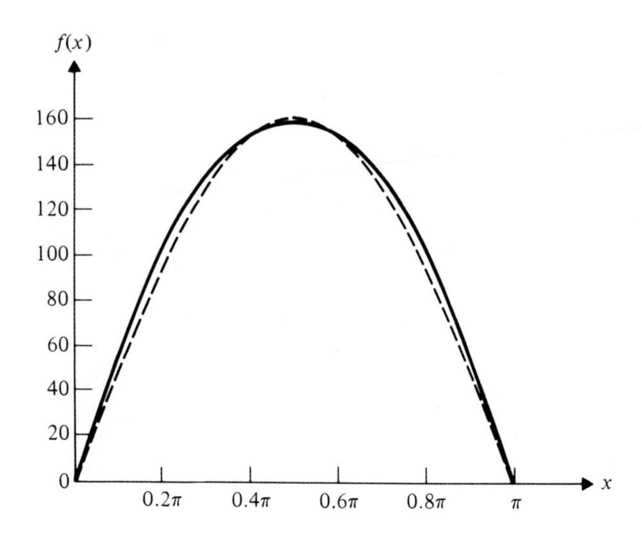

Figure 4.4-3
One-term approximation to the temperature distribution (Example 4.4-1).

Section 4.3. Thus (Exercise 2)

$$b_n = \frac{2}{\pi}\int_0^\pi 64x(\pi - x)\sin nx\, dx$$
$$= \begin{cases} 0, & \text{if } n \text{ is even,} \\ \dfrac{512}{\pi n^3}, & \text{if } n \text{ is odd.} \end{cases}$$

The complete solution is*

$$u(x, y) = \frac{512}{\pi}\sum_{m=1} \frac{\sin (2m - 1)x \sinh (2m - 1)(b - y)}{(2m - 1)^3 \sinh (2m - 1)b}, \tag{4.4-2}$$

where we have replaced n by $2m - 1$. ■

We call attention to the fact that the analytic solution in Eq. (4.4-2) may be used to evaluate $u(x, y)$ at various points. The computational work is not as forbidding as one would expect because the Fourier sine series representation of the function $64x(\pi - x)$ is a rapidly converging series. In Fig. 4.4-3 we show the function together with *one term* of its Fourier series representation to show the remarkable resemblance. In Fig. 4.4-4 we show some of the **isothermal curves** in the plate. Note that the temperature at the points (0, 0)

*In the remainder of this section we omit the upper limit on infinite series.

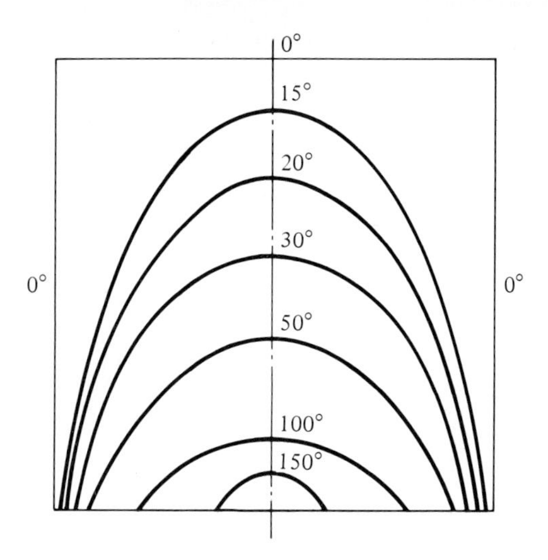

Figure 4.4–4
Isothermal curves (Example 4.4–1).

and $(\pi, 0)$ is zero because these are points at which the odd periodic extension of the function being represented is continuous.

In the next example we illustrate the use of the **principle of superposition** for solving a boundary-value problem.

EXAMPLE 4.4–2 Solve the following problem.

P.D.E.: $u_{xx} + u_{yy} = 0, \quad 0 < x < 1, \quad 0 < y < 2;$
B.C.: $u(0, y) = u(1, y) = 0, \quad 0 < y < 2,$
$u(x, 0) = u(x, 2) = 64x(1 - x), \quad 0 < x < 1.$

Solution This problem is similar to the one in Example 4.4–1, so we can use the results obtained there. As we are working on a different interval here, a few obvious changes will have to be made.

First, we solve the following similar problem:

P.D.E.: $u_{xx} + u_{yy} = 0, \quad 0 < x < 1, \quad 0 < y < 2;$
B.C.: $u(0, y) = u(1, y) = 0, \quad 0 < y < 2,$
$u(x, 2) = 0, \quad u(x, 0) = 64x(1 - x), \quad 0 < x < 1.$

The solution is (Exercise 3)

$$u(x, y) = \frac{512}{\pi^3} \sum_{m=1} \frac{\sin \pi x(2m - 1) \sinh \pi(2 - y)(2m - 1)}{(2m - 1)^3 \sinh 2\pi(2m - 1)}. \quad \textbf{(4.4–3)}$$

Next we solve the following problem:

$$\begin{aligned}
&\text{P.D.E.:} \quad u_{xx} + u_{yy} = 0, \quad 0 < x < 1, \quad 0 < y < 2;\\
&\text{B.C.:} \quad u(0, y) = u(1, y) = 0, \quad 0 < y < 2,\\
&\qquad\quad u(x, 0) = 0, \quad u(x, 2) = 64x(1 - x), \quad 0 < x < 1.
\end{aligned}$$

The solution is (Exercise 4)

$$u(x, y) = \frac{512}{\pi^3} \sum_{m=1} \frac{\sin \pi x(2m - 1) \sinh \pi y(2m - 1)}{(2m - 1)^3 \sinh 2\pi(2m - 1)}. \tag{4.4-4}$$

Finally, we add Eqs. (4.4-3) and (4.4-4), and the following sum is the solution to the original problem.

$$u(x, y) = \frac{512}{\pi^3} \sum_{m=1} \frac{\sin \pi x(2m - 1)}{(2m - 1)^3 \sinh 2\pi(2m - 1)}[\sinh \pi y(2m - 1) + \sinh \pi(2 - y)(2m - 1)]. \tag{4.4-5}$$

Some of the isothermals are shown in Fig. 4.4-5. Note that there are no discontinuities in contrast with the next example. ■

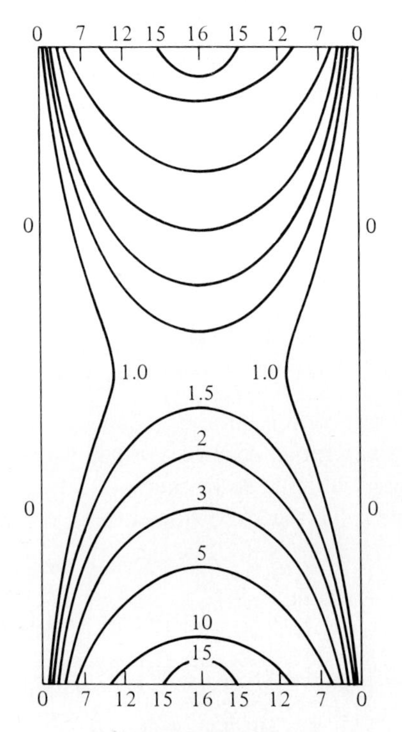

Figure 4.4-5
Isothermal curves (Example 4.4-2).

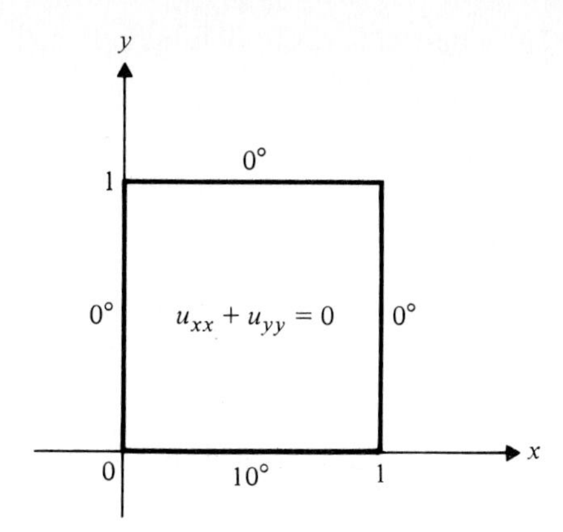

Figure 4.4-6
Boundary-value problem (Example 4.4-3).

EXAMPLE 4.4-3 Solve the following problem.

$$\begin{aligned} \text{P.D.E.:} \quad & u_{xx} + u_{yy} = 0, \quad 0 < x < 1, \quad 0 < y < 1; \\ \text{B.C.:} \quad & u(0, y) = u(1, y) = 0, \quad 0 < y < 1, \\ & u(x, 1) = 0, \; u(x, 0) = 10, \quad 0 < x < 1. \end{aligned}$$

Solution The **Dirichlet problem** (see Section 3.1) is shown diagrammatically

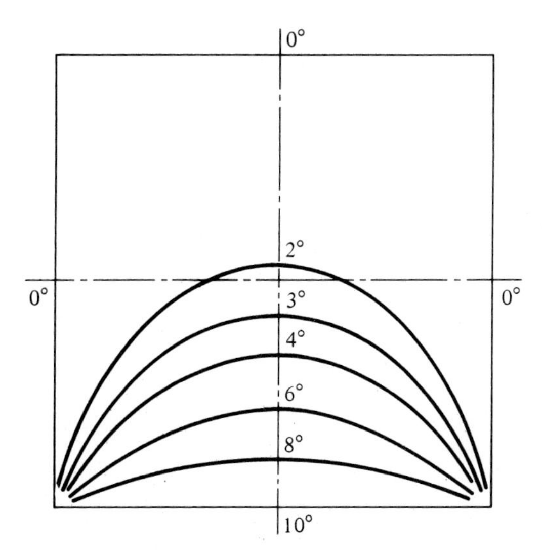

Figure 4.4-7
Isothermal curves (Example 4.4-3).

in Fig. 4.4–6. The solution is easily obtained (Exercise 5) as

$$u(x, y) = \frac{40}{\pi} \sum_{m=1} \frac{\sin (2m - 1)\pi x \sinh \pi(2m - 1)(1 - y)}{(2m - 1) \sinh \pi(2m - 1)}. \quad \textbf{(4.4–6)}$$

Some of the isothermals* are shown in Fig. 4.4–7. This time the points (0, 0) and (1, 0) are points of discontinuity, since we cannot specify the temperatures at those points to be both 0° and 10°. According to Theorem 4.2–1, however, the temperature at these points is 5°, that is, the average of the two values at the discontinuities. Figure 4.4–7 shows that this average value is not realistic either. In fact, the problem as stated is not physically realizable in the neighborhoods of the vertices of the square. ■

We conclude this section with a medical application of Fourier analysis.

The use of high-frequency sound waves has become common in medical diagnosis. Sound waves do not appear to cause any undesirable side effects and, unlike X-rays, can be safely used to examine the position and development of a fetus. The human heart can be examined by sending short bursts of sound through the chest wall and recording the echoes. Since various portions of the heart have different acoustic impedances, an **echocardiogram** is useful in diagnosis. If $f(t)$ denotes the time-varying amplitude of an echocardiogram waveform, then the *finite* Fourier series representation of $f(t)$ is given by (see Exercise 12)

$$f(t) = \sqrt{C_0} + \sum_{n=1}^{N} \sqrt{C_n} \sin (n\omega_0 t + \theta_n) + e(t).$$

Here ω_0, the fundamental frequency of the series, is 2π times the reciprocal of the heart period. The term $e(t)$ denotes the error resulting from using a finite value of N. The terms C_n and θ_n are the "power" and "phase angle," respectively, of the nth harmonic. Raeside, Chu, and Chandraratna have shown† that a knowledge of only C_0 and C_1 was sufficient to distinguish between those patients with a normal heart and those with three different cardiac conditions.

Key Words and Phrases

isothermal curves
principle of superposition
Dirichlet problem
equipotential curves
echocardiogram

*In case u represents *potential* instead of *temperature*, the isothermals become **equipotential curves**.

†D. E. Raeside, W. K. Chu, and P. A. N. Chandraratna, "Medical Applications of Fourier Analysis," *SIAM Review 20*, No. 4 (1978), pp. 850–854.

4.4 Exercises

• **1.** Verify that the functions $u_n(x, y)$ in Eq. (4.4–1) satisfy the partial differential equation and the three homogeneous boundary conditions of Example 4.4–1.

2. Find the Fourier sine coefficients b_n in Example 4.4–1.

3. Verify that the solution $u(x, y)$ in Eq. (4.4–3) of the first simplified problem of Example 4.4–2 satisfies all the boundary conditions.

4. Verify that the solution $u(x, y)$ in Eq. (4.4–4) of the second simplified problem of Example 4.4–2 satisfies all the boundary conditions.

5. Obtain the solution $u(x, y)$ in Eq. (4.4–6) for the problem of Example 4.4–3.

6. Show that the Fourier series for $f(t)$ in the medical application can be written as shown in the text. (*Hint*: Use the trigonometric identity for sin $(A + B)$.)

•• **7.** Solve the following problem:

$$\begin{aligned} &\text{P.D.E.:} \quad V_{xx} + V_{yy} = 0, \quad 0 < x < 1, \quad 0 < y < 1; \\ &\text{B.C.:} \quad V(0, y) = V(1, y) = 0, \quad 0 < y < 1, \\ &\qquad\quad V(x, 1) = 0, \quad V(x, 0) = \sin \pi x, \quad 0 < x < 1. \end{aligned}$$

8. Give a physical interpretation of the problem in Exercise 7.

***9.** Graph some of the curves $V(x, y) = C$ where C is a constant, for the problem of Exercise 7.

10. A string of unit length is fastened at both ends. The string is given an initial displacement $f(x) = 0.01x$ and released from rest.

(a) Write the partial differential equation satisfied by the displacement and all boundary and initial conditions.

(b) Solve the problem in part (a).

11. Consider the following problem:

$$\begin{aligned} &\text{P.D.E.:} \quad y_{tt} = a^2 y_{xx}, \quad 0 < x < 1, \quad t > 0; \\ &\text{B.C.:} \quad y(0, t) = y(1, t) = 0, \quad t > 0; \\ &\text{I.C.:} \quad \left.\begin{aligned} y(x, 0) &= 0 \\ y_t(x, 0) &= 0.01x \end{aligned}\right\} \quad 0 < x < 1. \end{aligned}$$

(a) Interpret the problem in physical terms.

(b) Solve the problem.

(c) Verify that the solution satisfies all boundary and initial conditions.

***(d)** Compute $y(1/2, t)$ for $t = 0, 0.1, 0.2, \ldots, 1.0$, carrying the work to three decimals.

(e) Graph the results of part (d) to show a time history of the midpoint of the string.

••• **12.** Given the following problem:

$$\begin{aligned} &\text{P.D.E.:} \quad y_{tt} = a^2 y_{xx}, \quad 0 < x < \pi, \quad t > 0; \\ &\text{B.C.:} \quad \left.\begin{aligned} y(0, t) &= 0 \\ y_x(\pi, t) &= 0 \end{aligned}\right\} \quad t > 0; \\ &\text{I.C.:} \quad \left.\begin{aligned} y(x, 0) &= f(x) \\ y_t(x, 0) &= 0 \end{aligned}\right\} \quad 0 < x < \pi. \end{aligned}$$

(a) Explain why the second boundary condition can be interpreted that the end of the string at $x = \pi$ is *free*.

(b) Show that the free end boundary condition can be generalized to $y_x(\pi, t) = g(t)$, $t > 0$.

(c) What restrictions must be placed on the function $g(t)$ in part (b)?

4.5 CONVERGENCE

In this section we discuss the convergence of Fourier series and the closely related topics of differentiation and integration of Fourier series. These subjects are somewhat theoretical in nature, so we will only give them a broad-brush treatment. For further details the references at the end of this chapter may be consulted.

It will be convenient in what follows to consider functions $f(x)$ defined on $[0, 2\pi]$. This is no restriction, since we saw in Section 4.2 (see especially Exercise 16 there) that a scale change can be made to transform to some other interval. We will thus be dealing with series of the form

$$\frac{1}{2}a_0 + \sum_{n=0}^{\infty} a_n \cos nx + b_n \sin nx, \tag{4.5–1}$$

where

$$a_n = \frac{1}{\pi}\int_0^{2\pi} f(x) \cos nx\, dx, \qquad n = 0, 1, 2, \ldots, \tag{4.5–2}$$

and

$$b_n = \frac{1}{\pi}\int_0^{2\pi} f(x) \sin nx\, dx, \qquad n = 1, 2, \ldots. \tag{4.5–3}$$

We will also use the finite sum

$$S_N(x) = \frac{1}{2}a_0 + \sum_{n=1}^{N} a_n \cos nx + b_n \sin nx. \tag{4.5–4}$$

First, we consider an example that shows the convergence of a Fourier series at its best.

EXAMPLE 4.5–1 Obtain the Fourier series of the function

$$f(x) = \sin\frac{x}{2}, \qquad 0 \le x \le 2\pi,$$
$$f(x + 2\pi) = f(x),$$

and analyze the convergence of the series.

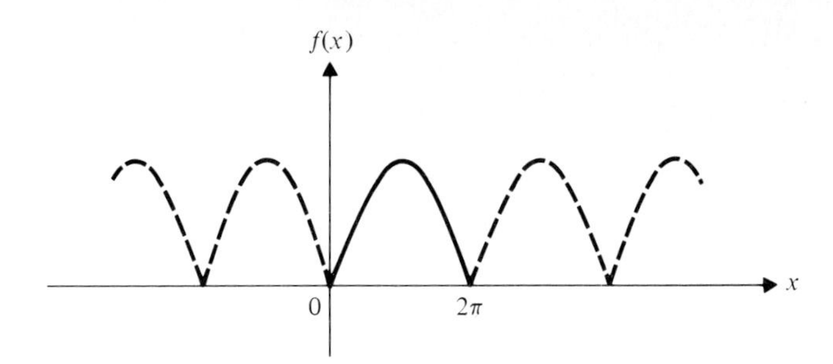

Figure 4.5-1
The periodic function of Example 4.5-1.

Solution We see in Fig. 4.5-1 that the given periodic function is even; hence

$$b_n \equiv 0, \qquad n = 1, 2, \ldots,$$

$$a_0 = \frac{1}{\pi}\int_0^{2\pi} \sin\frac{x}{2}\,dx = \frac{4}{\pi},$$

$$a_n = \frac{1}{\pi}\int_0^{2\pi} \sin\frac{x}{2}\cos\frac{nx}{2}\,dx$$

$$= \begin{cases} 0, & \text{when } n \text{ is odd,} \\ \dfrac{4}{\pi(1-n^2)}, & \text{when } n \text{ is even.} \end{cases}$$

(See Exercise 1.) Thus

$$f(x) = \frac{2}{\pi} - \frac{4}{\pi}\sum_{m=1} \frac{\cos mx}{4m^2 - 1} \tag{4.5-5}$$

is the Fourier series representation of $f(x)$. Now

$$\left|\frac{\cos mx}{4m^2 - 1}\right| \le \frac{1}{4m^2 - 1} < \frac{1}{m^2},$$

and since (see Exercise 19 in Section 4.2)

$$\sum_{m=1} \frac{1}{m^2} = \frac{\pi^2}{6},$$

it follows that the series in Eq. (4.5-5) is *absolutely* and *uniformly convergent* for all x. This, in turn, means that given $\epsilon > 0$, we can choose a value of N that depends only on ϵ such that

$$|f(x) - S_N(x)| < \epsilon$$

as shown in Fig. 4.5-2. ■

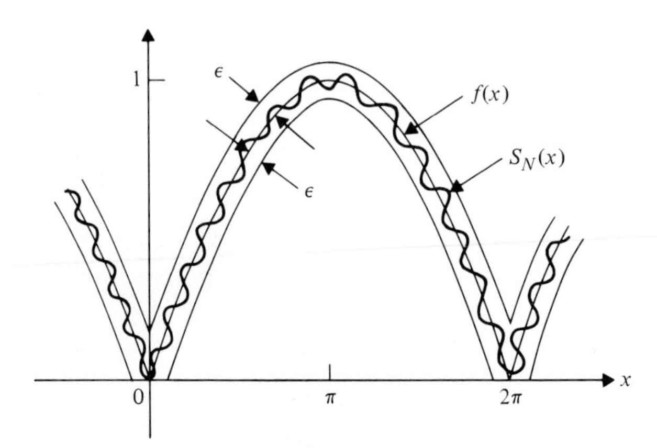

Figure 4.5-2
Uniform convergence.

In Example 4.5-1 we have an illustration of the following theorem, whose proof we omit.*

THEOREM 4.5-1
If $f(x)$ is continuous and has a piecewise continuous derivative in $0 \leq x \leq 2\pi$, if $f(x)$ has period 2π, and if $f(0) = f(2\pi)$, then the Fourier series representation of $f(x)$ converges everywhere uniformly and absolutely to $f(x)$.

If $f(x)$ does not meet the strict requirements of Theorem 4.5-1, then Theorem 4.2-1 may be used. It is very important, however, that all the hypotheses of theorems be satisfied in a particular case. For instance, in 1910, Fejér† gave an example‡ of a *continuous* function whose Fourier series representation *diverged* for a certain value of x.

Convergence of Fourier series can be examined further by considering $S_N(x)$, the Nth partial sum, given in Eq. (4.5-4). By substituting the values of a_n and b_n from Eqs. (4.5-2) and (4.5-3) we obtain (Exercise 2)

$$S_N(x) = \frac{1}{2\pi}\int_0^{2\pi} f(s)\,ds + \frac{1}{\pi}\sum_{n=1}^{N}\left[\cos nx \int_0^{2\pi} f(s)\cos ns\,ds + \sin nx \int_0^{2\pi} f(s)\sin ns\,ds\right]$$

$$= \frac{1}{\pi}\int_0^{2\pi}\left[\frac{1}{2} + \sum_{n=1}^{N}\cos n(x-s)\right] f(s)\,ds. \qquad \textbf{(4.5-6)}$$

*For a proof, see Hans Sagan, *Boundary and Eigenvalue Problems in Mathematical Physics* (New York: Wiley, 1961), p. 120 ff.

†Lipot (Leopold) Fejér (1880–1959), a Hungarian mathematician.

‡See Jane Cronin-Scanlon, *Advanced Calculus* (Boston: D. C. Heath, 1967), p. 231 ff.

In order to simplify the integrand in Eq. (4.5–6) we consider the expression

$$\frac{1}{2} + \sum_{n=1}^{N} \cos n(x - s), \tag{4.5–7}$$

which can be written (Exercise 3)

$$\frac{1}{2} \frac{\sin [(N + \frac{1}{2})(x - s)]}{\sin \frac{1}{2}(x - s)}, \tag{4.5–8}$$

provided that

$$\tfrac{1}{2}(x - s) \neq k\pi, \qquad k = 0, \pm 1, \pm 2, \ldots .$$

The function

$$D_N(t) = \begin{cases} \dfrac{\sin (N + \frac{1}{2})t}{2 \sin \frac{1}{2} t}, & t \neq 2k\pi, \quad k = 0, \pm 1, \pm 2, \ldots, \\ N + \frac{1}{2}, & t = 2k\pi, \quad k = 0, \pm 1, \pm 2, \ldots, \end{cases}$$

is called the **Dirichlet kernel**. This kernel can be used to express the sum of the first N terms of a Fourier series in closed form. From Eq. (4.5–6) we have

$$S_N(x) = \frac{1}{\pi} \int_0^{2\pi} f(s)\, D_N(x - s)\, ds. \tag{4.5–9}$$

In Fig. 4.5–3 we show the Dirichlet kernel $D_N(t)$ for $N = 8$. Except at the origin and at 2π this function remains finite no matter how large N becomes.

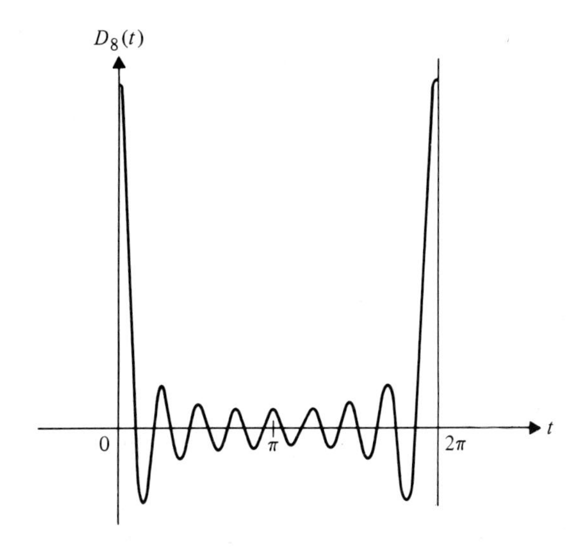

Figure 4.5–3
The Dirichlet kernel $D_8(t)$.

The function does not approach any limit, however, as N increases because it oscillates. As a result, it is difficult to use Eq. (4.5–9) for an estimation of the error between $f(x)$ and its Fourier series.

If $f(x)$ is periodic of period 2π, then we can replace $x - s$ by t in Eq. (4.5–9) to obtain (Exercise 5)

$$\begin{aligned} S_N(x) &= \frac{1}{\pi}\int_0^{2\pi} f(x - t)D_N(t)\,dt \\ &= \frac{1}{\pi}\int_0^{\pi} f(x - t)D_N(t)\,dt + \frac{1}{\pi}\int_{\pi}^{2\pi} f(x - t)D_N(t)\,dt \\ &= \frac{1}{\pi}\int_0^{\pi} [f(x + t) + f(x - t)]D_N(t)\,dt. \end{aligned} \tag{4.5–10}$$

We can now state the following theorem.

THEOREM 4.5–2

Let f be periodic of period 2π and integrable on $[0, 2\pi)$. In order that, for a given x_0, the sequence of partial sums $\{S_N(x_0)\}$ should converge to a limit $S(x_0)$, it is necessary and sufficient that

$$\lim_{N\to\infty} \frac{1}{\pi}\int_0^{\pi} [f(x_0 + t) + f(x_0 - t) - 2S(x_0)]D_N(t)\,dt = 0.$$

Proof *The result follows from Eq. (4.5–10) and from the relations (Exercise 6)*

$$\frac{1}{\pi}\int_0^{2\pi} D_N(t)\,dt = \frac{2}{\pi}\int_0^{\pi} D_N(t)\,dt = 1$$

and

$$\frac{1}{\pi}\int_0^{\pi} 2S(x_0)D_N(t)\,dt = S(x_0). \qquad \square$$

We consider next two problems closely related to convergence of Fourier series, namely, differentiation and integration of such series. What we are interested in first is the following problem. If $f(x)$ has a Fourier series representation, then under what conditions will $f'(x)$ have a representation that corresponds to the differentiated series?

Differentiation of a Fourier Series

In Example 4.3–1 we obtained the representation

$$f(x) \sim \frac{\pi}{2} - \frac{4}{\pi}\sum_{m=1} \frac{\cos(2m - 1)x}{(2m - 1)^2} \tag{4.5–11}$$

of the function $f(x) = x$ for $0 \le x \le \pi$. If we differentiate both members of (4.5–11), we obtain

$$f'(x) \sim \frac{4}{\pi} \sum_{m=1} \frac{\sin (2m - 1)x}{(2m - 1)}. \quad \textbf{(4.5–12)}$$

It can be shown (Exercise 7) that the above is the representation of $f'(x) = 1$ on $0 < x < \pi$ and that $f'(0) = f'(\pi) = 0$ as expected. On the other hand, if we begin with (4.3–7), then differentiation produces

$$f'(x) \sim 2 \sum_{n=1} (-1)^{n+1} \cos nx,$$

which converges *nowhere*, since the limit of the nth term does not approach zero as $n \to \infty$, a necessary condition for convergence. Clearly, the conditions under which a Fourier series representation may be differentiated term by term must be examined. Sufficient conditions are given in the following theorem, which we state without proof.

THEOREM 4.5–3

Let $f(x)$ have a Fourier series representation for $0 \le x \le 2\pi$ given by

$$f(x) \sim \frac{1}{2} a_0 + \sum_{n=1} (a_n \cos nx + b_n \sin nx)$$

with

$$a_n = \frac{1}{\pi} \int_0^{2\pi} f(s) \cos ns\, ds, \qquad n = 0, 1, 2, \ldots,$$

and

$$b_n = \frac{1}{\pi} \int_0^{2\pi} f(s) \sin ns\, ds, \qquad n = 1, 2, \ldots.$$

Then the Fourier series representation is differentiable at each point where $f''(x)$ exists, provided that $f(x)$ is continuous and $f'(x)$ is piecewise continuous on $0 \le x \le 2\pi$ and that $f(0) = f(2\pi)$. Moreover,

$$f'(x) \sim \sum_{n=1} n(-a_n \sin nx + b_n \cos nx), \qquad 0 < x < 2\pi. \quad \textbf{(4.5–13)}$$

Integration of a Fourier Series

Integration of a Fourier series is a much simpler matter. This is to be expected, since integration is a "smoothing" process that tends to eliminate discontinuities, whereas the process of differentiation has the opposite effect. This is shown dramatically in Fig. 4.5–4 using a graph of data taken by students on a subway trip. The acceleration of the subway car is shown in Fig. 4.5–4(a), and the velocity and position obtained by calculation are shown in Figs. 4.5–4(b) and 4.5–4(c), respectively.

The following theorem applies to integration of a Fourier series.

THEOREM 4.5–4

Let $f(x)$ be piecewise continuous on $0 < x < 2\pi$ and have a Fourier series representation

$$f(x) \sim \frac{1}{2} a_0 + \sum_{n=1} (a_n \cos nx + b_n \sin nx)$$

with a_n and b_n as before. Then

$$\int_0^x f(s)\, ds = \frac{1}{2} a_0 x + \sum_{n=1} \frac{1}{n} [a_n \sin nx - b_n (\cos nx - 1)] \quad \textbf{(4.5–14)}$$

for $0 \le x \le 2\pi$.

EXAMPLE 4.5–2 Obtain the Fourier series representation of x^2 on $[-\pi, \pi]$.

Solution We have, from (4.3–7),

$$x \sim 2 \sum_{n=1} \frac{(-1)^{n+1} \sin nx}{n},$$

a representation that is valid on $(-\pi, \pi)$. Using Theorem 4.5–4, we have

$$\frac{x^2}{2} = -2 \sum_{n=1} \frac{(-1)^{n+1}}{n^2} (\cos nx - 1)$$

$$= -2 \sum_{n=1} \frac{(-1)^{n+1}}{n^2} \cos nx + 2 \sum_{n=1} \frac{(-1)^{n+1}}{n^2}.$$

Figure 4.5–4
Data from a subway ride. (From Solomon Garfunkel, "A Laboratory and Computer Based Approach to Calculus," *MAA Monthly* 79, p. 282.) (a) Acceleration of subway car. (b) Speed of subway car (integrated from acceleration data). (c) Position of subway car (integrated from acceleration data).

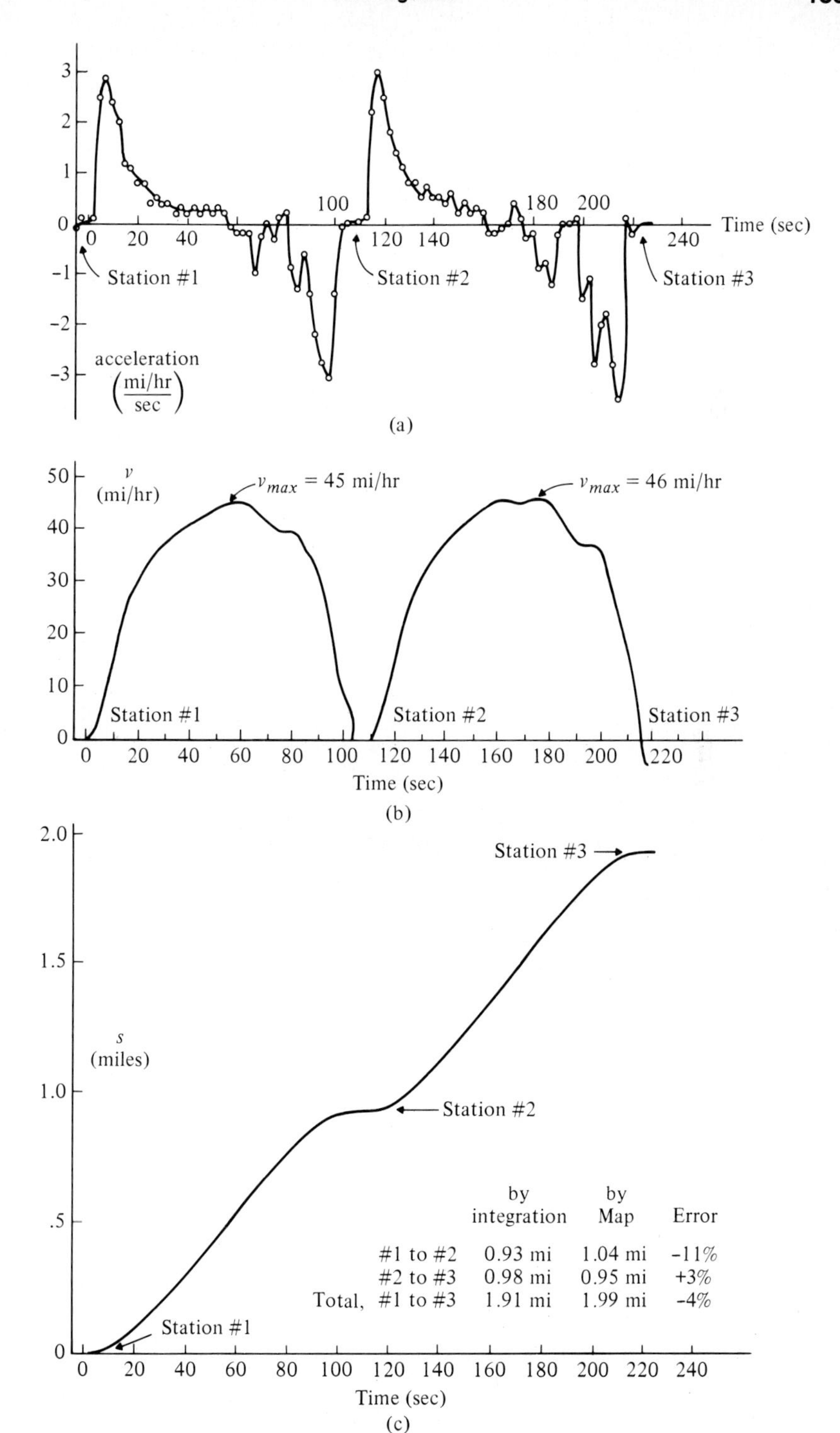
3
2
1
0
-1
-2
-3
acceleration
(mi/hr / sec)
0 20 40 100 120 140 180 200 240
Time (sec)
Station #1
Station #2
Station #3
(a)
50
40
30
20
10
0
v
(mi/hr)
v_{max} = 45 mi/hr
v_{max} = 46 mi/hr
Station #1
Station #2
Station #3
0 20 40 60 80 100 120 140 160 180 200 220
Time (sec)
(b)
2.0
1.5
1.0
.5
0
s
(miles)
Station #3
Station #2
Station #1
by integration
by Map
Error
#1 to #2 0.93 mi 1.04 mi -11%
#2 to #3 0.98 mi 0.95 mi +3%
Total, #1 to #3 1.91 mi 1.99 mi -4%
0 20 40 60 80 100 120 140 160 180 200 220 240
Time (sec)
(c)

But the last term is $\pi^2/6$ (see Exercise 38 in Section 4.3); hence

$$x^2 = \frac{\pi^2}{3} - 4 \sum_{n=1} \frac{(-1)^{n+1}}{n^2} \cos nx. \tag{4.5–15}$$

We leave it as an exercise to show that the last result is valid for all x (see Exercise 8). ■

Key Words and Phrases

Dirichlet kernel   Fejér kernel

4.5 Exercises

• 1. Carry out the computational details in Example 4.5–1.

2. Obtain the expression for $S_N(x)$ shown in Eq. (4.5–6).

3. Show that

$$2 \sin \frac{t}{2}\left[\frac{1}{2} + \sum_{n=1}^{N} \cos nt\right] = \sin\left(N + \frac{1}{2}t\right).$$

(*Hint*: Use the trigonometric identity

$$2 \sin \frac{t}{2} \cos nt = \sin\left(n + \frac{1}{2}\right)t - \sin\left(n - \frac{1}{2}\right)t\Big).$$

4. Demonstrate that

$$D_N(t) = N + \frac{1}{2}$$

if $t = 2k\pi$, $k = 0, \pm 1, \pm 2, \ldots$.

5. Show that $S_N(x)$ can be written in the form shown in Eq. (4.5–10).

6. Obtain the results

$$\frac{1}{\pi}\int_0^{2\pi} D_N(t)\, dt = \frac{2}{\pi}\int_0^{\pi} D_N(t)\, dt = 1.$$

7. Show that the representation in (4.5–12) is for a square wave of amplitude 1 and period 2π.

8. Verify that Eq. (4.5–15) is valid for all x.

•• 9. Explain how the representation (4.5–11) meets the hypotheses of Theorem 4.5–3.

10. Explain how

$$f(x) \sim 2 \sum_{n=1} \frac{(-1)^{n+1} \sin nx}{n}$$

fails to meet the conditions of Theorem 4.5–3.

11. Obtain a Fourier series representation for $f(x) = x$ on $0 \le x < 2\pi$ by differentiating an appropriate representation of x^2. For what values of x is your result valid?

12. **(a)** Obtain the result

$$\sin x \sim \frac{2}{\pi} + \frac{4}{\pi} \sum_{n=1} \frac{\cos 2nx}{(1 - 4n^2)}, \qquad 0 < x < \pi.$$

(b) Show that the result in part (a) meets the hypotheses of Theorem 4.5–3.

(c) Obtain the Fourier series representation for $\cos x$, $0 < x < \pi$, from the result in part (a).

13. **(a)** Obtain the Fourier series representation of the function

$$f(x) = \begin{cases} -1, & \text{for } -\pi < x < 0, \\ 1, & \text{for } 0 < x < \pi, \end{cases}$$

$$f(x + 2\pi) = f(x).$$

(b) Show that the representation obtained in part (a) meets the hypotheses of Theorem 4.5–3.

(c) Obtain the Fourier series representation of the function

$$g(x) = |x| \qquad \text{for } 0 < x < \pi, \qquad g(x + 2\pi) = g(x).$$

(d) Discuss the convergence of the representation in parts (a) and (c).

14. **(a)** Obtain the Fourier series representation of the function

$$f(x) = \frac{1}{2}(\pi - x), \qquad 0 < x < 2\pi, \qquad f(x + 2\pi) = f(x).$$

(b) Show that the representation in part (a) satisfies the hypotheses of Theorem 4.5–4.

(c) Obtain the representation of the function

$$g(x) = \frac{x}{4}(2\pi - x), \qquad 0 \le x \le 2\pi.$$

(d) Discuss the convergence of the representations in parts (a) and (c).

15. From the result in Exercise 14(d) show that

$$\frac{1}{8\pi} \int_0^{2\pi} x(2\pi - x)\, dx = \sum_{n=1} \frac{1}{n^2},$$

and thus find the sum of the series.

16. Show that

$$\int_0^{\pi/2} \frac{\sin (2n + 1)t}{\sin t}\, dt = \frac{\pi}{2}.$$

17. Obtain the result

$$S_N(x) = \frac{1}{2\pi} \int_{-\pi}^{\pi} f(t) \sum_{n=-N}^{N} \exp [in(x - t)]\, dt,$$

where f is periodic of period 2π.

18. **(a)** Show that $D_N(t)$ is an even function.
(b) Show that $D_N(t)$ is periodic of period 2π.

••• **19.** In Eq. (4.5-4), define $S_0(x)$ as

$$\frac{1}{2} a_0 = \frac{1}{2\pi} \int_0^{2\pi} f(x)\, dx.$$

Define

$$\sigma_N = \frac{S_0 + S_1 + \cdots + S_N}{N},$$

which is the *arithmetic mean* of the first N partial sums.

(a) The **Fejér kernel** $F_N(x)$ is defined as follows:

$$F_N(x) = \frac{1}{N} \sum_{j=0}^{N-1} D_j(x).$$

Show that

$$F_N(x) = \frac{1}{N} \sum_{j=0}^{N-1} \frac{\sin (j + \frac{1}{2})x}{2 \sin (x/2)}.$$

(b) Obtain the alternate form of $F_N(x)$:

$$F_N(x) = \frac{1}{2N} \left[\frac{\sin (Nx/2)}{\sin (x/2)} \right]^2.$$

(*Hint*: Express the sum in part (a) in terms of exponentials, and then sum the resulting geometric series.)

(c) Obtain the relation

$$S_N(x) = \frac{1}{\pi} \int_0^{2\pi} f(x + t) D_N(t)\, dt,$$

where $f(x)$ is periodic of period 2π.

(d) Using the result in part (c) and assuming that

$$\lim_{N \to \infty} S_N(x) = f(x),$$

show that

$$\sigma_N(x) = \frac{1}{\pi}\int_0^{2\pi} f(x + t)F_N(t)\,dt.$$

***20.** Show that using three terms of the series in Eq. (4.5–5) is enough to give $f(\pi)$ to within 1.2 percent.

21. Read the article "Pointwise Convergence of Fourier Series" by Paul R. Chernoff in *Amer. Math. Monthly* (*87*), 5, May 1980, pp. 399–400.

REFERENCES

Asplund, Edgar, and L. Bungart, *A First Course in Integration*. New York: Holt, Rinehart and Winston, 1966.
An advanced coverage of Lebesgue integration and its application to the theory of Fourier series.

Broman, Arne, *Introduction to Partial Differential Equations from Fourier Series to Boundary-Value Problems*. Reading, Mass.: Addison-Wesley, 1970.
An intermediate-level treatment of Fourier series and orthogonal systems.

Churchill, Ruel V., and J. W. Brown, *Fourier Series and Boundary Value Problems*, 3rd ed. New York: McGraw-Hill, 1978.
A good coverage of the topics at an intermediate level.

Cronin-Scanlon, Jane, *Advanced Calculus*. Boston: D. C. Heath, 1967.
Contains the theoretical topics necessary for an understanding of convergence of Fourier series written at an intermediate level.

Farlow, Stanley J., *Partial Differential Equations for Scientists & Engineers*. New York: John Wiley, 1982.
A good nontheoretical treatment written in the form of 47 lessons, each one beginning with a list of objectives. Includes many helpful diagrams and figures.

Gustafson, Karl E., *Introduction to Partial Differential Equations and Hilbert Space Methods*. New York: John Wiley, 1980.
The last half of the book contains an advanced treatment of Fourier series. Included are some modern applications involving nonlinear equations having soliton solutions. Contains detailed hints and solutions to some exercises.

Hanna, J. Ray, *Fourier Series and Integrals of Boundary Value Problems*. New York: John Wiley, 1982.
An elementary treatment of the subjects.

Jackson, Dunham, *Fourier Series and Orthogonal Polynomials*, Carus Mathematical Monograph No. 6. Washington, D.C.: Math. Assoc. of America, 1941.
An excellent theoretical coverage of Fourier series and their convergence.

Rudin, Walter, *Principles of Mathematical Analysis*, 3rd ed. New York: McGraw-Hill, 1976.
An unusually complete treatment of uniform convergence and Lebesgue integration theory at an intermediate level.

5 Fourier Integrals and Transforms

5.1 THE FOURIER INTEGRAL

The following extension of Fourier series to the Fourier integral is intended to be a *plausible* rather than a *mathematical* development. Our justification for this lies in the fact that a rigorous treatment such as that found in I. N. Sneddon's *Fourier Transforms* (New York: McGraw-Hill, 1951) would lead us too far afield.

We have seen that a (periodic) function $f(x)$ that satisfies the Dirichlet conditions, Theorem 4.2–1,* may be represented by a Fourier series. The Dirichlet conditions are *sufficient* but not *necessary*. The "representation" is in the sense that the series will converge to the average or mean value of $f(x)$ at points where the function has a jump discontinuity.

As an example, the function defined by

$$\begin{aligned} f(x) &= e^{-|x|}, \qquad -L < x < L, \\ f(x + 2L) &= f(x) \end{aligned} \tag{5.1–1}$$

has a Fourier series representation

$$f(x) \sim \frac{1}{2} a_0 + \sum_{n=1}^{\infty} \left(a_n \cos \frac{n\pi x}{L} + b_n \sin \frac{n\pi x}{L} \right), \tag{5.1–2}$$

where

$$a_n = \frac{1}{L} \int_{-L}^{L} f(s) \cos \frac{n\pi s}{L}\, ds, \qquad n = 0, 1, 2, \ldots,$$

$$b_n = \frac{1}{L} \int_{-L}^{L} f(s) \sin \frac{n\pi s}{L}\, ds, \qquad n = 1, 2, \ldots.$$

*See also Exercise 23 in Section 4.2.

We are using s as the dummy variable of integration here. For the example given in Eqs. (5.1–1) we would have $b_n = 0$ for all n and a simpler formula for a_n, since $f(x)$ is an even function.

If we denote a function such as the one defined in Eqs. (5.1–1) by $f_L(x)$ to highlight the fact that this is a *periodic* function with half-period equal to L, then we can make an obvious observation. The Fourier series representation given by (5.1–2) is valid no matter how large L is so long as it remains finite. This then leads us naturally to consider the function $f(x)$ defined as

$$f(x) = \lim_{L\to\infty} f_L(x).$$

The two functions $f(x)$ and $f_L(x)$ are shown in Figure 5.1–1. Now $f(x)$ is no longer a periodic function, although it is still piecewise smooth. We will impose one more condition on $f(x)$, namely, that it be **absolutely integrable** on the real line, that is, that the improper integral

$$\int_{-\infty}^{\infty} |f(x)|\, dx$$

be finite. In the case of our sample function we have

$$\int_{-\infty}^{\infty} |e^{-|x|}|\, dx = 2\int_0^{\infty} e^{-x}\, dx = 2 \lim_{L\to\infty} \int_0^L e^{-x}\, dx.$$

We can easily verify that the value of the integral is 2 (Exercise 1). Following the usual custom, we write

$$\int_0^{\infty} f(x)\, dx \qquad \text{instead of} \qquad \lim_{L\to\infty} \int_0^L f(x)\, dx.$$

Now we make the substitution $\alpha_n = n\pi/L$ and replace a_n and b_n by their

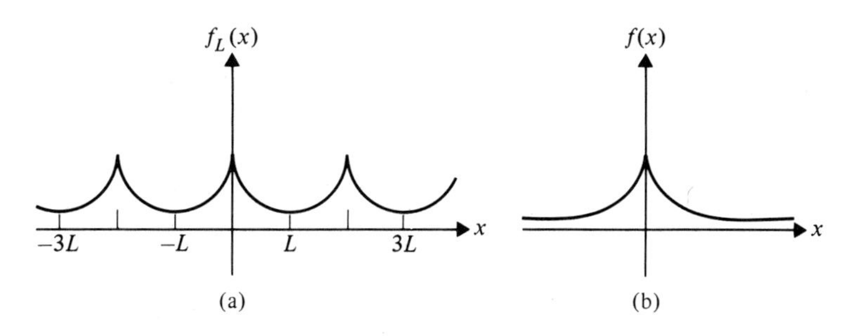

Figure 5.1–1
(a) Periodic function. (b) Nonperiodic function.

values in (5.1-2). Then

$$f_L(x) \sim \frac{1}{2L}\int_{-L}^{L} f_L(s)\,ds + \frac{1}{L}\sum_{n=1}^{\infty}\left[\cos(\alpha_n x)\int_{-L}^{L} f_L(s)\cos(\alpha_n s)\,ds\right.$$

$$\left. + \sin(\alpha_n x)\int_{-L}^{L} f_L(s)\sin(\alpha_n s)\,ds\right]. \tag{5.1-3}$$

Let

$$\Delta\alpha = \alpha_{n+1} - \alpha_n = \frac{(n+1)\pi}{L} - \frac{n\pi}{L} = \frac{\pi}{L},$$

and write (5.1-3) as

$$f_L(x) \sim \frac{1}{2L}\int_{-L}^{L} f_L(s)\,ds + \frac{1}{\pi}\sum_{n=1}^{\infty}\left[\cos(\alpha_n x)\,\Delta\alpha\int_{-L}^{L} f_L(s)\cos(\alpha_n s)\,ds\right.$$

$$\left. + \sin(\alpha_n x)\,\Delta\alpha\int_{-L}^{L} f_L(s)\sin(\alpha_n s)\,ds\right]. \tag{5.1-4}$$

This last representation of the periodic, piecewise smooth function $f_L(x)$ is valid for any *finite* L.

We now let L approach infinity. Then the first integral on the right of (5.1-4) approaches zero because $f(x)$ is absolutely integrable.* Moreover, it seems plausible that the infinite series becomes an integral from 0 to ∞. Hence

$$f(x) = \frac{1}{\pi}\int_{0}^{\infty}\left[\cos(\alpha x)\int_{-\infty}^{\infty} f(s)\cos(\alpha s)\,ds\right.$$

$$\left. + \sin(\alpha x)\int_{-\infty}^{\infty} f(s)\sin(\alpha s)\,ds\right]d\alpha, \tag{5.1-5}$$

which is the **Fourier integral representation** of $f(x)$. Equation (5.1-5) is often written in the form

$$f(x) = \int_{0}^{\infty}[A(\alpha)\cos\alpha x + B(\alpha)\sin\alpha x]\,d\alpha, \qquad -\infty < x < \infty, \tag{5.1-6}$$

where the **Fourier integral coefficients** are functions given by

$$A(\alpha) \doteq \frac{1}{\pi}\int_{-\infty}^{\infty} f(s)\cos\alpha s\,ds$$

and (5.1-7)

$$B(\alpha) = \frac{1}{\pi}\int_{-\infty}^{\infty} f(s)\sin\alpha s\,ds$$

*Recall that an improper integral is integrable if it is absolutely integrable.

The "plausible" part of the above argument stems from the definition

$$\alpha_n = \frac{n\pi}{L}.$$

As L becomes larger and larger, the numbers α_n become more dense on the real half-line $[0, \infty)$. Hence it is natural to replace the *discrete* numbers α_n by the *continuous* real variable α in the limit. At the same time, of course, the infinite series is replaced by the definite integral from zero to infinity.

Sufficient conditions for the validity of Eq. (5.1–5) can be stated in the form of the following theorem.

THEOREM 5.1-1

If $f(x)$ is piecewise smooth and absolutely integrable on the real line, then $f(x)$ can be represented by a Fourier integral. At a point where $f(x)$ is discontinuous, the representation gives the average of the left- and right-hand limits of $f(x)$ at that point.

The Fourier integral can be written in a more compact form. Going back to Eq. (5.1–5), we recognize that the terms $\cos(\alpha x)$ and $\sin(\alpha x)$ do not depend on s; hence these terms may be put *inside* the integrals. Then we have

$$f(x) = \frac{1}{\pi}\int_0^\infty \int_{-\infty}^\infty f(s)\,[\cos\alpha x \cos\alpha s + \sin\alpha x \sin\alpha s]\,ds\,d\alpha \quad \textbf{(5.1–8)}$$

$$= \frac{1}{\pi}\int_0^\infty \int_{-\infty}^\infty f(s)\cos\alpha(s - x)\,ds\,d\alpha. \quad \textbf{(5.1–9)}$$

If $f(x)$ is an *even* function, then $f(s)\sin\alpha s$ is an odd function of s, and Eq. (5.1–8) reduces to

$$f(x) = \frac{2}{\pi}\int_0^\infty \int_0^\infty f(s)\cos\alpha x\cos\alpha s\,ds\,d\alpha. \quad \textbf{(5.1–10)}$$

If $f(x)$ is an *odd* function, then $f(x)\cos\alpha s$ is an odd function of s, and Eq. (5.1–8) reduces to

$$f(x) = \frac{2}{\pi}\int_0^\infty \int_0^\infty f(s)\sin\alpha x\sin\alpha s\,ds\,d\alpha. \quad \textbf{(5.1–11)}$$

We call Eqs. (5.1–10) and (5.1–11) the **Fourier cosine integral** and **Fourier sine integral** representations of $f(x)$, respectively. In case $f(x)$ is defined only on $(0, \infty)$, then we can make either an even or an odd extension of the function and use Eq. (5.1–10) or Eq. (5.1–11), respectively.

From Eq. (5.1–9) we see that the inner integral is an even function of α; hence we can write

$$f(x) = \frac{1}{2\pi}\int_{-\infty}^\infty \int_{-\infty}^\infty f(s)\cos\alpha(s - x)\,ds\,d\alpha. \quad \textbf{(5.1–12)}$$

On the other hand,

$$\frac{i}{2\pi}\int_{-\infty}^{\infty}\int_{-\infty}^{\infty} f(s)\sin\alpha(s - x)\,ds\,d\alpha = 0 \tag{5.1-13}$$

because the inner integral is an odd function of α. Thus, adding Eqs. (5.1-12) and (5.1-13), we obtain

$$f(x) = \frac{1}{2\pi}\int_{-\infty}^{\infty}\int_{-\infty}^{\infty} f(s)\,e^{i\alpha(s-x)}\,ds\,d\alpha, \tag{5.1-14}$$

which is the **complex form of the Fourier integral**. This can also be written in the equivalent form

$$f(x) = \frac{1}{2\pi}\int_{-\infty}^{\infty} C(\alpha)\,e^{-i\alpha x}\,d\alpha,$$

where

$$C(\alpha) = \int_{-\infty}^{\infty} f(s)\,e^{i\alpha s}\,ds,$$

which is called the **complex Fourier integral coefficient.**

Observe that when the requirement that a piecewise smooth function be *periodic* is replaced by the requirement that it be *absolutely integrable*, then the Fourier *series* representation is replaced by the Fourier *integral* representation. Note the similarities and the differences in the two types of representations.

EXAMPLE 5.1-1 Obtain the Fourier integral representation of the function $f(x) = \exp(-|x|)$, $-\infty < x < \infty$.

Solution The given function is absolutely integrable (Exercise 1) and is an even function (Fig. 5.1-1); hence we have $B(\alpha) \equiv 0$ in Eqs. (5.1-7). Then

$$\begin{aligned} A(\alpha) &= \frac{1}{\pi}\int_{-\infty}^{\infty} e^{-|s|}\cos\alpha s\,ds \\ &= \frac{2}{\pi}\int_{0}^{\infty} e^{-s}\cos\alpha s\,ds \\ &= \frac{2}{\pi(1 + \alpha^2)}, \end{aligned}$$

a result that can be found in a table of definite integrals. Using Eq. (5.1-6) we have

$$e^{-|x|} = \frac{2}{\pi}\int_{0}^{\infty} \frac{\cos\alpha x}{1 + \alpha^2}\,d\alpha, \tag{5.1-15}$$

which is the required Fourier integral representation. ■

We remark that the integral representation in Eq. (5.1–15) is valid for all values of x. We cannot, however, "evaluate" the integral except for *particular* values of x. In this respect also we have an analogy to Fourier series.

EXAMPLE 5.1–2 Find the Fourier sine integral representation of the function

$$f(x) = \begin{cases} \cos x, & \text{for } 0 < x < \pi/2, \\ 0, & \text{for } x > \pi/2. \end{cases}$$

Solution For a Fourier sine integral representation we need to start with an *odd* function. Hence we make an odd extension of the given function as shown in Fig. 5.1–2. Then (Exercise 6)

$$\begin{aligned} B(\alpha) &= \frac{2}{\pi} \int_0^{\pi/2} \cos s \sin \alpha s \, ds \\ &= \frac{2}{\pi} \left[\frac{\alpha - \sin(\alpha\pi/2)}{\alpha^2 - 1} \right]. \end{aligned}$$

Thus

$$f(x) = \frac{2}{\pi} \int_0^\infty \frac{[\alpha - \sin(\alpha\pi/2)] \sin \alpha x}{\alpha^2 - 1} \, d\alpha. \quad \blacksquare \tag{5.1–16}$$

Note that although the integrand in Eq. (5.1–16) appears to be discontinuous at $\alpha = 1$, it can be shown (Exercise 7) that this is not the case.

From the result in Eq. (5.1–16) and using Theorem 5.1–1, we can obtain

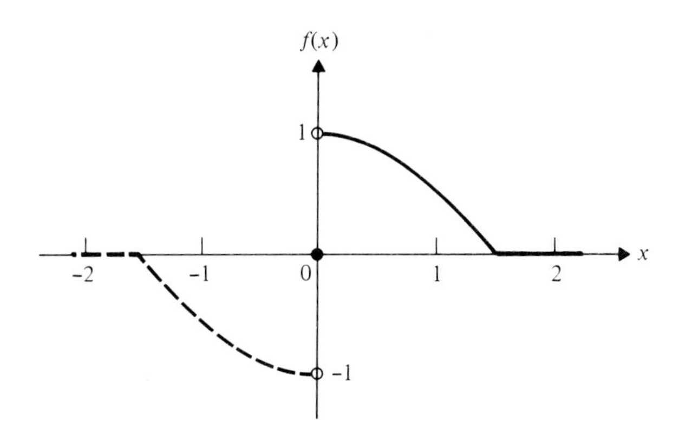

Figure 5.1–2
Odd extension (Example 5.1–2).

the following results.

$$\int_0^\infty \frac{[u - \sin(\pi u/2)]\sin(\pi u/2)}{u^2 - 1}\,du = 0 \tag{5.1-17}$$

$$\int_0^\infty \frac{[u - \sin(\pi u/2)]\sin(\pi u/4)}{u^2 - 1}\,du = \frac{\pi\sqrt{2}}{4} \tag{5.1-18}$$

Next we illustrate the use of the complex form of the Fourier integral.

EXAMPLE 5.1-3 Use Eq. (5.1-14) to obtain the Fourier integral representation of the function

$$f(x) = \begin{cases} \sin x, & \text{for } -\pi < x < \pi, \\ 0, & \text{elsewhere.} \end{cases}$$

Solution The function $f(x)$ is shown in Fig. 5.1-3. We have from Eq. (5.1-14)

$$\begin{aligned} f(x) &= \frac{1}{2\pi}\int_{-\infty}^{\infty}\int_{-\pi}^{\pi} \sin s\, e^{i\alpha(s-x)}\,ds\,d\alpha \\ &= \frac{1}{2\pi}\int_{-\infty}^{\infty}\int_{-\pi}^{\pi} \sin s\,[\cos\alpha(s - x) + i\sin\alpha(s - x)]\,ds\,d\alpha \\ &= \frac{1}{\pi}\int_0^{\infty}\int_{-\pi}^{\pi} \sin s\cos\alpha(s - x)\,ds\,d\alpha \\ &= \frac{2}{\pi}\int_0^{\infty}\int_0^{\pi} \sin s\sin\alpha s\sin\alpha x\,ds\,d\alpha \\ &= \frac{1}{\pi}\int_0^{\infty} \sin\alpha x\left[\frac{\sin\alpha\pi}{1 - \alpha} + \frac{\sin\alpha\pi}{1 + \alpha}\right]d\alpha \\ &= \frac{2}{\pi}\int_0^{\infty} \frac{\sin\alpha\pi}{1 - \alpha^2}\sin\alpha x\,d\alpha. \end{aligned} \tag{5.1-19}$$

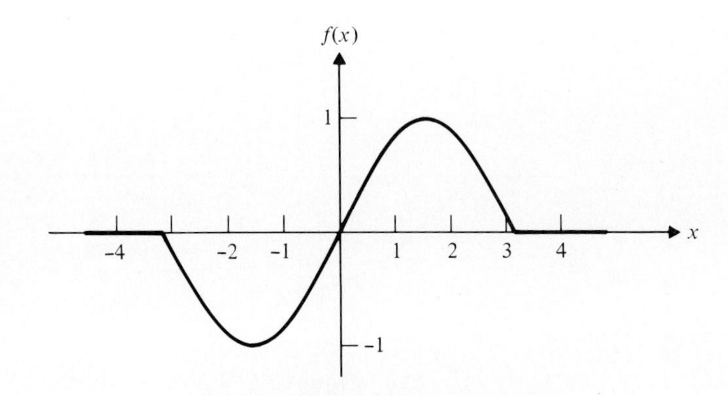

Figure 5.1-3
The function in Example 5.1-3.

Observe that the result is the Fourier *sine* integral representation of $f(x)$, which is reasonable, since $f(x)$ is an *odd* function. Accordingly, the above could have been simplified by starting with Eq. (5.1–11). ■

In the next section we will examine the complex coefficient $C(\alpha)$ in greater detail. It will be renamed the Fourier transform and will provide another useful tool for solving boundary-value problems.

Key Words and Phrases

absolutely integrable function
Fourier integral representation
Fourier integral coefficients
Fourier cosine integral
Fourier sine integral
complex form of the Fourier integral
complex Fourier integral coefficient

5.1 Exercises

• **1.** Show that

$$\int_{-\infty}^{\infty} e^{-|x|}\, dx = 2.$$

2. Explain why

$$\lim_{L\to\infty} \frac{1}{2L} \int_{-L}^{L} f_L(s)\, ds = 0$$

in (5.1–4).

3. Carry out the details in obtaining $A(\alpha)$ in Example 5.1–1.

4. Verify that Eq. (5.1–15) gives the correct result when $x = 0$.

5. Graph the function $A(\alpha)$ of Example 5.1–1.

6. Carry out the details in obtaining $B(\alpha)$ in Example 5.1–2.

7. Show that

$$\lim_{\alpha\to 1} \frac{\alpha - \sin(\alpha\pi/2)}{\alpha^2 - 1}$$

is finite.

8. Obtain each of the following results.
(a) Eq. (5.1–17) (b) Eq. (5.1–18)

9. In Example 5.1–3, fill in the missing details using the properties of odd and even functions.

10. Examine the integrand of (5.1–19) at $\alpha = 1$.

•• **11.** Determine which of the following functions is absolutely integrable on the real line.

(a) $f(x) = |1 - x|, \quad -1 \le x \le 1$

(b) $f(x) = \sin \pi x$

(c) $f(x) = x^{1/3}$

12. From Eqs. (5.1–6) and (5.1–7), show that if

$$f(x) = \begin{cases} 1, & \text{for } |x| < 1, \\ 0, & \text{for } |x| > 1, \\ \frac{1}{2}, & \text{for } |x| = 1, \end{cases}$$

then $f(x)$ satisfies the conditions of Theorem 5.1–1 and hence has a valid Fourier integral representation for all x given by

$$f(x) = \frac{2}{\pi}\int_0^\infty \frac{\sin \alpha \cos \alpha x}{\alpha}\, d\alpha, \qquad -\infty < x < \infty.$$

13. Use the result of Exercise 12 to show that

$$\int_0^\infty \frac{\sin 2\alpha}{\alpha}\, d\alpha = \frac{\pi}{2}.$$

14. Obtain the result

$$\int_0^\infty \frac{\sin ax}{x}\, dx = \frac{\pi}{2}, \qquad \text{if } a > 0.$$

(*Hint*: Make a change of variable in Exercise 13.)

15. Given the function

$$f(x) = \begin{cases} e^{-x}, & \text{for } x > 0, \\ 0, & \text{for } x < 0, \\ \frac{1}{2}, & \text{for } x = 0, \end{cases}$$

show that $f(x)$ satisfies the conditions of Theorem 5.1–1 and hence has a valid Fourier integral representation for all x given by

$$f(x) = \frac{1}{\pi}\int_0^\infty \frac{\cos \alpha x + \alpha \sin \alpha x}{1 + \alpha^2}\, d\alpha, \qquad -\infty < x < \infty.$$

16. Obtain the result

$$\int_0^\infty \frac{1}{1 + \alpha^2}\, d\alpha = \frac{\pi}{2}$$

from Exercise 15.

17. Prove that the function

$$f(x) = \begin{cases} 0, & \text{for } x \le 0, \\ \sin x, & \text{for } 0 \le x \le \pi \\ 0, & \text{for } x \ge \pi, \end{cases}$$

has a Fourier integral representation given by

$$f(x) = \frac{1}{\pi}\int_0^\infty \frac{\cos \alpha x + \cos \alpha(\pi - x)}{1 - \alpha^2}\, d\alpha, \qquad -\infty < x < \infty.$$

18. Use the result of Exercise 17 to show that

$$\int_0^\infty \frac{\cos(\pi\alpha/2)}{1-\alpha^2}\,d\alpha = \pi/2.$$

19. Obtain the Fourier sine integral representation of each of the following functions:

(a) $f(x) = e^{-x}, \quad x > 0$

(b) $f(x) = \begin{cases} h, & \text{for } 0 < x < L,\ h > 0, \\ 0, & \text{for } x > L \end{cases}$

(c) $f(x) = \begin{cases} x - 1, & \text{for } 0 < x < 1, \\ 0, & \text{for } x > 1 \end{cases}$

(d) $f(x) = \begin{cases} \sin \pi x, & \text{for } 0 < x < 1, \\ 0, & \text{for } x > 1 \end{cases}$

(e) $f(x) = \begin{cases} x^2, & \text{for } 0 < x < 1, \\ 0, & \text{for } x > 1 \end{cases}$

20. Obtain the Fourier cosine integral representation of each of the following functions:

(a) $f(x) = \begin{cases} \sin x, & \text{for } 0 < x < \pi, \\ 0, & \text{for } x > \pi \end{cases}$

(b) $f(x) = \begin{cases} h, & \text{for } 0 < x < L,\ h > 0, \\ 0, & \text{for } x > L \end{cases}$

(c) $f(x) = \begin{cases} 1 - x, & \text{for } 0 < x < 1, \\ 0, & \text{for } x > 1 \end{cases}$

(d) $f(x) = \begin{cases} 2, & \text{for } 0 < x < 1, \\ 1, & \text{for } 1 < x < 2, \\ 0, & \text{for } x > 2 \end{cases}$

(e) $f(x) = \begin{cases} x^2, & \text{for } 0 < x < 1, \\ 0, & \text{for } x > 1 \end{cases}$

21. Obtain the complex Fourier integral representation of each of the following functions:

(a) $f(x) = \begin{cases} \cos x, & \text{for } 0 < x < \pi, \\ 0, & \text{elsewhere} \end{cases}$

(b) $f(x) = \begin{cases} h, & \text{for } |x| < L,\ h > 0, \\ 0, & \text{for } |x| > L \end{cases}$

(c) $f(x) = \begin{cases} e^{-x}, & \text{for } x > 0, \\ 0, & \text{for } x < 0 \end{cases}$

22. Sketch the Fourier integral representation obtained in each of the following exercises:

(a) Exercise 19(b)
(b) Exercise 19(d)
(c) Exercise 20(b)
(d) Exercise 20(c)
(e) Exercise 21(b)

23. Verify that the following Fourier integral representations give the correct value for the indicated value of x:

(a) Exercise 20(b) at $x = 0$

(b) Exercise 20(b) at $x = L$
(c) Exercise 20(d) at $x = 0$
(d) Exercise 21(c) at $x = 0$

••• **24.** Obtain the Fourier integral representation of the function

$$f(x) = \frac{1}{1 + x^2}.$$

(*Hint*: Observe that $\sin \alpha x/(1 + x^2)$ is an odd function of x; then use the result in Eq. (5.1–15) with α and x interchanged.)

25. Find the Fourier integral representation of the function

$$f(x) = \frac{\sin x}{x}.$$

26. Prove that the function of Exercise 25 is *not* absolutely integrable. Nevertheless, the Fourier integral representation obtained there is valid. Does this violate Theorem 5.1–1? Explain.

27. Find the Fourier integral representation of the function

$$f(x) = \exp(-a^2x^2).$$

28. Show that in each case the integrands are defined when $\alpha = 1$.
(a) Exercise 17 **(b)** Exercise 18

5.2 FOURIER TRANSFORMS

The Fourier transform is obtained from Eq. (5.1–14) as follows. Rewrite Eq. (5.1–14) as an *iterated integral*,*

$$f(x) = \frac{1}{2\pi} \int_{-\infty}^{\infty} f(s)\, e^{i\alpha s}\, ds \int_{-\infty}^{\infty} e^{-i\alpha x}\, d\alpha.$$

Next, note that the variable s is a *dummy variable* and may be replaced by any other, say, x, since the integral is a function of α. Thus we have the pair of equations

$$\mathscr{F}\{f(x)\} = \bar{f}(\alpha) = \int_{-\infty}^{\infty} f(x)\, e^{i\alpha x}\, dx$$
$$\mathscr{F}^{-1}\{\bar{f}(\alpha)\} = f(x) = \frac{1}{2\pi} \int_{-\infty}^{\infty} \bar{f}(\alpha)\, e^{-i\alpha x}\, d\alpha. \qquad \textbf{(5.2–1)}$$

Such a pair is called a **Fourier transform pair.** We call $\bar{f}(\alpha)$ the **Fourier transform** of $f(x)$. The second equation in (5.2–1) defines $f(x)$, which is the **in-**

*This is possible because of a Fubini theorem. See B. D. Craven, *Functions of Several Variables* (London: Chapman and Hall, 1981), p. 73.

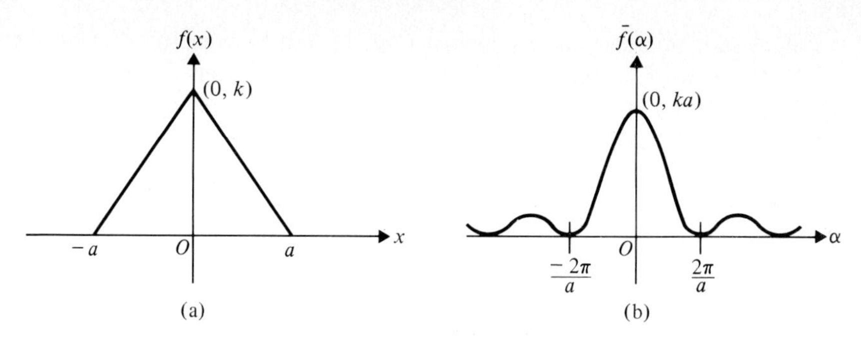

Figure 5.2-1
A Fourier transform pair.

verse Fourier transform of $\bar{f}(\alpha)$. An example of a simple function and its Fourier transform is shown in Fig. 5.2-1.

Before we give an example of the use of the Fourier transform, a few words of caution are in order. Some authors define a Fourier transform pair as

$$\bar{f}(\alpha) = \frac{1}{\sqrt{2\pi}} \int_{-\infty}^{\infty} f(x)\, e^{i\alpha x}\, dx,$$

$$f(x) = \frac{1}{\sqrt{2\pi}} \int_{-\infty}^{\infty} \bar{f}(\alpha)\, e^{-i\alpha x}\, d\alpha$$

in order to keep as much symmetry as possible in the pair. Others interchange the negative signs in the exponents, while still others use combinations of these two variations. Because of the negative sign in *one* of the exponents, true symmetry is impossible to achieve. The important thing is that any given pair should reduce to Eq. (5.1-14) or its equivalent. Thus in applications it does not matter if the factor $1/2\pi$ is inserted when the function is transformed or when the inverse transform is found. You should keep this fact in mind when consulting tables of Fourier transforms. It is possible to eliminate the $1/2\pi$ factor as a coefficient entirely and write the Fourier transform pair as (Exercise 1)

$$\bar{f}(\alpha) = \int_{-\infty}^{\infty} f(x)\, e^{2\pi i\alpha x}\, dx,$$

$$f(x) = \int_{-\infty}^{\infty} \bar{f}(\alpha)\, e^{-2\pi i\alpha x}\, d\alpha. \tag{5.2-2}$$

This last form has the advantage that $2\pi\alpha$ can be replaced by ω, the angular frequency, producing even more compact expressions.

Transforms are common in mathematics, since they are useful in converting many problems into more easily solvable ones. For example, logarithms are useful for converting multiplication into addition, the Laplace transform is useful for converting certain ordinary differential equations into algebraic equations, and both the Laplace and Fourier transforms are useful for converting certain partial differential equations into ordinary differential equations. In each case an inverse process is necessary.

In preparation for the next example we compute the Fourier transform of du/dx and d^2u/dx^2. We *assume* that u and du/dx both approach zero as $x \to \pm\infty$, that u is piecewise smooth, and that u is absolutely integrable on the real line. We have, after integrating by parts,

$$\int_{-\infty}^{\infty} \frac{du}{dx} e^{i\alpha x}\, dx = ue^{i\alpha x}\Big|_{-\infty}^{\infty} - i\alpha \int_{-\infty}^{\infty} ue^{i\alpha x}\, dx$$

$$= -i\alpha \bar{u}(\alpha), \tag{5.2-3}$$

using the assumptions. For the second derivative we have, after integrating by parts,

$$\int_{-\infty}^{\infty} \frac{d^2u}{dx^2} e^{i\alpha x}\, dx = \frac{du}{dx} e^{i\alpha x}\Big|_{-\infty}^{\infty} - \alpha i \int_{-\infty}^{\infty} \frac{du}{dx} e^{i\alpha x}\, dx$$

$$= -i\alpha[-i\alpha \bar{u}(\alpha)] = -\alpha^2 \bar{u}(\alpha),$$

using the result of Eq. (5.2–3). We can also write the above as

$$\mathscr{F}\left\{\frac{d^2u}{dx^2}\right\} = -\alpha^2 \bar{u}(\alpha), \tag{5.2-4}$$

where $\mathscr{F}$ indicates the Fourier transform defined in Eq. (5.2–1).

EXAMPLE 5.2–1 Solve the following one-dimensional diffusion equation:

$$u_t(x, t) = u_{xx}(x, t), \qquad -\infty < x < \infty, \qquad 0 < t,$$

given that $u(x, 0) = f(x)$ and $|u(x, t)| < \infty$.

Solution We assume that $u(x, t)$ and $u_x(x, t)$ approach zero as $x \to \pm\infty$ and that $f(x)$ is piecewise smooth and absolutely integrable on the real line. Then $f(x)$ has a Fourier transform

$$\bar{f}(\alpha) = \int_{-\infty}^{\infty} f(x)\, e^{i\alpha x}\, dx,$$

$$= \int_{-\infty}^{\infty} f(s)\, e^{i\alpha s}\, ds$$

if we change the dummy variable of integration in order to prevent confusion later. The Fourier transform of $u_t(x, t)$ is

$$\int_{-\infty}^{\infty} \frac{\partial u(x, t)}{\partial t} e^{i\alpha x}\, dx = \frac{\partial}{\partial t} \int_{-\infty}^{\infty} u(x, t) e^{i\alpha x}\, dx$$

$$= \frac{\partial \bar{u}(\alpha, t)}{\partial t} = \frac{d\bar{u}(\alpha, t)}{dt},$$

where we have assumed that differentiation and integration may be interchanged (see Theorem 1.6–1) and where we have noted that α plays the role of a parameter in obtaining the transform.

Transforming the given partial differential equation and the initial condition produces

$$\frac{d\bar{u}}{dt} + \alpha^2\bar{u} = 0,$$

$$\bar{u}(\alpha, 0) = \bar{f}(\alpha).$$

The solution to this problem is readily found (Exercise 2) to be

$$\bar{u}(\alpha, t) = \bar{f}(\alpha)e^{-\alpha^2 t}.$$

Using the inversion formula (5.2–1) to obtain $u(x, t)$, we have

$$\begin{aligned} u(x, t) &= \frac{1}{2\pi}\int_{-\infty}^{\infty} \bar{u}(\alpha, t)e^{-i\alpha x}\, d\alpha = \frac{1}{2\pi}\int_{-\infty}^{\infty} \bar{f}(\alpha)e^{-\alpha^2 t}e^{-i\alpha x}d\alpha \\ &= \frac{1}{2\pi}\int_{-\infty}^{\infty}\int_{-\infty}^{\infty} f(s)e^{i\alpha s}e^{-\alpha^2 t}e^{-i\alpha x}d\alpha\, ds \\ &= \frac{1}{2\pi}\int_{-\infty}^{\infty}\int_{-\infty}^{\infty} f(s)e^{i\alpha(s-x)}e^{-\alpha^2 t}d\alpha\, ds. \end{aligned}$$

Now

$$e^{-\alpha^2 t}e^{i\alpha(s-x)} = e^{-\alpha^2 t}\,[\cos \alpha(s - x) + i \sin \alpha(s - x)],$$

and the first term in this sum is an even function of α, while the second term is an odd function of α. Hence

$$\begin{aligned} u(x, t) &= \frac{1}{\pi}\int_{-\infty}^{\infty}\int_{0}^{\infty} f(s) \cos \alpha(s - x)\, e^{-\alpha^2 t}d\alpha\, ds \\ &= \frac{1}{2\sqrt{\pi t}}\int_{-\infty}^{\infty} f(s) \exp\,[-(s - x)^2/4t]\, ds, \end{aligned} \qquad \textbf{(5.2–5)}$$

using a result found in published tables of definite integrals, namely,

$$\int_0^{\infty} \cos bx \exp\,(-a^2x^2)\, dx = \frac{\sqrt{\pi}}{2a} \exp\,(-b^2/4a^2), \qquad \text{if } a > 0. \qquad \blacksquare$$

From a practical viewpoint the evaluation of $u(x, t)$ from Eq. (5.2–5) for various values of x and t may be done numerically. The use of the **fast Fourier transform (FFT)** technique is recommended in this case.*

*See Anthony Ralston and P. Rabinowitz, *A First Course In Numerical Analysis* (New York: McGraw-Hill, 1978), p. 263 ff.

Fourier Sine Transform

If $f(x)$ is an odd function, then we define the **Fourier sine transform** pair

$$\mathscr{F}_s\{f(x)\} = \bar{f}_s(\alpha) = \int_0^\infty f(x) \sin \alpha x \, dx,$$
$$f(x) = \frac{2}{\pi} \int_0^\infty \bar{f}_s(\alpha) \sin \alpha x \, d\alpha. \tag{5.2-6}$$

We use repeated integration by parts to compute the Fourier sine transform of d^2u/dx^2 to obtain

$$\begin{aligned} \int_0^\infty \frac{d^2u}{dx^2} \sin \alpha x \, dx &= \frac{du}{dx} \sin \alpha x \bigg|_0^\infty - \alpha \int_0^\infty \frac{du}{dx} \cos \alpha x \, dx \\ &= -\alpha \int_0^\infty \frac{du}{dx} \cos \alpha x \, dx \\ &= -\alpha u \cos \alpha x \bigg|_0^\infty - \alpha^2 \int_0^\infty u \sin \alpha x \, dx. \end{aligned}$$

Hence

$$\mathscr{F}_s\left\{\frac{d^2u}{dx^2}\right\} = \alpha u(0) - \alpha^2 \bar{u}_s(\alpha). \tag{5.2-7}$$

In the above computation we have made the assumptions that u and du/dx approach zero as $x \to \infty$ and that

$$\int_0^\infty |u| \, dx$$

is finite.

Fourier Cosine Transform

If $f(x)$ is an even function, then we define the **Fourier cosine transform** pair

$$\mathscr{F}_c\{f(x)\} = \bar{f}_c(\alpha) = \int_0^\infty f(x) \cos \alpha x \, dx,$$
$$f(x) = \frac{2}{\pi} \int_0^\infty \bar{f}_c(\alpha) \cos \alpha x \, d\alpha. \tag{5.2-8}$$

In the same manner as before, we can compute the Fourier cosine transform of d^2u/dx^2 to obtain (Exercise 3)

$$\mathscr{F}_c\left\{\frac{d^2u}{dx^2}\right\} = -u'(0) - \alpha^2 \bar{u}_c(\alpha). \tag{5.2-9}$$

under the previous assumptions. When the meaning is clear, we will omit the subscripts "s" and "c" in order to simplify the notation.

EXAMPLE 5.2–2 Solve the following boundary-value problem (Fig. 5.2–2):

$$\begin{aligned} &\text{P.D.E.:} && u_t = ku_{xx}, \quad 0 < x < \infty, \quad 0 < t; \\ &\text{B.C.:} && u_x(0, t) = 0, \quad 0 < t; \\ &\text{I.C.:} && u(x, 0) = f(x), \quad 0 < x < \infty. \end{aligned}$$

Solution Since $0 < x < \infty$, we can make either an odd or an even extension of the given function $f(x)$. In other words, it "appears" that we may solve the problem by means of either the Fourier sine or the cosine transform. We note, however, that in the boundary condition we are given the value of $u_x(x, t)$ at $x = 0$. For this reason we choose the cosine transform.

A few words of explanation are in order here regarding the physical significance of the boundary condition $u_x(0, t) = 0$. We saw in Section 3.4 that the heat flux (the amount of heat per unit area per unit time) across a surface is proportional to the thermal conductivity and the change of temperature in the direction of the outward normal to the surface. In the present example the change of temperature in the direction of the outward normal is $-u_x(0, t)$ at $x = 0$. Ordinarily, $-u_x(0, t) = g(t)$, that is, a function of t. In the case of **perfect insulation**, however, we take $g(t) \equiv 0$, since there can be no heat flux across the surface.

We assume that u and du/dt approach zero as $x \to \infty$ and that both $u(x, t)$ and $f(x)$ are absolutely integrable on $0 < x < \infty$. Then

$$u(x, t) = \frac{2}{\pi}\int_0^\infty \bar{u}(\alpha, t) \cos \alpha x \, d\alpha,$$

and if we use Eq. (5.2–9), the partial differential equation and initial condition are transformed into

$$\frac{d\bar{u}}{dt} + \alpha^2 k \bar{u} = 0, \qquad \bar{u}(0) = \bar{f}(\alpha).$$

Hence (see Example 5.2–1)

$$\bar{u}(\alpha, t) = \bar{f}(\alpha)e^{-\alpha^2 kt},$$

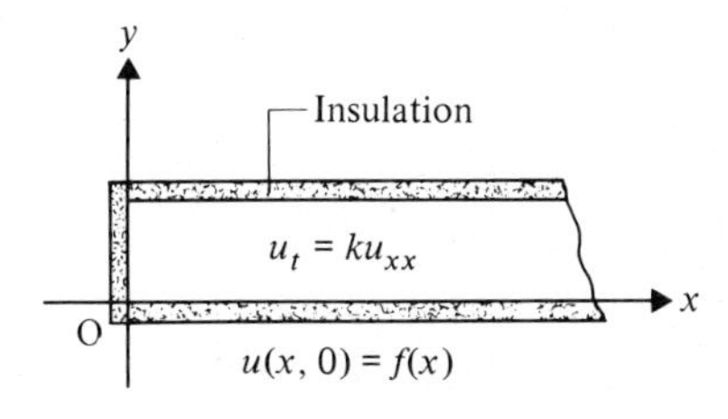

Figure 5.2–2
The problem of Example 5.2–2.

and, using the inverse cosine transform,

$$u(x, t) = \frac{2}{\pi}\int_0^\infty \bar{f}(\alpha)e^{-\alpha^2kt}\cos\alpha x\,d\alpha.$$

But

$$\bar{f}(\alpha) = \int_0^\infty f(x)\cos\alpha x\,dx$$

$$= \int_0^\infty f(s)\cos\alpha s\,ds.$$

Thus

$$u(x, t) = \frac{2}{\pi}\int_0^\infty f(s)\cos\alpha s\,ds\int_0^\infty e^{-\alpha^2kt}\cos\alpha x\,d\alpha. \quad \blacksquare$$

EXAMPLE 5.2–3 Solve the following boundary-value problem:

$$\begin{aligned}
\text{P.D.E.:}\quad & u_{tt} = a^2u_{xx}, && -\infty < x < \infty, \quad 0 < t;\\
\text{I.C.:}\quad & u(x, 0) = f(x), && -\infty < x < \infty,\\
& u_t(x, 0) = g(x), && -\infty < x < \infty.
\end{aligned}$$

Solution This problem can be solved by a method similar to the one we used in solving the heat equation in Example 5.2–1. We assume that u, $f(x)$, and $g(x)$ all have Fourier transforms. Then the partial differential equation is transformed into

$$\frac{d^2\bar{u}}{dt^2} + \alpha^2a^2\bar{u} = 0$$

with initial conditions

$$\bar{u}(0) = \bar{f}(\alpha) \qquad \text{and} \qquad \bar{u}'(0) = \bar{g}(\alpha).$$

Hence

$$\bar{u}(\alpha, t) = c_1(\alpha)\cos\alpha at + c_2(\alpha)\sin\alpha at,$$

with $c_1(\alpha) = \bar{f}(\alpha)$. Then

$$\bar{u}'(\alpha, t) = -\alpha a\bar{f}(\alpha)\sin\alpha at + c_2(\alpha)\,\alpha a\cos\alpha at,$$

and the condition

$$\bar{u}'(0) = \bar{g}(\alpha)$$

implies that

$$c_2(\alpha) = \bar{g}(\alpha)/\alpha a.$$

Thus

$$\bar{u}(\alpha, t) = \bar{f}(\alpha)\cos\alpha at + \frac{\bar{g}(\alpha)\sin\alpha at}{\alpha a}$$

and

$$u(x, t) = \frac{1}{2\pi} \int_{-\infty}^{\infty} \left[\bar{f}(\alpha) \cos \alpha at + \frac{\bar{g}(\alpha) \sin \alpha at}{\alpha a} \right] e^{-i\alpha x} \, d\alpha.$$

The last expression can be put into a more easily recognizable form if we make use of the connection between the hyperbolic functions of complex quantities and the circular functions, namely,

$$\sin x = -\frac{1}{2} i(e^{ix} - e^{-ix}) = -i \sinh ix$$

and

$$\cos x = \frac{1}{2}(e^{ix} + e^{-ix}) = \cosh ix.$$

Using these, we can write the solution to the problem as

$$u(x, t) = \frac{1}{2\pi} \int_{-\infty}^{\infty} \bar{f}(\alpha) \frac{e^{-i\alpha(x-at)} + e^{-i\alpha(x+at)}}{2} \, d\alpha$$

$$+ \frac{1}{2\pi} \int_{-\infty}^{\infty} \bar{g}(\alpha) \frac{e^{-i\alpha(x-at)} + e^{-i\alpha(x+at)}}{2\alpha ai} \, d\alpha.$$

But

$$f(x - at) = \frac{1}{2\pi} \int_{-\infty}^{\infty} \bar{f}(\alpha) e^{-i\alpha(x-at)} \, d\alpha,$$

so that the first integral in the solution is

$$\frac{1}{2} [f(x - at) + f(x + at)].$$

Further, from

$$g(x) = \frac{1}{2\pi} \int_{-\infty}^{\infty} \bar{g}(\alpha) e^{-i\alpha x} \, d\alpha,$$

we obtain for arbitrary c and d

$$\int_c^d g(x) \, dx = \frac{1}{2\pi} \int_{-\infty}^{\infty} \bar{g}(\alpha) \, d\alpha \int_c^d e^{-i\alpha x} \, dx$$

$$= \frac{1}{2\pi} \int_{-\infty}^{\infty} \bar{g}(\alpha) \, d\alpha \left(\frac{e^{-i\alpha x}}{-i\alpha} \right) \Bigg|_c^d$$

$$= \frac{1}{2\pi} \int_{-\infty}^{\infty} \bar{g}(\alpha) \frac{e^{-i\alpha c} - e^{-i\alpha d}}{i\alpha} \, d\alpha,$$

assuming that we can change the order of integration. Hence

$$\frac{1}{2\pi} \int_{-\infty}^{\infty} \bar{g}(\alpha) \frac{e^{-i\alpha(x-at)} - e^{-i\alpha(x+at)}}{2ai\alpha} \, d\alpha = \frac{1}{2a} \int_{x-at}^{x+at} g(s) \, ds,$$

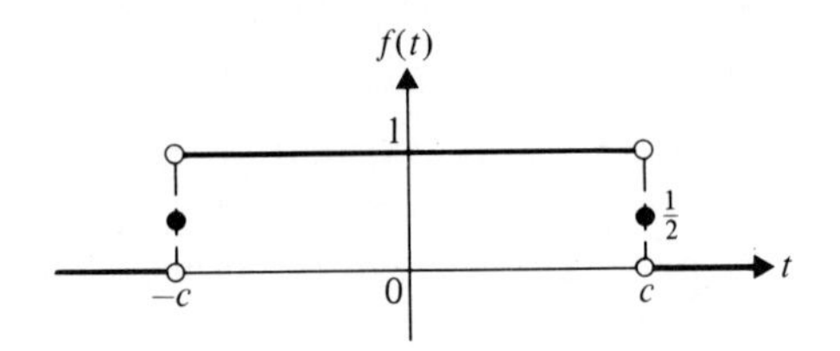

Figure 5.2-3
A rectangular pulse.

and the final solution to our problem takes on the familiar form of D'Alembert's solution

$$u(x, t) = \frac{1}{2}[f(x + at) + f(x - at)] + \frac{1}{2a}\int_{x-at}^{x+at} g(s)\,ds \qquad \textbf{(5.2-10)}$$

(see Eq. (3.3-8)). ■

An important special application of the Fourier transform will be given next. Suppose that we have a rectangular unit pulse of duration $2c$ defined by

$$f(t) = \begin{cases} 1, & \text{when } |t| < c, \\ 0, & \text{when } |t| > c, \\ 1/2, & \text{when } |t| = c. \end{cases}$$

This pulse is shown in Fig. 5.2-3.

Since $f(t)$ is piecewise smooth and absolutely integrable, we can compute its Fourier transform (also called the **spectrum of the function**) as follows:

$$\bar{f}(\alpha) = \int_{-c}^{c} f(t)\, e^{i\alpha t}\,dt = \int_{-c}^{c} e^{i\alpha t}\,dt$$

$$= \left.\frac{e^{i\alpha t}}{i\alpha}\right|_{-c}^{c} = \frac{1}{i\alpha}(e^{i\alpha c} - e^{-i\alpha c}).$$

Then, using the relation

$$\frac{e^{ix} - e^{-ix}}{2i} = \sin x,$$

we obtain

$$\bar{f}(\alpha) = \frac{2 \sin \alpha c}{\alpha}.$$

The graph of $\bar{f}(\alpha)$, the spectrum of $f(t)$, is shown in Fig. 5.2-4.

Note that we can evaluate $\bar{f}(0)$ by means of the well-known limit

$$\lim_{\theta \to 0} \frac{\sin \theta}{\theta} = 1.$$

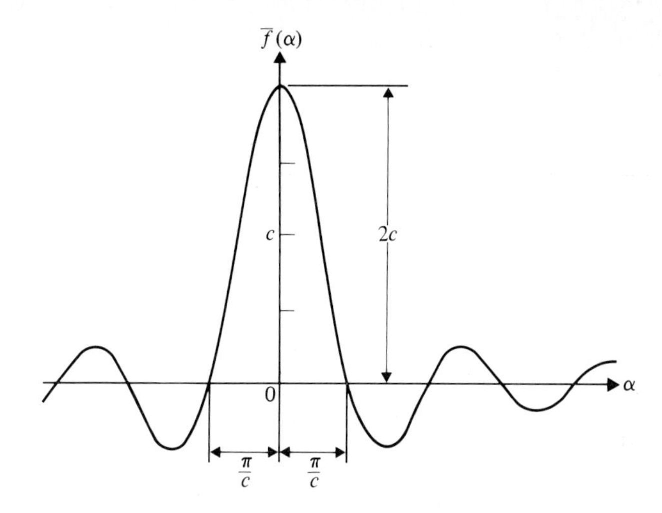

Figure 5.2-4
The spectrum of a rectangular pulse.

Hence $\bar{f}(0) = 2c$, as shown in Fig. 5.2-4. As $c \to \infty$, the pulse $f(t)$ becomes increasingly broad in time. On the other hand, the central peak of $\bar{f}(\alpha)$ becomes increasingly higher and narrower. Thus most of the energy of the pulse lies within this central peak of width $2\pi/c$. Therefore the longer the pulse, the more narrow the **spectral bandwidth** into which its energy is concentrated.

If we think of α as angular frequency ($\alpha = 2\pi f$) and let $\Delta\alpha$ be the angular frequency separating the maximum of $\bar{f}(\alpha)$ at $\alpha = 0$ from the first zero at $\alpha = \pi/c$, then $\Delta\alpha = \pi/c$. However, c represents the corresponding duration of the pulse in time, call it Δt; hence $(\Delta\alpha)(\Delta t) = \pi$ or $2\pi\ \Delta f\ \Delta t = \pi$ or $\Delta f\ \Delta t = 1/2$. Thus there is a *constant* relation between the time duration of the pulse and its frequency bandwidth. In other words, the shape of a pulse in the time domain and the shape of its amplitude spectrum in the frequency domain are not independent. A relation of this kind forms the basis of the **Heisenberg* uncertainty principle** in quantum mechanics manifested in the uncertainty relation between position and momentum measurements.†

A visual formulation of the foregoing relation between the time and frequency domains can be obtained. Imagine that all the harmonic components indicated by $\bar{f}(\alpha)$ are graphed as functions of *time*. Now, if the ordinates are added, the result is the single step $f(t)$. Stated in another way, since each harmonic component extends from $t = -\infty$ to $t = \infty$, the linear superposition of harmonic components results in a complete cancellation for $|t| > c$, in unity for $|t| < c$, and in one-half for $|t| = c$. Still another, and

*Werner (Carl) Heisenberg (1901–1976), a German physicist.

†See Kurt B. Wolf, *Integral Transforms in Science and Engineering* (New York: Plenum Press, 1979), Chapter 7.

more mathematical, way of expressing the above is in the form

$$f(t) = \frac{1}{2\pi}\int_{-\infty}^{\infty} \bar{f}(\alpha)e^{-i\alpha t}\, d\alpha.$$

We mention one important offshoot of the rectangular pulse example. If we let c become infinite, then the energy of the pulse is confined to a *zero* bandwidth. Under this limiting condition, $\bar{f}(\alpha)$ becomes the so-called **Dirac delta function*** $\delta(\alpha)$, which has the unusual properties

$$\delta(\alpha) = 0 \qquad \text{if } \alpha \neq 0 \tag{5.2–11}$$

and

$$\int_{-\infty}^{\infty} \delta(\alpha)\, d\alpha = 1. \tag{5.2–12}$$

The Dirac delta function plays an important role in ordinary differential equations when the forcing functions are impulses. In connection with Fourier transforms, the delta function broadens the theory by admitting periodic functions that do not possess Fourier transforms.†

Although the Dirac delta function defined above is not a function in the mathematical sense, it is a useful entity. This "pseudofunction" belongs to a class of generalized functions, and its use as a function may be studied by "distribution theory."‡ One of the most useful properties of $\delta(\alpha)$ is its *sifting property*, that is, its ability to pick (or sift) out the value of a function at a particular point. If $f(t)$ is continuous on $t \geq 0$, then

$$\int_0^{\infty} f(t)\, \delta(t)\, dt = f(0) \tag{5.2–13}$$

and

$$\int_0^{\infty} f(t)\, \delta(t - t_0) = f(t_0). \tag{5.2–14}$$

Key Words and Phrases

- Fourier transform pair
- Fourier transform
- inverse Fourier transform
- fast Fourier transform (FFT)
- Fourier sine transform
- Fourier cosine transform
- perfect insulation
- spectrum of a function
- spectral bandwidth
- Heisenberg uncertainty principle
- Dirac delta function

---

*After Paul A. M. Dirac (1902–), a British theoretical physicist known for his work in quantum mechanics.

†See R. N. Bracewell, *The Fourier Transform and Its Applications*, 2nd ed. (New York: McGraw-Hill, 1978).

‡See Arthur E. Danese, *Advanced Calculus: An Introduction to Applied Mathematics*, Vol. 2 (Boston: Allyn and Bacon, 1965), Chapter 24.

5.2 Exercises

• **1.** Obtain the Fourier transform pair Eqs. (5.2–2) from Eqs. (5.2–1). (*Hint*: Replace α by $2\pi\lambda$ in the second of Eqs. (5.2–1).)

2. Verify that

$$\bar{u}(\alpha, t) = \bar{f}(\alpha) \exp(-\alpha^2 t)$$

is the unique solution of the initial-value problem in Example 5.2–1.

3. Obtain the result shown in Eq. (5.2–9).

4. Solve Example 5.2–1 if $f(x)$ is defined as follows (see Fig. 5.2–1)

$$f(x) = \begin{cases} 1 + x, & \text{for } -1 < x \le 0, \\ 1 - x, & \text{for } 0 \le x < 1, \\ 0, & \text{otherwise.} \end{cases}$$

5. Solve Example 5.2–2 if $f(x)$ is defined as follows.

$$f(x) = \begin{cases} b, & \text{for } 0 < x < c,\ b > 0, \\ 0, & \text{for } x > c. \end{cases}$$

6. Carry out the details required to obtain $\bar{u}(\alpha, t)$ in Example 5.2–3.

•• **7.** What is the solution of the problem in Example 5.2–1 if $f(x) = \exp(-|x|)$?

8. Solve Example 5.2–2 if the boundary condition $u_x(0, t) = 0$ is replaced by $u(0, t) = 0$.

9. If $f(x) = \exp(-x)$, solve each of the following problems:
(a) Example 5.2–2
(b) Exercise 8

10. **(a)** Obtain the spectrum of the function

$$f(t) = \begin{cases} \cos t, & \text{for } -\pi/2 \le t \le \pi/2, \\ 0, & \text{elsewhere} \end{cases}$$

(b) Graph the function and its spectrum.

11. Find the Fourier transform of each of the following functions defined on $(-\infty, \infty)$.

(a) $g(x) = \begin{cases} \sin \pi x, & \text{for } 0 < x < 1, \\ 0, & \text{otherwise} \end{cases}$

(b) $f(t) = \begin{cases} |\cos \pi t|, & \text{for } 0 < t < 1, \\ 0, & \text{otherwise} \end{cases}$

(c) $f(x) = \begin{cases} \exp(-x), & \text{for } 0 < x < 1, \\ 0, & \text{otherwise} \end{cases}$

(d) $g(t) = \begin{cases} t^2, & \text{for } |t| < 1, \\ 0, & \text{for } |t| > 1 \end{cases}$

(e) $F(x) = \begin{cases} -h, & \text{for } -c < x < 0,\ h > 0, \\ h, & \text{for } 0 < x < c, \\ 0, & \text{otherwise} \end{cases}$

12. Find the Fourier *sine* transform of each of the following functions defined on $(0, \infty)$.

(a) $\phi(x) = \begin{cases} h, & \text{for } 0 < x < \dfrac{1}{h}, \\ 0, & \text{for } x > \dfrac{1}{h}, \end{cases}$

(b) $\psi(t) = \begin{cases} \sin \pi t, & \text{for } 0 < t < 1, \\ 0, & \text{for } t > 1 \end{cases}$

(c) $F(x) = \begin{cases} 1 - x, & \text{for } 0 < x < 1, \\ 0, & \text{for } x > 1 \end{cases}$

(d) $G(t) = \begin{cases} \exp t, & \text{for } 0 < t < 2, \\ 0, & \text{for } t > 2 \end{cases}$

13. Find the Fourier *cosine* transform of each of the following functions defined on $[0, \infty)$.

(a) $f(x) = \begin{cases} \sin x, & \text{for } 0 \le x < \pi, \\ 0, & \text{for } x > \pi \end{cases}$

(b) $\phi(x) = \begin{cases} h, & \text{for } 0 \le x < \dfrac{1}{h}, \\ 0, & \text{for } x > \dfrac{1}{h} \end{cases}$

(c) $F(t) = \begin{cases} |\cos \pi t|, & \text{for } 0 \le t < 1, \\ 0, & \text{for } t > 1 \end{cases}$

(d) $G(x) = \begin{cases} 0, & \text{for } 0 \le x < \dfrac{\pi}{2}, \\ \cos x, & \text{for } \dfrac{\pi}{2} < x < \dfrac{3\pi}{2}, \\ 0, & \text{for } x > \dfrac{3\pi}{2} \end{cases}$

14. Let $f(x)$ be absolutely integrable on $-\infty < x < \infty$ and have Fourier transform $\bar{f}(\alpha)$. Show the following for $a > 0$.

(a) $\dfrac{1}{a} f(x/a)$ has transform $\bar{f}(a\alpha)$.

(b) $f(ax)$ has transform $\dfrac{1}{a}\bar{f}(\alpha/a)$.

(c) $\bar{f}(x)$ has transform $2\pi f(-\alpha)$.

15. Let $f(x)$ be absolutely integrable on $-\infty < x < \infty$ and have Fourier transform $\bar{f}(\alpha)$. Show each of the following for any real b.

(a) $f(x - b)$ has transform $e^{i\alpha b}\bar{f}(\alpha)$.

(b) $\frac{1}{2}[f(x - b) + f(x + b)]$ has transform $(\cos \alpha b)\bar{f}(\alpha)$.

16. Prove that finding the Fourier transform of a function is a linear operation.

17. Use the definition of Fourier transform,

$$\bar{f}(\alpha) = \frac{1}{\sqrt{2\pi}} \int_{-\infty}^{\infty} f(x)\, e^{i\alpha x}\, dx,$$

to show that $f(x) = \exp(-x^2/2)$ is its own Fourier transform.

18. Explain why the negative sign in the exponent can be removed from the second equation of (5.2–1) and placed in the first equation.

••• 19. The function (compare Fig. 5.2–4)

$$\operatorname{sinc} x = \frac{\sin \pi x}{\pi x}$$

is useful as a filtering or interpolating function. Prove the following properties of this function.

(a) $\operatorname{sinc} 0 = 1$ **(b)** $\operatorname{sinc} n = 0$, n a nonzero integer

(c) $\int_{-\infty}^{\infty} \operatorname{sinc} x \, dx = 1$

20. The function (compare Exercise 19) $\operatorname{sinc}^2 x$ represents the power radiation pattern of a uniformly excited antenna or the intensity of light in the Fraunhofer diffraction pattern of a slit. Graph this function.

21. Find the Fourier transform of $\operatorname{sinc}^2 x$ and compare it with $\bar{f}(\alpha)$, the transform of $f(x)$ in Exercise 4.

22. Prove the following properties of the Fourier transform for real functions $f(x)$:
(a) If $f(x)$ is even, then $\bar{f}(\alpha)$ is real and even.
(b) If $f(x)$ is odd, then $\bar{f}(\alpha)$ is imaginary and odd.
(c) If $f(x)$ is neither even nor odd, then $\bar{f}(\alpha)$ has an even real part and an odd imaginary part.

23. Illustrate the properties in Exercise 22 by using the results of Exercises 11(d), 11(e), and 11(b), respectively.

24. Find the Fourier transform of the *finite wave train*

$$f(t) = \begin{cases} \sin t, & \text{for } |t| \le 6\pi, \\ 0, & \text{otherwise.} \end{cases}$$

5.3 APPLICATIONS

We saw in Examples 5.2–1 to 5.2–3 how the Fourier transform can be used to solve boundary-value problems involving the diffusion and wave equations. In this section we give applications of the transform to the potential equation and to certain linear ordinary differential equations. Our first example also illustrates an application of the Fourier integral representation of a function.

EXAMPLE 5.3–1 Find a potential function $u(x, y)$ in the xy-plane that is bounded for $y \ge 0$ and for which

$$\lim_{y \to 0^+} u(x, y) = f(x).$$

Solution We solve Laplace's equation, also called the **potential equation,**

$$u_{xx} + u_{yy} = 0,$$

by separation of variables. This results in the two ordinary differential equations

$$X'' + \alpha^2 X = 0$$

and

$$Y'' - \alpha^2 Y = 0,$$

with solutions

$$X(x) = c_1 \cos \alpha x + c_2 \sin \alpha x$$

and

$$Y(y) = c_3 e^{\alpha y} + c_4 e^{-\alpha y},$$

respectively. We choose $c_3 = 0$ and $\alpha > 0$, since the solution must be bounded for $y \geq 0$. Thus the solution has the form*

$$u(x, y) = \int_0^\infty e^{-\alpha y}(A \cos \alpha x + B \sin \alpha x)\, d\alpha,$$

and we can satisfy the given boundary condition by considering A and B to be functions of α. Then

$$u(x, y) = \int_0^\infty e^{-\alpha y}[A(\alpha) \cos \alpha x + B(\alpha) \sin \alpha x]\, d\alpha,$$

and

$$f(x) = \int_0^\infty [A(\alpha) \cos \alpha x + B(\alpha) \sin \alpha x]\, d\alpha.$$

If we refer to Eq. (5.1–6), the last equation can be satisfied if we take

$$A(\alpha) \cos \alpha x + B(\alpha) \sin \alpha x = \frac{1}{\pi} \int_{-\infty}^\infty f(s) \cos \alpha(s - x)\, ds,$$

that is, if we use the Fourier integral representation of $f(x)$. Then

$$u(x, y) = \frac{1}{\pi} \int_0^\infty \left[\int_{-\infty}^\infty f(s) \cos \alpha(s - x)\, ds\right] e^{-\alpha y}\, d\alpha.$$

Assuming that we can reverse the order of integration, we have

$$u(x, y) = \frac{1}{\pi} \int_0^\infty \left[\int_{-\infty}^\infty e^{-\alpha y} \cos \alpha(s - x)\, d\alpha\right] f(s)\, ds.$$

*Note that an *integral* is necessary, since the eigenvalues α^2 are continuous here.

Finally, evaluating the inner integral (Exercise 1) gives us

$$u(x, y) = \frac{1}{\pi} \int_{-\infty}^{\infty} \frac{yf(s)}{y^2 + (s - x)^2}\, ds. \qquad \blacksquare \tag{5.3–1}$$

Since $-\infty < x < \infty$ in Example 5.3–1, it is possible to use the Fourier exponential transform as an alternative method of obtaining the solution. Transforming x, we have the problem

$$\frac{d^2\bar{u}(\alpha, y)}{dy^2} - \alpha^2\bar{u}(\alpha, y) = 0,$$
$$\bar{u}(\alpha, 0) = \bar{f}(\alpha), \tag{5.3–2}$$

assuming that $f(x)$ is absolutely integrable on the real line. The *bounded* solution of this problem can be written as (Exercise 19)

$$\bar{u}(\alpha, y) = \begin{cases} \bar{f}(\alpha)e^{\alpha y}, & \text{for } \alpha < 0, \\ \bar{f}(\alpha)e^{-\alpha y}, & \text{for } \alpha > 0, \end{cases} \tag{5.3–3}$$

or, more compactly, as

$$\bar{u}(\alpha, y) = \bar{f}(\alpha)e^{-|\alpha|y}. \tag{5.3–4}$$

The Fourier inverse transform of this is

$$\begin{aligned} u(x, y) &= \frac{1}{2\pi} \int_{-\infty}^{\infty} \bar{f}(\alpha)e^{-|\alpha|y}e^{-i\alpha x}\, d\alpha \\ &= \frac{1}{2\pi} \int_{-\infty}^{\infty} \int_{-\infty}^{\infty} f(s)e^{-|\alpha|y}e^{-i\alpha s}e^{-i\alpha x}\, ds\, d\alpha \\ &= \frac{1}{2\pi} \int_{-\infty}^{\infty} f(s)\, ds \int_{-\infty}^{\infty} e^{-|\alpha|y}[\cos \alpha(s - x) + i \sin \alpha(s - x)]\, d\alpha \\ &= \frac{1}{\pi} \int_{-\infty}^{\infty} f(s)\, ds \int_{0}^{\infty} e^{-\alpha y}\cos \alpha(s - x)\, d\alpha \\ &= \frac{1}{\pi} \int_{-\infty}^{\infty} \frac{yf(s)}{y^2 + (s - x)^2}\, ds, \end{aligned}$$

which agrees with the previous result. We point out that the two approaches used are equivalent, since in both cases we used the Fourier integral representation of the function $f(x)$.

A question naturally arises at this point, namely, could the problem be solved by using the Fourier transform to transform y instead of x? Since the range of y is $0 < y < \infty$, the Fourier sine and cosine transforms appear to be applicable. Recall, however, that Eq. (5.2–7) shows that the sine transform requires that $u(x, 0)$ be known, whereas Eq. (5.2–9) shows that the cosine transform requires knowledge of $u_y(x, 0)$. Thus it appears that the sine transform is the one to use.

Transforming y by the sine transform converts the potential equation into

$$\frac{d^2\bar{u}_s(x, \alpha)}{dx^2} - \alpha^2\bar{u}_s(x, \alpha) = -\alpha f(x), \tag{5.3–5}$$

which is a *nonhomogeneous* ordinary differential equation. The complementary solution of Eq. (5.3–5) is a linear combination of cosh αx and sinh αx, both of which are *unbounded* for $0 < \alpha < \infty$ and $-\infty < x < \infty$. This difficulty can be overcome by taking the two *arbitrary* constants to be zero, that is, by using $\bar{u}_s(x, \alpha) = 0$ as the complementary solution of Eq. (5.3–5). Accordingly, we need only a particular integral of Eq. (5.3–5). It should be clear now that there are difficulties with the method being used that were not present in the previous ones.

The next example shows an application of the Fourier transform to certain ordinary linear differential equations.

EXAMPLE 5.3–2 Obtain a solution of*

$$x\frac{d^2y}{dx^2} + \frac{dy}{dx} - xy = 0.$$

Solution In order to obtain the Fourier transform of the three terms on the left, we need to observe that from the definition

$$\bar{y}(\alpha) = \int_{-\infty}^{\infty} ye^{i\alpha x}\,dx$$

we can obtain

$$\frac{d\bar{y}}{d\alpha} = i\int_{-\infty}^{\infty} xye^{i\alpha x}\,dx.$$

Hence the transform of xy is given by

$$\int_{-\infty}^{\infty} xye^{i\alpha x}\,dx = -i\frac{d\bar{y}}{d\alpha}.$$

Integration by parts produces (Exercise 2)

$$\int_{-\infty}^{\infty} \frac{dy}{dx}e^{i\alpha x}\,dx = -i\alpha\bar{y}(\alpha),$$

*The equation is the **modified Bessel equation** of order zero (see Section 7.2).

and using the above results and integrating by parts twice, we find that (Exercise 3)

$$\int_{-\infty}^{\infty} x \frac{d^2y}{dx^2} e^{i\alpha x}\, dx = 2i\alpha \bar{y} + \alpha^2 i \frac{d\bar{y}}{d\alpha}.$$

In obtaining the transforms above, we have made the assumptions that y and dy/dx are absolutely integrable. Thus the differential equation is transformed into (Exercise 4).

$$(\alpha^2 + 1) \frac{d\bar{y}}{d\alpha} + \alpha \bar{y} = 0,$$

which is easily solved by separating the variables. Hence (Exercise 5)

$$\bar{y}(\alpha) = \frac{C}{\sqrt{\alpha^2 + 1}},$$

where C is an arbitrary constant. Using the inverse transform, we obtain (Exercise 6)

$$y(x) = \frac{C}{2\pi} \int_{-\infty}^{\infty} \frac{e^{-i\alpha x}}{\sqrt{\alpha^2 + 1}}\, d\alpha = \frac{C}{\pi} \int_0^{\infty} \frac{\cos \alpha x\, d\alpha}{\sqrt{\alpha^2 + 1}} = cK_0(|x|),$$

where $K_0(x)$ is the **modified Bessel function of the second kind** of order zero. It is one of the two linearly independent solutions of the given differential equation, the other being $I_0(x)$, the **modified Bessel function of the first kind** of order zero. ■

Additional examples of the use of transform methods for solving boundary-value problems will be given in Section 6.4.

Key Words and Phrases

potential equation
modified Bessel equation
modified Bessel function of the second kind
modified Bessel function of the first kind

5.3 Exercises

• **1.** In Example 5.3–1, evaluate

$$\int_0^{\infty} e^{-\alpha y} \cos \alpha(s - x)\, d\alpha,$$

and verify the result obtained in the text.

2. Assuming that $y \to 0$ as $x \to \pm\infty$, show that

$$\int_{-\infty}^{\infty} \frac{dy}{dx} e^{i\alpha x}\, dx = -i\alpha \bar{y}(\alpha).$$

3. Use integration by parts to obtain each of the following results. Assume that $y \to 0$ as $x \to \pm\infty$ and $dy/dx \to 0$ as $x \to \pm\infty$.

(a) $$\int_{-\infty}^{\infty} x \frac{d^2y}{dx^2} e^{i\alpha x}\, dx = i\alpha \bar{y}(\alpha) - i\alpha \int_{-\infty}^{\infty} x \frac{dy}{dx} e^{i\alpha x}\, dx$$

(b) $$\int_{-\infty}^{\infty} x \frac{dy}{dx} e^{i\alpha x}\, dx = \bar{y}(\alpha) - \alpha \frac{d\bar{y}}{d\alpha}$$

4. Use the results in Exercise 3 and the text to show that the Fourier transform of

$$x \frac{d^2y}{dx^2} + \frac{dy}{dx} - xy = 0$$

is

$$(\alpha^2 + 1) \frac{d\bar{y}}{d\alpha} + \alpha \bar{y} = 0.$$

5. Solve the equation

$$(\alpha^2 + 1) \frac{d\bar{y}}{d\alpha} + \alpha \bar{y} = 0.$$

6. Find the Fourier inverse transform of

$$\bar{y}(\alpha) = \frac{C}{\sqrt{\alpha^2 + 1}},$$

Where C is an arbitrary constant.

•• **7.** Find the temperature $u(x, t)$ in a semi-infinite rod, initially at temperature zero, given that one end is kept at constant temperature u_0. State any assumptions that must be made to solve this problem by using the Fourier sine transform.

8. A semi-infinite rod is insulated at one end and has an initial temperature distribution given by e^{-ax}, $a > 0$. Find the temperature $u(x, t)$ by:

(a) using separation of variables;

(b) using the Fourier cosine transform.

(c) Verify that the results in parts (a) and (b) are the same.

9. A semi-infinite rod has one end at temperature zero and has an initial temperature distribution given by $f(x)$. Solve this problem given that

$$f(x) = \begin{cases} u_0, & \text{for } 0 < x < L, \\ 0, & \text{otherwise.} \end{cases}$$

10. Solve the following problem:

$$\begin{aligned} &\text{P.D.E.:} && u_t = k u_{xx}, && x > 0, && t > 0; \\ &\text{B.C.:} && u_x(0, t) = -\beta, && t > 0, && \beta > 0; \\ &\text{I.C.:} && u(x, 0) = 0, && x > 0. \end{aligned}$$

11. Use the Fourier cosine transform to show that the steady-state temperatures in the semi-infinite slab $y > 0$, when the temperature on the edge $y = 0$ is kept at unity over the interval $|x| < c$ and at zero outside this interval, is given by

$$u(x, y) = \frac{1}{\pi}\left[\arctan\left(\frac{c + x}{y}\right) + \arctan\left(\frac{c - x}{y}\right)\right].$$

(*Hint*: The result

$$\int_0^\infty e^{-ax}x^{-1} \sin bx \, dx = \arctan \frac{b}{a},$$

$a > 0$, $b > 0$, may be useful.)

12. **(a)** Use the method of Example 5.3–2 to obtain a particular solution of the equation

$$x \frac{d^2y}{dx^2} + \frac{dy}{dx} + xy = 0.$$

(b) Compare the result in part (a) with that found in Example 1.5–5. Comment.

13. Explain how it is known that the separation constant must be chosen as shown in Example 5.3–1.

14. **(a)** Find the solution to the problem of Example 5.3–1 if $f(x)$ is given by

$$f(x) = \begin{cases} 1, & 0 < x < c, \\ 0, & \text{otherwise.} \end{cases}$$

(b) From the result in part (a), compute $u(0, 1)$, $u(c/2, 1)$, and $u(c, 1)$. Are your results reasonable?

15. **(a)** Show that the solution to the problem of Example 5.3–1 if $f(x)$ has the values

$$f(x) = \begin{cases} 0, & x < 0 \\ 1/2, & x = 0 \\ 1, & x > 0 \end{cases}$$

is given by

$$u(x, y) = \frac{1}{\pi}\left(\frac{\pi}{2} + \arctan \frac{x}{y}\right)$$

(b) Prove that the solution in part (a) is bounded for $y \geq 0$.
(c) Verify that the solution satisfies Laplace's equation if $y > 0$.
(d) Verify that

$$\lim_{y \to 0^+} u(x, y) = f(x).$$

(e) Is $f(x)$ absolutely integrable? Does this violate any theorem in Section 5.1?

16. Obtain a particular solution of each of the following differential equations by using the Fourier transform:

(a) $x \dfrac{d^2y}{dx^2} + \dfrac{dy}{dx} + y = 0$

(b) $\dfrac{d^2y}{dx^2} + x \dfrac{dy}{dx} + y = 0$

(c) $\dfrac{d^2y}{dx^2} + \dfrac{dy}{dx} + xy = 0$

(d) $\dfrac{d^2y}{dx^2} + x\dfrac{dy}{dx} - xy = 0$

17. Show that the Fourier transform of

$$\frac{1}{\sqrt{4\pi a}} \exp\,(-x^2/4a)$$

is $\exp\,(-a\alpha^2)$ for $a > 0$.

••• **18.** An important application of the Fourier transform is the resolution of a *finite* sinusoidal wave into an *infinite* one. If $\sin \omega_0 t$ is clipped so that N cycles remain, then

$$f(t) = \begin{cases} \sin \omega_0 t, & \text{for } |t| < \dfrac{N\pi}{\omega_0}, \\ 0, & \text{elsewhere.} \end{cases}$$

(a) Use the Fourier sine transform to show that

$$\bar{f}_s(\alpha) = \frac{2}{\pi}\int_0^{N\pi/\omega_0} \sin \omega_0 t \sin \alpha t \, dt.$$

(b) Compute $\bar{f}(\omega_0)$, and explain the significance of this value.

19. Show that *bounded* solutions of

$$\frac{d^2\bar{u}(\alpha, y)}{dy^2} - \alpha^2\bar{u}(\alpha, y) = 0, \qquad \bar{u}(\alpha, 0) = \bar{f}(\alpha)$$

are given by

$$\bar{u}(\alpha, y) = \begin{cases} \bar{f}(\alpha)e^{\alpha y}, & \text{for } \alpha < 0, \\ \bar{f}(\alpha)e^{-\alpha y}, & \text{for } \alpha > 0, \end{cases}$$

or, more compactly, by

$$\bar{u}(\alpha, y) = \bar{f}(\alpha)e^{-|\alpha|y}.$$

20. Discuss ways in which a particular solution of Eq. (5.3–5) might be found.

21. **(a)** Use the Fourier transform to find a solution of the equation

$$3y'' + 2y' + xy = 0.$$

(b) Verify the solution obtained in part (a). (*Hint*: Show that after substituting into the given differential equation and collecting the terms, the result can be written

$$\int_{-\infty}^{\infty} \exp\,(-iu)\, du.$$

Then use Cauchy's theorem—see Section 10.5 of the author's *Advanced Engineering Mathematics* (Reading, Mass.: Addison-Wesley, 1982).)

22. Read the article "Computerized Tomography: The New Medical X-Ray Technology," by L. A. Shepp and J. B. Kruskal in *Amer. Math. Monthly 85*, No. 6, June–July 1978, pp. 420–439.

REFERENCES

Bracewell, Ronald N., *The Fourier Transform and its Applications*, 2nd ed. New York: McGraw-Hill, 1978.
An excellent book that is especially suited for applied mathematicians and engineers.

Davies, B. *Integral Transforms and Their Applications.* New York: Springer-Verlag, 1978.
An advanced treatment of the Fourier transform that includes transforms in two or more variables.

Erdelyi, A., Magnus, W., Oberhettinger, F., and Tricomi, F. G., *Tables of Integral Transforms (Bateman Manuscript Project).* New York: McGraw-Hill, 1954.
An excellent set of tables of Fourier and Laplace transforms.

Kraut, Edgar A., *Fundamentals of Mathematical Physics.* New York: McGraw-Hill, 1967.
An excellent coverage of integral transforms and partial differential equations at an intermediate level.

Sneddon, Ian N., *Fourier Transforms.* New York: McGraw-Hill, 1951.
This is a complete theoretical and applied treatment of the subject at an advanced level.

Sneddon, Ian N., *The Use of Integral Transforms.* New York: McGraw-Hill, 1972.
This advanced theoretical coverage is very complete and includes tables of Fourier transforms.

Weinberger, H. F., *A First Course in Partial Differential Equations with Complex Variables and Transform Methods.* Waltham, Mass.: Blaisdell, 1965.
A somewhat advanced and thorough treatment of the Fourier transform and other topics.

Wolf, Kurt B., *Integral Transforms in Science and Engineering.* New York: Plenum Press, 1979.
An advanced coverage of the Fourier transform but one that contains many excellent computer-generated graphs.

6 Boundary-Value Problems in Rectangular Coordinates

6.1 LAPLACE'S EQUATION

In Section 3.2 we indicated that one of the most common second-order partial differential equations is Laplace's equation,

$$u_{xx} + u_{yy} + u_{zz} = 0. \qquad \textbf{(6.1-1)}$$

Up to this point we have dealt with this equation in one and two dimensions. We now present an example of a boundary-value problem in *three* dimensions in order to illustrate **double Fourier series**.

EXAMPLE 6.1–1 Solve the following boundary-value problem:

P.D.E.: $u_{xx} + u_{yy} + u_{zz} = 0, \quad 0 < x < a, \quad 0 < y < b, \quad 0 < z < c;$

B.C.: $u(0, y, z) = u(a, y, z) = 0, \quad 0 < y < b, \quad 0 < z < c,$

$u(x, 0, z) = u(x, b, z) = 0, \quad 0 < x < a, \quad 0 < z < c,$

$u(x, y, c) = 0, \quad u(x, y, 0) = f(x, y), \quad 0 < x < a, \quad 0 < y < b.$

Solution This problem could arise in attempting to find the potential function inside a rectangular parallelepiped in which four lateral faces and the top are at potential zero and the potential on the bottom is a given function of x and y (see Fig. 6.1–1). We shall later specify the properties that this function should have.

We solve the problem by the method of separation of variables. Let

$$u(x, y, z) = X(x)Y(y)Z(z),$$

differentiate, and substitute into Eq. (6.1–1) to obtain

$$X''YZ + XY''Z + XYZ'' = 0.$$

The primes denote *ordinary* derivatives with respect to the arguments of the

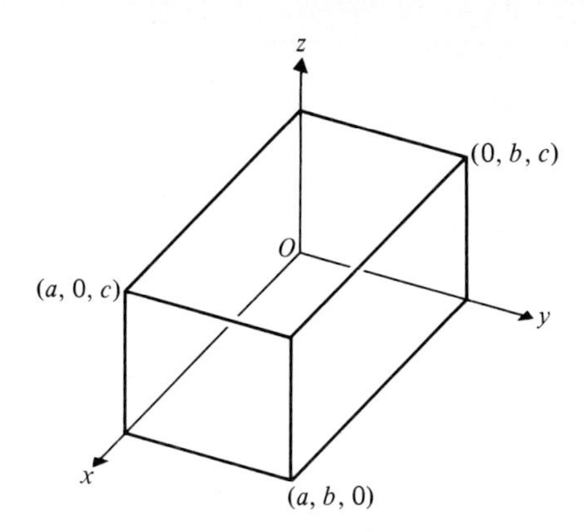

Figure 6.1-1
A Dirichlet problem in $\mathbb{R}^3$.

functions. As usual, we are interested in finding a nontrivial solution; hence the last equation can be divided by the product XYZ. Then

$$\frac{Y''}{Y} + \frac{Z''}{Z} = -\frac{X''}{X} = \lambda, \qquad \textbf{(6.1-2)}$$

where λ is a separation constant whose exact nature will be determined by the boundary conditions. Note that although the separation of variables is not complete, the left-hand side of Eq. (6.1-2) is free of x, while the remaining term contains only x. This condition can exist only if both terms are constant as shown.

The two-point boundary-value problem in X can be written as

$$X'' + \lambda X = 0, \qquad X(0) = 0, \qquad X(a) = 0. \qquad \textbf{(6.1-3)}$$

We leave it for the exercises (see Exercise 1) to show that $\lambda = 0$ and $\lambda < 0$ lead to trivial solutions. Thus

$$X(x) = c_1 \cos(\sqrt{\lambda}x) + c_2 \sin(\sqrt{\lambda}x),$$

and the condition $X(0) = 0$ implies that $c_1 = 0$, whereas $X(a) = 0$ implies that $\sqrt{\lambda} = n\pi/a$, $n = 1, 2, \ldots$. Hence the eigenvalues of the Sturm–Liouville problem (6.1-3) are

$$\lambda = n^2\pi^2/a^2, \qquad n = 1, 2, \ldots,$$

and the corresponding eigenfunctions are

$$X_n(x) = \sin\left(\frac{n\pi}{a}x\right), \qquad n = 1, 2, \ldots.$$

We have suppressed the arbitrary constant, since *any* constant multiple of the above eigenfunctions are also solutions of the homogeneous boundary-value problem in (6.1-3).

A second separation of variables now produces

$$\frac{Z''}{Z} - \frac{n^2\pi^2}{a^2} = -\frac{Y''}{Y} = \mu.$$

The two-point boundary-value problem in Y has exactly the same form as (6.1–3); hence

$$\mu = m^2\pi^2/b^2, \qquad m = 1, 2, \ldots,$$

and the corresponding eigenfunctions are

$$Y_m(y) = \sin\left(\frac{m\pi}{b}y\right), \qquad m = 1, 2, \ldots.$$

Although both separation constants are functions of the positive integers, they are *independent* of each other.

The problem in Z can be written as

$$Z'' - \pi^2\left(\frac{n^2}{a^2} + \frac{m^2}{b^2}\right)Z = 0, \qquad Z(c) = 0,$$

or putting

$$\omega_{mn}^2 = \pi^2\left(\frac{n^2}{a^2} + \frac{m^2}{b^2}\right)$$

gives us

$$Z'' - \omega_{mn}^2 Z = 0, \qquad Z(c) = 0.$$

The solution to this problem for specified values of m and n is (see Exercise 2)

$$Z_{mn}(z) = B_{mn} \sinh \omega_{mn}(c - z),$$

where B_{mn} is a constant that depends on m and n. This constant will be evaluated when we apply the last (nonhomogeneous) boundary condition.

Since m and n are independent, we must take a linear combination of a linear combination of products for our final solution. This results in the **double infinite series**

$$u(x, y, z) = \sum_{n=1}^{\infty} \sum_{m=1}^{\infty} B_{mn} \sinh \omega_{mn}(c - z) \sin\left(\frac{m\pi}{b}y\right) \sin\left(\frac{n\pi}{a}x\right). \tag{6.1–4}$$

This is to be interpreted as follows: For each value of n, m takes on values $m = 1, 2, 3, \ldots$, which gives the *double* series.

Applying the final boundary condition, we have*

$$\sum_{n=1} \left[\sum_{m=1} B_{mn} \sinh (c\omega_{mn}) \sin\left(\frac{m\pi}{b}y\right)\right] \sin\left(\frac{n\pi}{a}x\right) = f(x, y), \tag{6.1–5}$$

*In the remainder of this chapter we omit the upper limit on infinite series.

and the brackets show that for each m we must have

$$\sum_{n=1} B_{mn} \sinh(c\omega_{mn}) \sin\left(\frac{m\pi}{b}y\right) = \frac{2}{a}\int_0^a f(s, y) \sin\left(\frac{n\pi s}{a}\right) ds. \tag{6.1-6}$$

In other words, for *each fixed* value of y $(0 < y < b)$, Eq. (6.1-5) shows that $f(x, y)$ must be represented as a Fourier sine series in x. On the other hand, the right-hand side of Eq. (6.1-6) is a function of y for each n—call it $F_n(y)$—and we can write

$$\sum_{n=1} B_{mn} \sinh(c\omega_{mn}) \sin\left(\frac{m\pi}{b}y\right) = F_n(y).$$

This shows that $F_n(y)$ is represented by a Fourier sine series in y so that the coefficients are given by

$$B_{mn} \sinh(c\omega_{mn}) = \frac{2}{b}\int_0^b F_n(t) \sin\left(\frac{m\pi}{b}t\right) dt.$$

Thus

$$B_{mn} = \frac{2}{b \sinh(c\omega_{mn})} \int_0^b F_n(t) \sin\left(\frac{m\pi}{b}t\right) dt$$

$$= \frac{4}{ab \sinh(c\omega_{mn})} \int_0^b \int_0^a f(s, t) \sin\left(\frac{n\pi}{a}s\right) \sin\left(\frac{m\pi}{b}t\right) ds\, dt. \tag{6.1-7}$$

Hence the formal solution to the problem is given by Eq. (6.1-4) with the B_{mn} given in Eq. (6.1-7) and ω_{mn} defined by

$$\omega_{mn} = \pi \sqrt{\frac{n^2}{a^2} + \frac{m^2}{b^2}}. \tag{6.1-8}$$

We observe that the function $f(x, y)$ of Example 6.1-1 should satisfy the Dirichlet conditions with respect to both variables; that is, for fixed $y = y_0$ $(0 < y_0 < b)$, $f(x, y_0)$ should be a piecewise smooth function of x $(0 < x < a)$. Similarly, for fixed $x = x_0$ $(0 < x_0 < a)$, $f(x_0, y)$ should be a piecewise smooth function of y $(0 < y < b)$. ■

In the next example we solve Laplace's equation over a semi-infinite domain.

EXAMPLE 6.1-2 Find the potential $V(x, y)$ at any point of a plate bounded by $x = 0$, $y = 0$, and $y = b$ if $V(0, y) = V(x, b) = 0$ and $V(x, 0) = f(x)$.

Solution We state the problem in mathematical terms, being careful to give the proper ranges of the variables (Fig. 6.1-2).

$$\begin{aligned}
\text{P.D.E.:}\quad & V_{xx} + V_{yy} = 0, \quad 0 < x < \infty, \quad 0 < y < b;\\
\text{B.C.:}\quad & V(0, y) = 0, \quad 0 < y < b,\\
& V(x, b) = 0, \quad 0 < x < \infty,\\
& V(x, 0) = f(x), \quad 0 < x < \infty.
\end{aligned}$$

We use the Fourier sine transform Eq. (5.2-6) to transform x because $V(0, y) = 0$ *and* $0 < x < \infty$. Then

$$\bar{V}(\alpha, y) = \int_0^\infty V(x, y) \sin \alpha x \, dx,$$

and if we use Eq. (5.2-7), the partial differential equation becomes

$$-\alpha^2 \bar{V}(\alpha, y) + \frac{d^2 \bar{V}(\alpha, y)}{dy^2} = 0.$$

The sine transform of $V(x, b) = 0$ is $\bar{V}(\alpha, b) = 0$, and that of $V(x, 0) = f(x)$ is $\bar{V}(\alpha, 0) = \bar{f}(\alpha)$. In order to make these transformations we must assume that

$$\lim_{x \to \infty} V(x, y) \quad \text{and} \quad \lim_{x \to \infty} V_x(x, y)$$

are both zero and that $f(x)$ is absolutely integrable on the real half-line, $0 < x < \infty$. (Why?)

The solution of the second-order, homogeneous, ordinary differential equation with constant coefficients is

$$\bar{V}(\alpha, y) = C_1(\alpha) \cosh (\alpha y) + C_2(\alpha) \sinh (\alpha y).$$

The condition $\bar{V}(\alpha, b) = 0$ yields

$$C_1 = -C_2 \frac{\sinh (\alpha b)}{\cosh (\alpha b)}$$

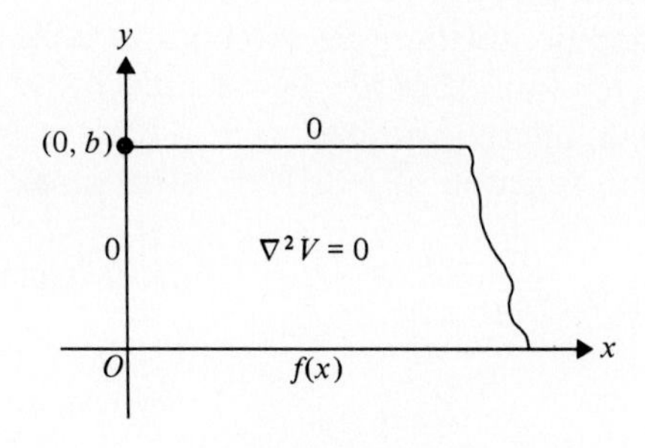

Figure 6.1-2

so that the updated solution becomes (Exercise 5)

$$\bar{V}(\alpha, y) = -C_2(\alpha)\frac{\sinh(\alpha b)}{\cosh(\alpha b)}\cosh(\alpha y) + C_2(\alpha)\sinh(\alpha y)$$

$$= \frac{C_2(\alpha)\sinh\alpha(y-b)}{\cosh(\alpha b)}.$$

From the condition $\bar{V}(\alpha, 0) = \bar{f}(\alpha)$ we now obtain

$$C_2(\alpha) = \frac{-\bar{f}(\alpha)\cosh(\alpha b)}{\sinh(\alpha b)}$$

and the updated solution,

$$\bar{V}(\alpha, y) = \frac{\bar{f}(\alpha)\sinh\alpha(b-y)}{\sinh(\alpha b)}.$$

Using the inverse transform (5.2-6), we get

$$V(x, y) = \frac{2}{\pi}\int_0^\infty \frac{\bar{f}(\alpha)\sinh\alpha(b-y)}{\sinh(\alpha b)}\sin(\alpha x)\,d\alpha$$

$$= \frac{2}{\pi}\int_0^\infty\int_0^\infty f(s)\sin\alpha s\,\frac{\sinh\alpha(b-y)}{\sinh(\alpha b)}\sin(\alpha x)\,ds\,d\alpha. \quad \textbf{(6.1-9)}$$

We note that no further simplification is possible unless $f(x)$ is known. (See Exercise 6.) ■

A function that satisfies Laplace's equation is said to be a **harmonic function**. Harmonic functions have some special properties, which we state in the following theorems.*

THEOREM 6.1-1
If a function f is harmonic in a bounded region and is zero everywhere on the boundary of the region, then f is identically zero throughout the region.

THEOREM 6.1-2
If a function f is harmonic in a bounded region and its normal derivative $\partial f/\partial n$ is zero everywhere on the boundary of the region, then f is a constant in the region.

A **Dirichlet problem** is defined as the problem of finding a function that is harmonic in a region and having prescribed values on the boundary of the region.

*Proofs can be found in Hans Sagan, *Boundary and Eigenvalue Problems in Mathematical Physics* (New York: John Wiley, 1961), Chapter 2.

THEOREM 6.1-3
If a Dirichlet problem for a bounded region has a solution, then that solution is unique.

A **Neumann problem** is defined as the problem of finding a function f that is harmonic in a region and having prescribed values of the normal derivative $\partial f/\partial n$ on the boundary of the region.

THEOREM 6.1-4
If a Neumann problem for a bounded region has a solution, then that solution is unique except possibly for an additive constant.

Dirichlet and Neumann problems arise naturally from the heat-conduction (or diffusion) equation when the steady-state solution is of interest. In two dimensions the heat-conduction equation is

$$u_t = k(u_{xx} + u_{yy}),$$

where u is the temperature and k is a constant called the **thermal diffusivity**. If we seek the **steady-state temperature**,* that is, the temperature after a long period of time has elapsed, then u is independent of t, and the above equation reduces to the two-dimensional Laplace equation. A similar situation holds in three dimensions.

EXAMPLE 6.1-3 Find the steady-state temperatures in a rectangular plate of length a and width b if the edges $x = 0$ and $x = a$ are perfectly insulated, the edge $y = b$ is kept at temperature zero, and the edge $y = 0$ has a temperature distribution given by $\sin(\pi x/a)$. See Fig. 6.1-3.

Solution Recall that **perfect insulation** of an interface means that there can be no flux across that interface. Consequently, the change of temperature in the direction of the outward normal to that interface must be zero. We also assume that the *faces* of the plate are perfectly insulated so that u does not change in the z-direction. Then we can state the following boundary-value problem:

$$\begin{aligned}
&\text{P.D.E.:} \quad && u_{xx} + u_{yy} = 0, \quad 0 < x < a, \quad 0 < y < b;\\
&\text{B.C.:} && \left.\begin{aligned} u_x(0, y) &= 0,\\ u_x(a, y) &= 0, \end{aligned}\right\} \quad 0 < y < b,\\
& && \left.\begin{aligned} u(x, b) &= 0,\\ u(x, 0) &= \sin\frac{\pi x}{a}, \end{aligned}\right\} \quad 0 < x < a.
\end{aligned}$$

*Some authors use the term *equilibrium temperature*.

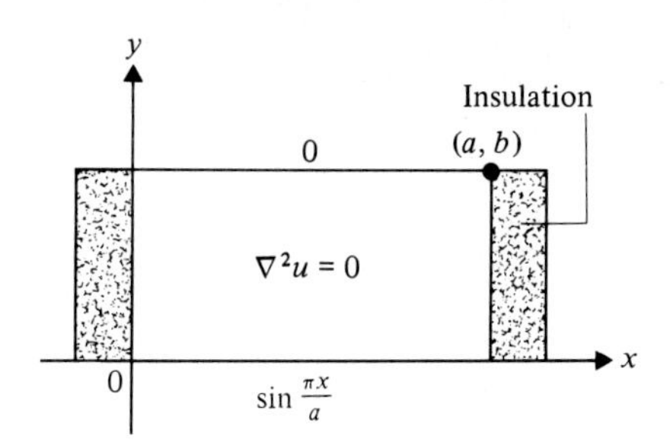

Figure 6.1-3

Separation of variables gives a Sturm–Liouville problem:

$$X'' + \lambda^2 X = 0, \qquad X'(0) = 0, \qquad X'(a) = 0.$$

The eigenfunctions are (Exercises 9, 10, and 11)

$$X_n(x) = \cos\left(\frac{n\pi}{a}x\right), \qquad n = 0, 1, 2, \ldots .$$

We also have

$$Y_n'' - \frac{n^2\pi^2}{a^2} Y_n = 0, \qquad Y_n(b) = 0,$$

with solutions (Exercises 12 and 13)

$$Y_n(y) = \frac{c_n}{\cosh\left(\frac{n\pi b}{a}\right)} \sinh \frac{n\pi}{a}(y - b), \qquad n = 1, 2, \ldots,$$

$$Y_0(y) = c_0(y - b).$$

Now that all the *homogeneous* boundary conditions are satisfied, we can form a linear combination of the eigenfunction products. Then we have

$$u(x, y) = c_0(y - b) + \sum_{n=1} c_n \cos\left(\frac{n\pi}{a}x\right) \frac{\sinh \frac{n\pi}{a}(y - b)}{\cosh\left(\frac{n\pi b}{a}\right)}.$$

Applying the fourth boundary condition, we get

$$-bc_0 + \sum_{n=1} c_n \left(\frac{-\sinh\left(\frac{n\pi b}{a}\right)}{\cosh\left(\frac{n\pi b}{a}\right)}\right) \cos\left(\frac{n\pi}{a}x\right) = \sin\left(\frac{\pi x}{a}\right)$$

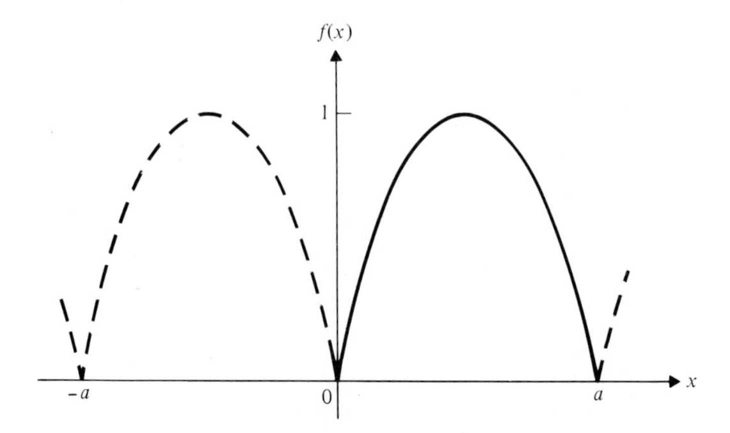

Figure 6.1-4
Even periodic extension of sin ($\pi x/a$).

or

$$-bc_0 + \sum_{n=1} a_n \cos\left(\frac{n\pi}{a}x\right) = \sin\left(\frac{\pi x}{a}\right).$$

This shows that we must make an even periodic extension of the given function sin ($\pi x/a$) in order to represent it as a Fourier *cosine* series (Fig. 6.1-4). Then,

$$a_0 = \frac{2}{a}\int_0^a \sin\frac{\pi s}{a}\,ds = \frac{4}{\pi};$$

hence $-bc_0 = 1/2\ a_0 = 2/\pi$, from which $c_0 = -2/\pi b$. We also have (Exercise 14)

$$\begin{aligned} a_n &= \frac{2}{a}\int_0^a \sin\frac{\pi s}{a}\cos\left(\frac{n\pi}{a}s\right)ds \\ &= \begin{cases} 0, & \text{if } n \text{ is odd,} \\ \dfrac{4}{\pi(1-n^2)}, & \text{if } n \text{ is even.} \end{cases} \end{aligned}$$

The solution can now be written as

$$u(x, y) = \frac{2}{\pi b}(b - y) + \frac{4}{\pi}\sum_{n=1} \frac{\cos\dfrac{2n\pi x}{a}\sinh\dfrac{2n\pi}{a}(b - y)}{(1 - 4n^2)\sinh\dfrac{2n\pi b}{a}}. \qquad \blacksquare$$

(6.1-10)

We emphasize that even though three of the four boundary conditions are homogeneous, the algebra involved in obtaining the solution can become quite complicated. For this reason the principle of superposition is recommended when more than one nonhomogeneous boundary condition is present. (See Section 3.2.)

It becomes a routine matter (Exercise 16) to verify that Eq. (6.1–10) satisfies the three homogeneous boundary conditions of the problem of Example 6.1–3. Further, we know from our study of Fourier series in Chapter 4 that the nonhomogeneous boundary condition is also satisfied. Thus the only remaining questions are what happens at the vertices of the rectangle, whether the solution is unique, and whether it satisfies Laplace's equation.

We have no difficulty verifying that $u(0, b) = u(a, b) = 0$. The results $u(a, 0) = u(0, 0) = 0$ are left for the exercises (see Exercise 17). Thus the boundary conditions can be written as

$$\begin{aligned} u_x(0, y) &= u_x(a, y) = 0, \qquad 0 < y < b, \\ u(x, b) &= 0, \qquad u(x, 0) = \sin(\pi x/a), \qquad 0 \le x \le a. \end{aligned}$$

In other words, at *each point* of the boundary, either u or its normal derivative is specified. A linear combination of these boundary conditions is known as **Robin's condition,*** and the following theorem applies.

THEOREM 6.1–5

Let $u(x, y)$ be harmonic in a bounded region R and satisfy Robin's condition

$$h\frac{\partial u}{\partial n} + ku = f(x, y), \qquad h \ge 0, \qquad k \ge 0, \tag{6.1–11}$$

on C, the boundary of R. Then $u(x, y)$ is the unique solution (except possibly for an additive constant) of $\nabla^2 u = 0$ in R satisfying the condition (6.1–11) on C. If $h = 0$ everywhere on C, then the additive constant is zero, but h and k cannot both be zero.

We omit the proof† but point out that Theorem 6.1–5 also includes Theorems 6.1–3 and 6.1–4 as special cases. When u is given on part of the boundary and its normal derivative is given on the remainder of the boundary, the conditions are called **mixed boundary conditions.**

The remaining question concerns the verification that the result in Eq. (6.1–10) is a harmonic function. We will postpone this question and consider it in Section 6.5 together with other related questions.

---

*After Victor G. Robin (1855–1897), a French mathematician.

†A proof of a somewhat more general theorem can be found in R. V. Churchill and J. W. Brown's *Fourier Series and Boundary Value Problems*, 3rd ed. (New York: McGraw-Hill, 1978), pp. 256 ff.

Key Words and Phrases

double Fourier series
double infinite series
harmonic function
Dirichlet problem
Neumann problem
thermal diffusivity
steady-state temperature
perfect insulation
Robin's condition
mixed boundary conditions

6.1 Exercises

• 1. Show that if $\lambda = 0$ or $\lambda < 0$, then the only solution to the boundary-value problem (6.1–3) is $X(x) = 0$.

2. Show that the solution to

$$Z'' - \omega_{mn}^2 Z = 0, \qquad Z(c) = 0,$$

is

$$B_{mn} \sinh \omega_{mn}(c - z),$$

where B_{mn} is an arbitrary constant whose value depends on both m and n.

3. Solve the problem of Example 6.1–1 if $a = b = c = \pi$.

4. Find B_{mn} in Example 6.1–1 if

$$f(x, y) = xy.$$

5. Obtain the result

$$\bar{V}(\alpha, y) = \frac{C_2(\alpha) \sinh \alpha(y - b)}{\cosh (\alpha b)}$$

from the previous step in Example 6.1–2.

6. Obtain the solution to Example 6.1–2 if $f(x) = e^{-x}$. Does this function meet the necessary requirements?

7. Obtain the solution to Example 6.1–2 if

$$f(x) = \begin{cases} \sin x, & \text{for } 0 < x < \pi, \\ 0, & \text{otherwise.} \end{cases}$$

8. Solve the problem of Example 6.1–2 if $b = 1$ and the condition $V(0, y) = 0$ is replaced by $V_x(0, y) = 0$.

9. Show that $X_0(x) = 1$ is an eigenfunction corresponding to the eigenvalue $\lambda = 0$ in Example 6.1–3.

10. In Example 6.1–3, show that the problem

$$X'' - \lambda^2 X = 0, \qquad X'(0) = X'(a) = 0,$$

has only the trivial solution.

11. Obtain the eigenfunctions

$$X_n(x) = \cos\left(\frac{n\pi}{a}x\right), \qquad n = 0, 1, 2, \ldots,$$

in Example 6.1–3.

12. Obtain the functions

$$Y_n(y) = \frac{c_n}{\cosh(n\pi b/a)} \sinh \frac{n\pi}{a}(y - b), \qquad n = 1, 2, \ldots,$$

in Example 6.1–3.

13. In Example 6.1–3, show that

$$Y_0(y) = c_0(y - b)$$

is the solution corresponding to the eigenvalue $n = 0$.

14. Show that

$$a_{2n} = \frac{4}{\pi(1 - 4n^2)}, \qquad n = 1, 2, \ldots,$$

in Example 6.1–3.

15. Solve the problem of Example 6.1–3 if $a = b = \pi$ and $u(x, 0) = \sin x$.

16. Verify that Eq. (6.1–10) satisfies the boundary conditions of Example 6.1–3.

17. Show that

$$u(a, 0) = u(0, 0) = 0$$

using Eq. (6.1–10). (*Hint*: Obtain S_N, the sum of the first N terms, by using partial fraction decomposition on $1/(1 - 4n^2)$, and then find $\lim S_N$.)

•• **18.** Solve the following boundary-value problem (Fig. 6.1–5):

P.D.E.: $u_{xx} + u_{yy} = 0, \quad 0 < x < 1, \quad y > 0;$

B.C.: $u(1, y) = 0, \quad y > 0,$

$u(0, y) = e^{-ay}, \quad a > 0, \quad y > 0,$

$u_y(x, 0) = 0, \quad 0 < x < 1.$

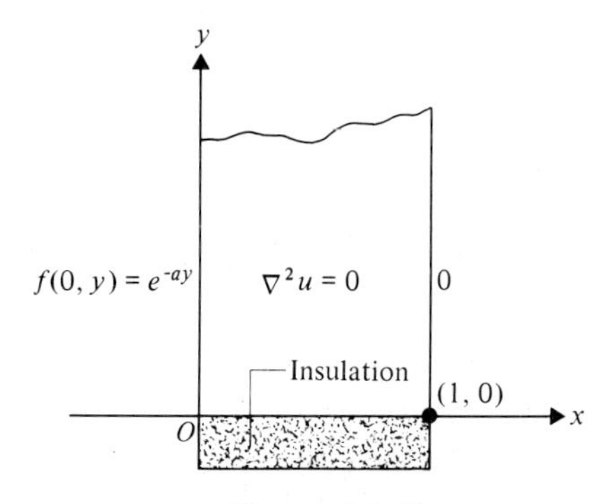

Figure 6.1–5

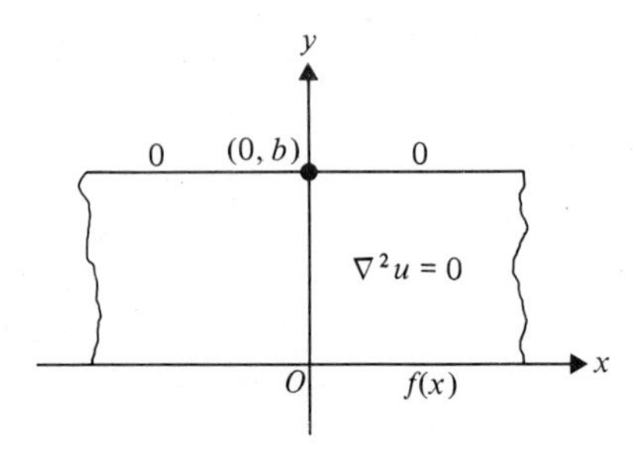

Figure 6.1-6

19. Solve Laplace's equation over the infinite strip $0 < y < b$, given that the potential is zero when $y = b$ and $f(x)$ when $y = 0$ (Fig. 6.1-6). State any other conditions that must be satisfied. (*Hint*: Use the Fourier transform.)

20. **(a)** Solve the Dirichlet problem:

$$\begin{aligned} \text{P.D.E.:} \quad & u_{xx} + u_{yy} = 0, \quad 0 < x < 1, \quad 0 < y < 1; \\ \text{B.C.:} \quad & u(0, y) = u(1, y) = 0, \quad 0 < y < 1, \\ & u(x, 0) = 0, \quad u(x, 1) = x, \quad 0 < x < 1. \end{aligned}$$

(b) Solve the Dirichlet problem:

$$\begin{aligned} \text{P.D.E.:} \quad & u_{xx} + u_{yy} = 0, \quad 0 < x < 1, \quad 0 < y < 1; \\ \text{B.C.:} \quad & u(x, 0) = u(x, 1) = 0, \quad 0 < x < 1, \\ & u(0, y) = 0, \quad u(1, y) = y, \quad 0 < y < 1. \end{aligned}$$

(c) Use superposition to obtain the solution of the Dirichlet problem:

$$\begin{aligned} \text{P.D.E.:} \quad & u_{xx} + u_{yy} = 0, \quad 0 < x < 1, \quad 0 < y < 1; \\ \text{B.C.:} \quad & u(x, 0) = 0, \quad u(x, 1) = x, \quad 0 < x < 1, \\ & u(0, y) = 0, \quad u(1, y) = y, \quad 0 < y < 1. \end{aligned}$$

21. A rectangular plate has its edges $y = 0$ and $y = b$ perfectly insulated, while its edges $x = 0$ and $x = a$ are kept at temperature zero. Use *separation of variables* to find the steady-state temperatures in the plate. Is the result in agreement with what you would expect in the light of Theorems 6.1–1 and 6.1–2?

22. Solve the problem of Example 6.1–1 given the condition that on the face $x = 0$, $u(0, y, z) = \sin(\pi y/b) \sin(\pi z/c)$, and on all other faces $u = 0$.

***23.** In Example 6.1–1, put $a = 1$, $b = 2$, and compute ω_{11}, ω_{12}, ω_{21}, ω_{22}.

••• **24.** Let $u(x, y) = f(\phi)$, where $\phi(x, y)$ is a nonconstant, harmonic function. Under what conditions is u harmonic?

25. Prove Theorem 6.1–3. (*Hint*: Suppose f and g are both solutions. Then consider the difference $f - g$, and use Theorem 6.1–1.)

26. Explain how the word "zero" in Theorem 6.1–1 can be replaced by "a constant K."

27. Solve the problem of Example 6.1–3 given the boundary condition

$$u(x, 0) = \frac{10}{a}(a - x), \qquad 0 < x < a,$$

all other conditions remaining unchanged.

28. **(a)** Verify that the solution to Exercise 27 satisfies the given boundary conditions.
(b) Compute $u(a, 0)$ and $u(0, 0)$.
(c) Explain the results obtained in part (b).

29. Solve the problem of Example 6.1–1 given that $u(x, y, c) = f(x, y)$ and all other faces are kept at zero.

30. Verify that Eq. (6.1–9) satisfies the boundary conditions in Example 6.1–2.

31. Assuming that differentiation and integration may be interchanged, verify that the function $V(x, y)$ given by Eq. (6.1–9) satisfies Laplace's equation.

32. **(a)** Show that $u(x, y) = xy$ is also a solution of the Dirichlet problem in Exercise 20(c).
(b) Use Theorem 6.1–3 to obtain a series expression for $f(x, y) = xy$.

6.2 THE WAVE EQUATION

Transverse Waves

In Section 3.3 we derived the vibrating string equation after making a number of simplifying assumptions. This derivation was then extended in a natural way to a **vibrating membrane** (such as a drumhead) to obtain the two-dimensional wave equation

$$u_{tt} = c^2(u_{xx} + u_{yy}). \tag{6.2–1}$$

We are calling the constant c here to distinguish it from the a in Section 3.3. It is defined as T_0g/w, where T_0 is the tension per unit length (constant in every direction), w is weight per unit area, and g is the acceleration due to gravity. A boundary-value problem involving this equation in a rectangular region will generally have four boundary conditions and two initial conditions specified. We illustrate with an example.

EXAMPLE 6.2–1 Solve the following problem.

$$\begin{aligned}
&\text{P.D.E.:} && u_{tt} = c^2(u_{xx} + u_{yy}), \quad 0 < x < a, \quad 0 < y < b, \quad t > 0;\\
&\text{B.C.:} && u(0, y, t) = u(a, y, t) = 0, \quad 0 < y < b, \quad t > 0,\\
& && u(x, 0, t) = u(x, b, t) = 0, \quad 0 < x < a, \quad t > 0;\\
&\text{I.C.:} && u_t(x, y, 0) = 0, \quad u(x, y, 0) = f(x, y), \quad 0 < x < a, \quad 0 < y < b.
\end{aligned}$$

Solution Using separation of variables in a different form from that used

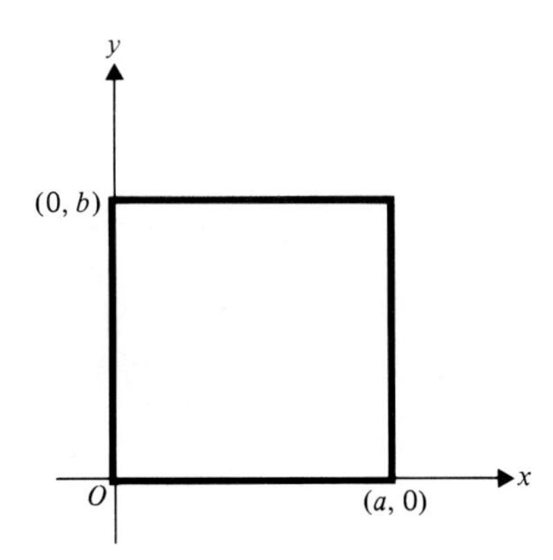

Figure 6.2-1

previously, let

$$u(x, y, t) = \Phi(x, y)T(t),$$

and substitute into the P.D.E. Then

$$\ddot{T}\Phi = c^2 T(\Phi_{xx} + \Phi_{yy}),$$

the dot denoting differentiation with respect to t. Dividing by $c^2\Phi T$ produces the desired separation,

$$\frac{\ddot{T}}{c^2 T} = \frac{(\Phi_{xx} + \Phi_{yy})}{\Phi} = -\lambda^2, \tag{6.2-2}$$

where $-\lambda^2$ is a negative constant.

The physical interpretation of this problem (see Fig. 6.2-1) is the following: A rectangular membrane is fastened along its four edges to a frame that is given an initial displacement perpendicular to the xy-plane (in the u-direction) defined by $f(x, y)$. Since no damping or other external forces are acting, we expect that interior points* of the membrane will vibrate indefinitely. For this reason the separation constant is chosen to be *negative*. In other words, we were motivated in choosing the above product form of $u(x, y, t)$ by the knowledge that u must be *periodic* in t.

From Eq. (6.2-2) we have

$$\ddot{T} + c^2\lambda^2 T = 0, \qquad \dot{T}(0) = 0,$$

using the homogeneous initial condition. The solution to this problem is (Exercise 1)

$$T(t) = \cos(c\lambda t),$$

where the nature of λ is still to be determined.

*Certain points, called **nodes**, may exhibit zero displacement.

In Eq. (6.2–2) we make the substitution $\Phi(x, y) = X(x)Y(y)$ to obtain

$$\frac{X''}{X} + \frac{Y''}{Y} = -\lambda^2.$$

But this is exactly the problem (including the boundary conditions) that we solved in Example 6.1–1 of the last section. Hence solutions of the present problem are products of

$$X_n(x) = \sin\left(\frac{n\pi}{a}x\right), \qquad n = 1, 2, \ldots,$$

$$Y_m(y) = \sin\left(\frac{m\pi}{b}y\right), \qquad m = 1, 2, \ldots,$$

$$T_{mn}(t) = \cos(c\omega_{mn}t),$$

where the m and n are independent and

$$\omega_{mn}^2 = \pi^2\left(\frac{n^2}{a^2} + \frac{m^2}{b^2}\right).$$

A linear combination of these products summed over m and n is an expression that still contains arbitrary constants. These constants can be found by imposing the final nonhomogeneous initial condition as in Example 6.1–1. Thus

$$u(x, y, t) = \sum_{m=1}^{\infty}\sum_{n=1}^{\infty} B_{mn} \sin\left(\frac{m\pi}{b}y\right) \sin\left(\frac{n\pi}{a}x\right) \cos(c\omega_{mn}t)$$

with

$$B_{mn} = \frac{4}{ab}\int_0^b \sin\left(\frac{m\pi}{b}y\right)\left[\int_0^a f(x, y) \sin\left(\frac{n\pi}{a}x\right) dx\right] dy.$$

We observe that $f(x, y)$, $f_x(x, y)$, and $f_y(x, y)$ should all be continuous on $0 < x < a$, $0 < y < b$, and vanish on the boundaries of the rectangle. Note that the angular frequency of the vibrating membrane ($c\omega_{mn}$) depends on both m and n and does not change by integral multiples of some fixed basic frequency. Consequently, the vibrating membrane does not produce a *musical* note as the vibrating string does. (See Exercise 3.) ■

Longitudinal Waves

The waves produced in a vibrating string are **transverse waves** (Fig. 6.2–2), that is, the direction of motion of the individual particles of the string is perpendicular to the direction of propagation of the waves (compare Section 3.3). In a solid metal bar, however, elastic waves that are **longitudinal waves** can occur, that is, the direction of motion of the individual particles is the same as the direction of propagation of the waves (Fig. 6.2–3).

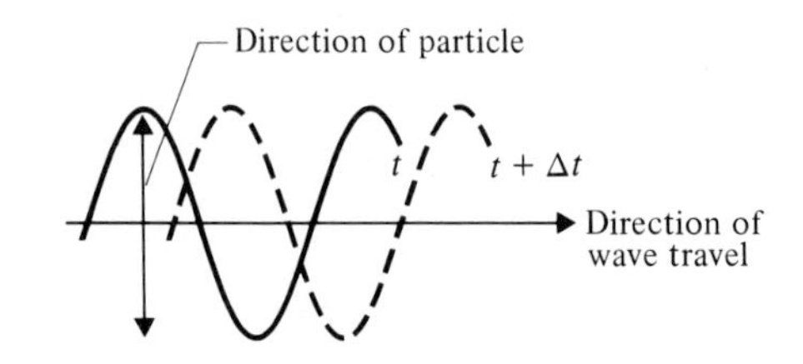

Figure 6.2-2
Transverse wave.

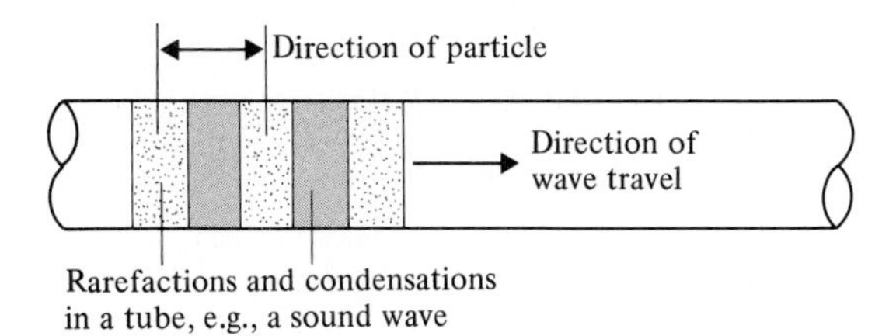

Figure 6.2-3
Longitudinal wave.

Consider a bar of uniform cross-section and density extending along the x-axis from the origin to the point $(0, L)$. We assume that the bar is perfectly elastic, meaning that if external forces are applied at the ends so that elongation takes place, tensile forces in the direction of the x-axis will result. If the external forces are now removed, the bar will vibrate longitudinally in accordance with the laws of elasticity. Suppose the bar has density ρ (mass per unit volume), cross-section A, and Young's modulus of elasticity E, and let a cross-section at x be displaced an amount u as shown in Fig. 6.2-4. We will also assume that A is small in comparison to L. From the definition of Young's modulus E, the force on the cross-section at x is given by

$$EA\,\frac{\partial u}{\partial x},$$

since $\partial u/\partial x$ represents the elongation per unit length. On the other hand, the force on this cross-section of length Δx is also given by

$$\rho A \Delta x\,\frac{\partial^2 u}{\partial t^2},$$

where $\partial^2 u/\partial t^2$ is evaluated at a point between x and $x + \Delta x$, say, at the center of mass of the element. The net force per unit length is

$$\frac{EA}{\Delta x}\left[\frac{\partial u(x + \Delta x, t)}{\partial x} - \frac{\partial u(x, t)}{\partial x}\right],$$

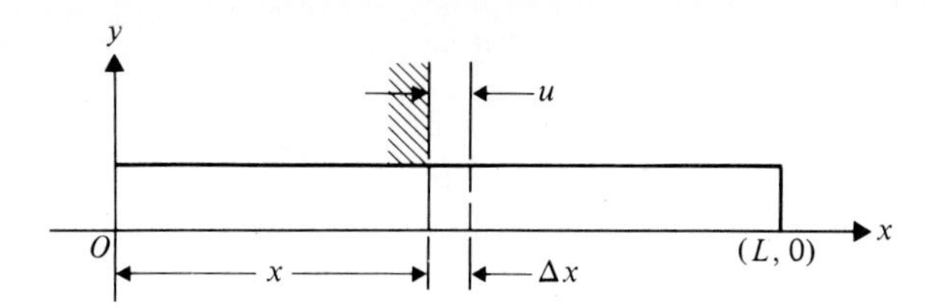

Figure 6.2-4
Longitudinal waves in a bar.

and on equating the two forces and taking the limit as $\Delta x \to 0$, we have

$$\frac{\partial^2 u}{\partial t^2} = \frac{E}{\rho}\frac{\partial^2 u}{\partial x^2} = c^2 \frac{\partial^2 u}{\partial x^2}. \qquad \textbf{(6.2-3)}$$

Thus the small longitudinal vibrations of an elastic rod satisfy the one-dimensional wave equation (Exercise 7).

EXAMPLE 6.2-2 The end $x = L$ of a long thin bar is kept fixed, and a constant compressive force of F_0 units per unit area is applied to the end $x = 0$ (Fig. 6.2-5). If the bar is initially at rest and unstrained, find the longitudinal displacement of an arbitrary cross-section at any time t.

Solution We need to solve the following boundary-value problem:

P.D.E.: $u_{tt} = c^2 u_{xx}$, $0 < x < L$, $t > 0$, $c^2 = E/\rho$;
B.C.: $u(L, t) = 0$, $t > 0$ (the end $x = L$ is kept fixed),
$Eu_x(0, t) = F_0$, $t > 0$ (a constant force is applied at $x = 0$);
I.C.: $u(x, 0) = 0$, $0 < x < L$ (the bar is initially unstrained),
$u_t(x, 0) = 0$, $0 < x < L$ (the bar is initially at rest).

We have indicated in the mathematical formulation of the problem how the *physical facts* are translated into *mathematical terms*. Since $u(x, t)$ represents the displacement of the bar, $u(L, t) = 0$ shows that the displacement is zero when $x = L$, that is, the end $x = L$ is kept fixed. As was stated earlier, the force on a cross-section at x is given by

$$EA\frac{\partial u}{\partial x},$$

and at the end $x = 0$ this force is given as $F_0 A$. From this it follows that

$$Eu_x(0, t) = F_0.$$

Since the bar is initially unstrained, it cannot have any displacement at $t = 0$, that is, $u(x, 0) = 0$. Finally, the bar is initially at rest, meaning that the velocity

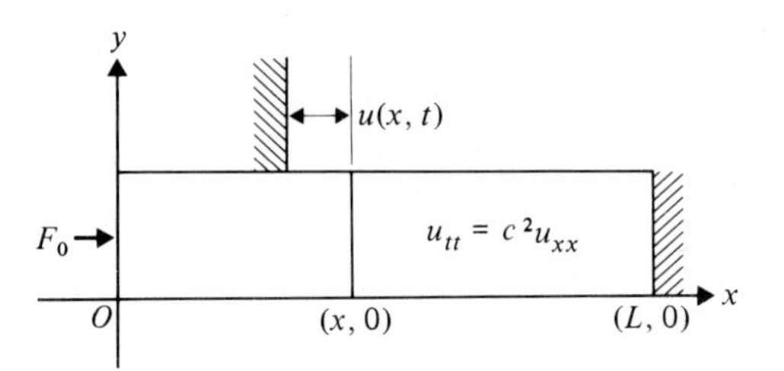

Figure 6.2-5

u_t is zero, at time $t = 0$. The ability to translate physical phenomena into mathematical symbolism is an invaluable aid in problem solving.

Although there are three homogeneous conditions, separation of variables will produce the result $u(x, t) = 0$, which is not a solution (Exercise 8). The difficulty is caused by the fact that the resulting Sturm–Liouville problem in T has a unique trivial solution. This situation could be remedied if we could somehow transfer the nonhomogeneous boundary condition into an initial condition. We can obtain a solution by making the change of variable

$$u(x, t) = U(x, t) - \phi(x),$$

where $\phi(x)$ is to be determined. Then the problem becomes the following one:

$$\begin{aligned}
&\text{P.D.E.:} \quad U_{tt} = c^2[U_{xx} - \phi''(x)], \quad 0 < x < L, \quad t > 0;\\
&\text{B.C.:} \quad \left.\begin{aligned} &U(L, t) - \phi(L) = 0,\\ &EU_x(0, t) - E\phi'(0) = F_0, \end{aligned}\right\} \quad t > 0;\\
&\text{I.C.:} \quad \left.\begin{aligned} &U(x, 0) - \phi(x) = 0,\\ &U_t(x, 0) = 0, \end{aligned}\right\} \quad 0 < x < L.
\end{aligned}$$

If we now set

$$\phi''(x) = 0, \qquad \phi(L) = 0, \qquad \phi'(0) = -F_0/E,$$

then $\phi(x) = (F_0/E)(L - x)$ (Exercise 9). Thus the problem is transformed into the following familiar one:

$$\begin{aligned}
&\text{P.D.E.:} \quad U_{tt} = c^2 U_{xx}, \quad 0 < x < L, \quad t > 0;\\
&\text{B.C.:} \quad \left.\begin{aligned} &U(L, t) = 0,\\ &U_x(0, t) = 0, \end{aligned}\right\} \quad t > 0;\\
&\text{I.C.:} \quad \left.\begin{aligned} &U(x, 0) = \frac{F_0}{E}(L - x),\\ &U_t(x, 0) = 0, \end{aligned}\right\} \quad 0 < x < L.
\end{aligned}$$

This problem *can* be solved by the method of separation of variables. We make the assumption that

$$U(x, t) = X(x)T(t)$$

and substitute into the partial differential equation. Then

$$X\ddot{T} = c^2X''T,$$

and when we divide both members by the nonzero (Why?) quantity c^2XT, we get

$$\frac{\ddot{T}}{c^2T} = \frac{X''}{X} = -\lambda^2. \tag{6.2–4}$$

We know that the separation constant must be negative for two reasons—first, because $T(t)$ must be periodic in t (Why?) and, second, because the problem in $X(x)$ has only the trivial solution if the separation constant is positive or zero (Exercise 10).

We leave it as an exercise (see Exercise 11) to show that the Sturm-Liouville problem

$$X'' + \lambda^2X = 0, \qquad X(L) = 0, \qquad X'(0) = 0, \tag{6.2–5}$$

has solutions

$$X_n(x) = \cos\,(2n - 1)\frac{\pi}{2L}x, \qquad n = 1, 2, \ldots .$$

It is important here to note that the eigenvalues of the problem (6.2–5) are

$$\lambda^2 = \left(\frac{2n - 1}{2L}\pi\right)^2, \qquad n = 1, 2, \ldots,$$

and that zero is *not* an eigenvalue. The second equation in (6.2–4) can be written

$$\ddot{T}_n + \left(\frac{2n - 1}{2L}\pi c\right)^2 T_n = 0, \qquad \dot{T}_n(0) = 0, \tag{6.2–6}$$

where the boundary condition comes from the homogeneous initial condition on $U(x, t)$. The solutions to (6.2–6) are (Exercise 12)

$$T_n(t) = \cos\left(\frac{2n - 1}{2L}\pi ct\right), \qquad n = 1, 2, \ldots .$$

Now we take a linear combination of products $X_n(x)T_n(t)$ so that

$$U(x, t) = \sum_{n=1} a_n \cos\,(2n - 1)\frac{\pi}{2L}x \cos\,(2n - 1)\frac{\pi c}{2L}t, \tag{6.2–7}$$

which satisfies all homogeneous conditions. When we impose the nonhomogeneous I.C., we get

$$U(x, 0) = \sum_{n=1} a_n \cos\,(2n - 1)\frac{\pi}{2L}x = \frac{F_0}{E}(L - x),$$

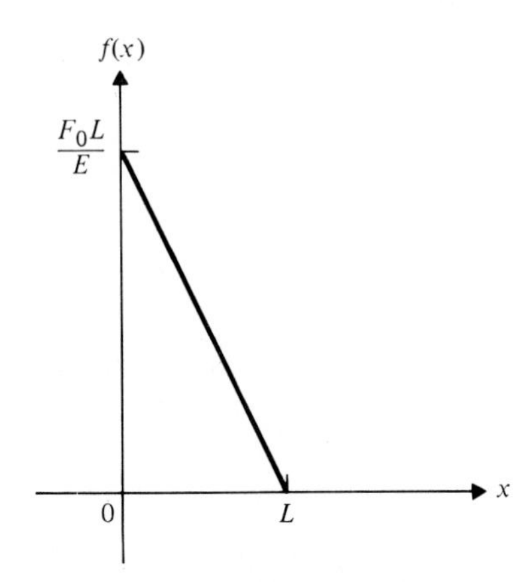

Figure 6.2-6

which is a Fourier cosine series representation of the function $(F_0/E)(L - x)$, shown in Fig. 6.2-6.

Note, however, that if we proceed as in Section 4.3 to make an *even periodic extension* of this function, we obtain the function shown in Fig. 6.2-7. This cannot be correct because such an extension would have a *constant* term $\frac{1}{2}a_0$. We have already established that zero is *not* an eigenvalue of the problem (6.2-5) or, stated another way, that $a_0 = 0$. Accordingly, the even periodic extension in this case must be the function $\Phi(x)$, shown in Fig. 6.2-8. We can now write the solution as

$$U(x, t) = \frac{1}{2}[\Phi[x + ct) + \Phi(x - ct)], \tag{6.2-8}$$

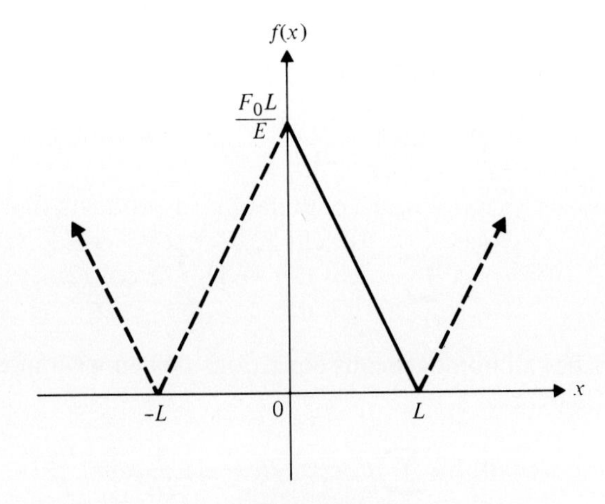

Figure 6.2-7
The even periodic extension of $(F_0/L)(L - x)$.

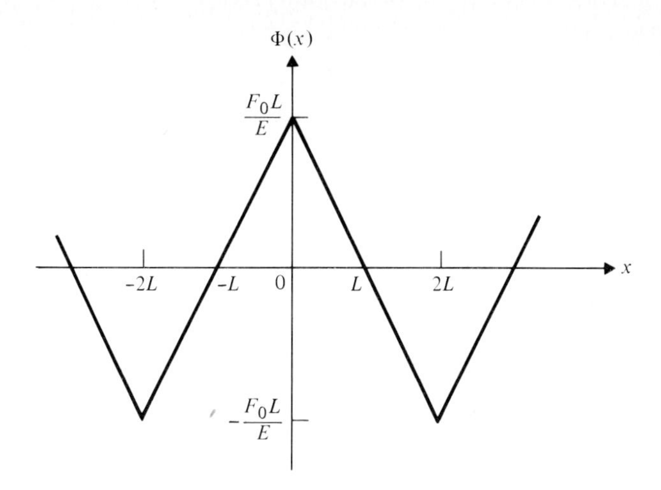

Figure 6.2-8
The function Φ(x).

following the method used to obtain Eq. (3.3-6). Alternatively, the solution can be written as in Eq. (6.2-7) with the coefficients a_n given by (Exercise 13)

$$a_n = \frac{F_0}{EL}\int_0^{2L} (L - x)\cos(2n-1)\frac{\pi}{2L}x\,dx$$

$$= \frac{8F_0L}{E\pi^2(2n-1)^2}, \qquad n = 1, 2, \ldots. \tag{6.2-9}$$

To see that Eqs. (6.2-7) and (6.2-8) are equivalent, we use the trigonometric identity

$$\cos\alpha\cos\beta = \frac{1}{2}[\cos(\alpha+\beta) + \cos(\alpha-\beta)]$$

in Eq. (6.2-7). Then we have

$$U(x, t) = \frac{1}{2}\sum_{n=1} a_n[\cos\omega_n(x + ct) + \cos\omega_n(x - ct)],$$

where

$$\omega_n = \frac{(2n-1)\pi}{2L}.$$

The function $\Phi(x)$, however, has the Fourier cosine series representation

$$\Phi(x) = \sum_{n=1} a_n\cos\omega_n x.$$

Thus

$$\frac{1}{2}\Phi(x + ct) = \frac{1}{2}\sum_{n=1} a_n \cos \omega_n(x + ct),$$

with a similar expression for $\frac{1}{2}\Phi(x - ct)$, and the equivalence of Eqs. (6.2–7) and (6.2–8) is established.

Accordingly, the solution to the original problem is

$$u(x, t) = \frac{F_0}{E}(x - L) + U(x, t) \qquad \textbf{(6.2–10)}$$

with $U(x, t)$ expressed in one of the two equivalent forms. We leave the full verification of these results for the exercises (see Exercises 14 and 15). ∎

In the problem of Example 6.2–2 we showed how a nonhomogeneous boundary condition can be transferred from one independent variable to another by a change in the dependent variable. Other examples of this technique will be given in Section 6.3.

Another problem in which the one-dimensional wave equation applies is considered next. Let θ be the angular displacement of a cross-section of a uniform circular shaft from some equilibrium position. If θ is small, then, according to the theory of elasticity, θ satisfies the partial differential equation

$$\theta_{tt} = c^2\theta_{xx}, \qquad \textbf{(6.2–11)}$$

with $c^2 = Gg/\rho$, where G is the shear modulus of elasticity, ρ is the density (mass per unit volume), and g is the acceleration due to gravity.

EXAMPLE 6.2–3 Suppose that a shaft of circular cross-section and length L is clamped at one end while the other end is twisted, then freed. As the shaft oscillates, the free end is clamped at time $t = t_0$. At this instant the angular velocity is $\omega_0 x/L$, and the angle θ is zero. State all conditions that apply, and solve the boundary-value problem.

Solution We take coordinates so that the shaft lies along the x-axis with the free end at $x = L$. Then we have the following.

P.D.E.: $\theta_{tt} = c^2\theta_{xx}, \quad 0 < x < L, \quad t > 0;$

B.C.: $\theta(0, t) = 0, \quad t > 0$ (clamped at one end),

$\theta(L, t) = 0, \quad t > t_0$ (free end clamped at $t = t_0$);

I.C.: $\left.\begin{aligned}\theta(x, t_0) &= 0,\\ \theta_t(x, t_0) &= \omega_0 x/L,\end{aligned}\right\} \quad 0 < x < L.$

We can simplify the problem considerably by taking $t_0 = 0$ so that the time coordinate* is shifted by an amount t_0. Then, using separation of variables, we find

*We observe that there is no loss of generality in taking the initial value of t as zero.

$$\frac{\ddot{T}}{c^2 T} = \frac{X''}{X} = -\lambda^2.$$

The separation constant must be *negative*, since we are dealing with *periodic* oscillations in time. Thus the two-point boundary-value problem

$$X'' + \lambda^2 X = 0, \qquad X(0) = 0, \qquad X(L) = 0,$$

has solutions (Exercise 16)

$$X_n(x) = \sin\left(\frac{n\pi}{L}x\right), \qquad n = 1, 2, \ldots,$$

while the initial-value problem

$$\ddot{T}_n + \frac{n^2\pi^2c^2}{L^2}T_n = 0, \qquad T_n(0) = 0,$$

has solutions (Exercise 17)

$$T_n(t) = \sin\left(\frac{n\pi c}{L}t\right), \qquad n = 1, 2, \ldots.$$

Hence

$$\theta(x, t) = \sum_{n=1}^{\infty} b_n \sin\left(\frac{n\pi}{L}x\right) \sin\left(\frac{n\pi c}{L}t\right)$$

with (Exercise 18)

$$\frac{n\pi c}{L} b_n = \frac{2}{L}\int_0^L \frac{\omega_0 s}{L} \sin\left(\frac{n\pi}{L}s\right) ds,$$

$$b_n = \frac{2\omega_0 L}{c\pi^2}\frac{(-1)^{n-1}}{n^2}. \qquad \blacksquare$$

Key Words and Phrases

vibrating membrane	transverse wave
node	longitudinal wave

6.2 Exercises

• **1.** Obtain the solution of the following problem.

$$\ddot{T} + \lambda^2 c^2 T = 0, \qquad \dot{T}(0) = 0$$

***2.** The frequencies ω_{mn} in Example 6.2–1 are called *characteristic frequencies*. List the first six characteristic frequencies of the vibrating rectangular membrane, that is, ω_{11}, ω_{12}, ω_{21}, ω_{22}, ω_{13}, ω_{31}, if $a = b = \pi$.

3. If in Example 6.2–1

$$u(x, y, 0) = k \sin \frac{\pi x}{a} \sin \frac{\pi y}{b},$$

where k is a constant, obtain the solution. Note that under this condition the vibrating membrane *does* produce a musical tone. What is the frequency of the tone?

4. Solve the problem of Example 6.2–1 given that $f(x, y) = xy(a - x)(b - y)$. Show that this function satisfies the required conditions in the example.

5. If the initial conditions of the problem of Example 6.2–1 are replaced by

$$u_t(x, y, 0) = g(x, y), \qquad u(x, y, 0) = 0,$$

what is the solution $u(x, y, t)$?

6. Use the principle of superposition to solve the problem of Example 6.2–1 if the initial conditions there are replaced by $u(x, y, 0) = f(x, y)$, $u_t(x, y, 0) = g(x, y)$.

7. Show that E/ρ in Eq. (6.2–3) has dimensions of (velocity)^2.

8. Use separation of variables in Example 6.2–2 to show that the equation in t has only the trivial solution $T(t) = 0$ regardless of the choice of the (real) separation constant.

9. Solve the two-point boundary-value problem

$$\phi''(x) = 0, \qquad \phi(L) = 0, \qquad \phi'(0) = -F_0/E.$$

10. Show that the problem

$$X''(x) - \lambda^2 X(x) = 0, \qquad X(L) = 0, \qquad X'(0) = 0,$$

has only the trivial solution when $\lambda \geq 0$.

11. Solve the boundary-value problem

$$X''(x) + \lambda^2 X(x) = 0, \qquad X(L) = 0, \qquad X'(0) = 0.$$

12. Solve the problem

$$\ddot{T}_n(t) + \lambda^2 T_n(t) = 0, \qquad \dot{T}_n(0) = 0,$$

where

$$\lambda = \frac{2n - 1}{2L}\pi, \qquad n = 1, 2, \ldots.$$

13. Carry out the necessary computations to arrive at the value of a_n given in Eq. (6.2–9).

14. Verify that

$$U(x, t) = \frac{1}{2}[\Phi(x + ct) + \Phi(x - ct)],$$

where $\Phi(x)$ is shown in Fig. 6.2–8, satisfies the P.D.E. and all conditions for the problem in $U(x, t)$ of Example 6.2–2.

15. Verify that $u(x, t)$ as given in Eq. (6.2–10) with $U(x, t)$ as in Exercise 14 satisfies all conditions for the problem of Example 6.2–2.

16. Find the eigenvalues and the corresponding eigenfunctions of the Sturm-Liouville problem

$$X'' + \lambda^2 X = 0, \qquad X(0) = 0, \qquad X(L) = 0.$$

17. Solve

$$\ddot{T} + \left(\frac{n\pi c}{L}\right)^2 T = 0, \qquad T(0) = 0.$$

18. Obtain the coefficients b_n in the problem of Example 6.2–3.

•• **19.** Deduce the dimensions of T_0 in Eq. (6.2–1) and of G in Eq. (6.2–10).

20. Solve Example 6.2–3 given that $\theta_t(x, t_0) = k$, a constant, all other conditions remaining the same.

21. Solve the vibrating string equation for the case of the plucked string of length L, that is, for initial conditions

$$u(x, 0) = \begin{cases} \dfrac{2h}{L}x, & \text{for } 0 \le x \le \dfrac{L}{2}, \\[2ex] \dfrac{2h}{L}(L - x), & \text{for } \dfrac{L}{2} \le x \le L, \qquad h > 0, \end{cases}$$

$$u_t(x, 0) = 0.$$

(See Fig. 6-2-9.)

22. Solve the vibrating string equation for the case of a string length of L and initial conditions

$$u(x, 0) = x(L - x), \qquad 0 \le x \le L,$$
$$u_t(x, 0) = 0.$$

***23.** If $L = 1$ and $c = 1$ in Exercise 22, compute
(a) $u(1/4, 2)$ **(b)** $u(1/2, 3/2)$

24. Deduce from Eqs. (6.2–7) and (6.2–9) that

$$\sum_{n=1} \frac{1}{(2n - 1)^2} = \frac{\pi^2}{8}.$$

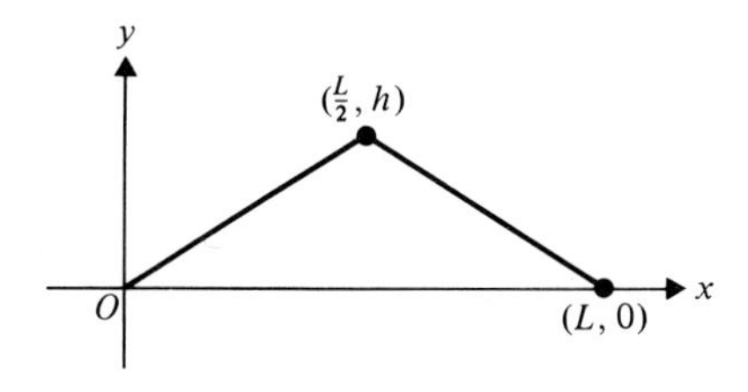

Figure 6.2-9

••• **25.** The *nonhomogeneous* wave equation leads to the following problem:

$$\begin{aligned} &\text{P.D.E.:} \quad u_{tt} = a^2u_{xx} + F(x, t), \quad -\infty < x < \infty, \quad t > 0; \\ &\text{I.C.:} \quad u(x, 0) = f(x), \quad u_t(x, 0) = g(x), \quad -\infty < x < \infty. \end{aligned}$$

(a) Give a physical interpretation of this problem.

(b) Verify that the solution is given by

$$u(x, t) = \frac{1}{2}[f(x + at) + f(x - at)] + \frac{1}{2a}\int_{x-at}^{x+at} g(s)\, ds + \frac{1}{2a}\int_0^t \int_{x-a(t-\tau)}^{x+a(t-\tau)} F(\xi, \tau)\, d\xi\, d\tau.$$

(c) Solve the problem if $F(x, t) = kg$, where k is a positive constant and g is the acceleration due to gravity.

(d) Solve the problem if $F(x, t) = kx^2$, where k is a positive constant.

26. Show that the wave equation is unchanged if x is replaced by $-\xi$ and hence that it follows that if $u(x, t)$ is a solution, then so is $u(-x, t)$. Show that a further consequence of this is that if the initial conditions are *both* odd (or even), then so is the solution.

27. **(a)** Referring to Exercise 21 and noting that the kinetic energy of an element ds of the vibrating string (see Section 3.3) is approximately $\rho\, dx\, u_t^2/2$, show that the total kinetic energy of motion of the nth mode is

$$K_n = \frac{\rho}{2}\int_0^L u_t^2\, dx = \frac{Mb_n^2\, \omega_n^2}{4} \sin^2 \omega_n t,$$

where M is the total mass of the string and

$$b_n = \frac{(-1)^{n+1}h}{\pi^2(2n - 1)^2}, \qquad \omega_n = \frac{(2n - 1)\pi c}{L}.$$

(b) Show that the net potential energy of the nth mode is given by

$$V_n = \frac{T_0}{2}\int_0^L u_x^2\, dx = \frac{Mb_n^2\, \omega_n^2}{4} \cos^2 \omega_n t.$$

(c) Obtain the total energy of the nth mode:

$$E_n = K_n + V_n = \frac{Mb_n^2\omega_n^2}{4}.$$

(d) Note that the energy in each of the excited modes is proportional to the square of the corresponding term in $u(x, t)$. Hence the energy ratios for the case of the plucked string are 1:1/81:1/625, which shows clearly that the energy in the system is dominated by the first or fundamental mode.

28. Solve the following problem:

$$\begin{aligned} &\text{P.D.E.:} \quad u_{tt} = u_{xx} + u, \quad 0 < x < \pi, \quad t > 0; \\ &\text{B.C.:} \quad u(0, t) = u(\pi, t) = 0, \quad t > 0; \\ &\text{I.C.:} \quad u(x, 0) = f(x), \quad u_t(x, 0) = 0. \end{aligned}$$

29. Solve each of the following nonhomogeneous problems:

(a) P.D.E.: $u_{tt} - c^2 u_{xx} = 1, \quad -\infty < x < \infty, \quad t > 0;$
I.C.: $u(x, 0) = \sin bx, \quad -\infty < x < \infty,$
$u_t(x, 0) = 0, \quad -\infty < x < \infty$

(b) P.D.E.: $u_{tt} - c^2 u_{xx} = 4x + t, \quad -\infty < x < \infty, \quad t > 0;$
I.C.: $u(x, 0) = 0, \quad -\infty < x < \infty,$
$u_t(x, 0) = \cosh bx, \quad -\infty < x < \infty$

30. Read the paper "Motivating Existence-Uniqueness Theory for Applications-Oriented Students," by A. D. Snider in *Amer. Math. Monthly*, December 1976, pp. 805–807.

6.3 THE DIFFUSION EQUATION

A classic example of a second-order partial differential equation of parabolic type is the diffusion equation. It has the form

$$u_t = k\nabla^2 u, \qquad k > 0, \tag{6.3–1}$$

where ∇^2 is the Laplacian operator

$$\frac{\partial^2}{\partial x^2} + \frac{\partial^2}{\partial y^2} + \frac{\partial^2}{\partial z^2}.$$

We show the operator in three dimensions and in rectangular coordinates here.

In this section we consider the diffusion equation together with various boundary conditions. It will be convenient to think of these problems as problems of **heat conduction** under various conditions. In some cases we will specify the temperatures at the boundaries; in other cases we will specify the flux of heat across the boundaries or the transfer of heat by **convection** into surroundings kept at constant temperature. This last case will involve **Newton's law of cooling** in the form

$$-\frac{du}{dn} = h(u - u_0), \qquad h > 0, \qquad u > u_0, \tag{6.3–2}$$

where du/dn is the (outward-pointing) normal derivative at the boundary, h is a positive constant, and u_0 is the constant temperature of the surroundings, that is, the **ambient temperature**.

At a point of the surface S we define the *flux* of heat Φ as the quantity of heat per unit area, per unit time, that is being conducted across S at the point. The flux Φ is proportional to the directional derivative of the temperature u in

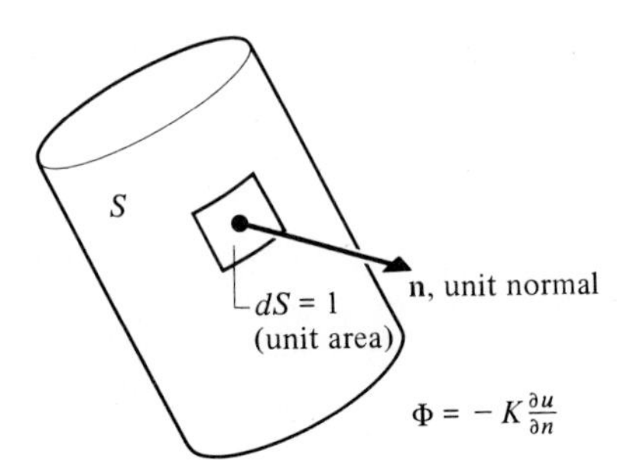

Figure 6.3-1
Flux.

a direction *normal* to the surface S, that is,

$$\Phi = -K\frac{\partial u}{\partial n}, \tag{6.3-3}$$

where the proportionality constant K (> 0) is the thermal conductivity and $\partial u/\partial n$ is the rate of change of temperature along the *outward* pointing normal (Fig. 6.3-1). The SI units* of flux are calories/cm²/sec.

In Examples 6.3-1, 6.3-2, and 6.3-3 we will illustrate how various boundary conditions can be treated. In these examples we will be considering a slender rod of constant cross-section encased in an insulating jacket so that heat conduction takes place in the x-direction only. We will thus be dealing with the *one-dimensional heat equation* in rectangular coordinates. The equation also applies to an infinite slab of finite width whose faces are insulated and whose edges are kept at constant temperatures or are insulated.

EXAMPLE 6.3-1 Solve the following problem.

$$\begin{aligned} &\text{P.D.E.:} \quad u_t = u_{xx}, \quad 0 < x < L, \quad t > 0; \\ &\text{B.C.:} \quad \left.\begin{aligned} u_x(0, t) &= 0, \\ u(L, t) &= 0, \end{aligned}\right\} \quad t > 0; \\ &\text{I.C.:} \quad u(x, 0) = f(x), \quad 0 < x < L. \end{aligned}$$

This is the problem of heat conduction in a bar whose left end is perfectly insulated, whose right end is kept at temperature zero, and along which there is an initial temperature distribution defined by $f(x)$. See Fig. 6.3-2.

Solution Using separation of variables, let $u(x, t) = X(x)T(t)$, and substitute into the partial differential equation. Then

$$X(x)\dot{T}(t) = X''(x)T(t),$$

*Système International d'Unites.

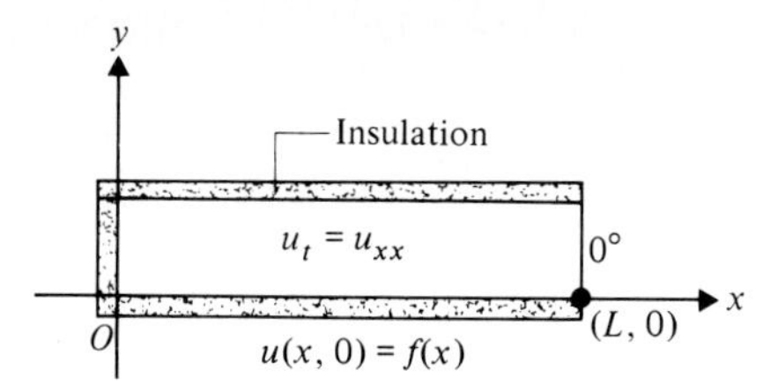

Figure 6.3-2

and dividing by $X(x)T(t)$, we get (Exercise 2)

$$\frac{\dot{T}(t)}{T(t)} = \frac{X''(x)}{X(x)} = -\lambda^2.$$

The Sturm–Liouville problem in $X(x)$ is

$$X'' + \lambda^2 X = 0, \qquad X'(0) = 0, \qquad X(L) = 0,$$

with solutions (Exercise 3)

$$X_n(x) = \cos\frac{(2n-1)}{L}\frac{\pi}{2}x, \qquad n = 1, 2, \ldots.$$

Since the first-order equations

$$\dot{T}_n(t) + \frac{(2n-1)^2\pi^2}{4L^2}T_n(t) = 0$$

have solutions

$$T_n(t) = \exp\left[\frac{-\pi^2(2n-1)^2}{4L^2}t\right],$$

we take for $u(x, t)$ the linear combination

$$u(x, t) = \sum_{n=1} a_{2n-1} \exp\left[\frac{-\pi^2(2n-1)^2}{4L^2}t\right]\cos\frac{(2n-1)\pi}{2L}x. \quad \textbf{(6.3-4)}$$

Then, applying the I.C., we have

$$\sum_{n=1} a_{2n-1} \cos\frac{(2n-1)\pi}{2L}x = f(x).$$

If $f(x)$ meets the required conditions, then it can be expanded in a Fourier cosine series so that

$$a_{2n-1} = \frac{2}{L}\int_0^L f(s)\cos\frac{(2n-1)\pi}{2L}s\,ds, \qquad n = 1, 2, \ldots. \quad \textbf{(6.3-5)}$$

The complete solution is given by Eq. (6.3-4) with a_{2n-1} defined in Eq. (6.3-5). ■

The next example illustrates how a **nonhomogeneous boundary condition** can be made homogeneous.

EXAMPLE 6.3–2 Solve the following problem. See Fig. 6.3–3.

$$\begin{aligned} &\text{P.D.E.:} && u_t = u_{xx}, \quad 0 < x < L, \quad t > 0; \\ &\text{B.C.:} && \left.\begin{aligned} u(0, t) &= 0, \\ u(L, t) &= u_0, \text{ a constant,} \end{aligned}\right\} \quad t > 0; \\ &\text{I.C.:} && u(x, 0) = 0, \quad 0 < x < L. \end{aligned}$$

Solution The method of separation of variables fails here because it leads to $u(x, t) = 0$, which is not a solution (Exercise 6). If we make a change of variable, however, the difficulty can be overcome. Accordingly, let

$$u(x, t) = U(x, t) + \phi(x),$$

so that the problem becomes the following one.

$$\begin{aligned} &\text{P.D.E.:} && U_t = U_{xx} + \phi''(x), \quad 0 < x < L, \quad t > 0; \\ &\text{B.C.:} && \left.\begin{aligned} U(0, t) + \phi(0) &= 0, \\ U(L, t) + \phi(L) &= u_0, \end{aligned}\right\} \quad t > 0; \\ &\text{I.C.:} && U(x, 0) + \phi(x) = 0, \quad 0 < x < L. \end{aligned}$$

Now if we choose $\phi(x)$ so that it satisfies

$$\phi''(x) = 0, \qquad \phi(0) = 0, \qquad \phi(L) = u_0, \tag{6.3–6}$$

then the resulting problem in $U(x, t)$ is one that can be solved by the method of separation of variables. The details are left for the exercises. (See Exercises 7 and 8.) ■

The technique used in solving the problem of Example 6.3–2 is an illustration of a powerful method in mathematics, that is, reducing a problem to one whose solution is known. Other examples of this technique occur in the integral calculus, where integration by substitution is used, and in ordinary differential equations, where certain second-order equations are reduced to first-order ones. See also Exercises 24 and 26.

We observe that an alternative way of looking at Example 6.3–2 is the following. One end of a bar is kept at temperature zero, and the other at

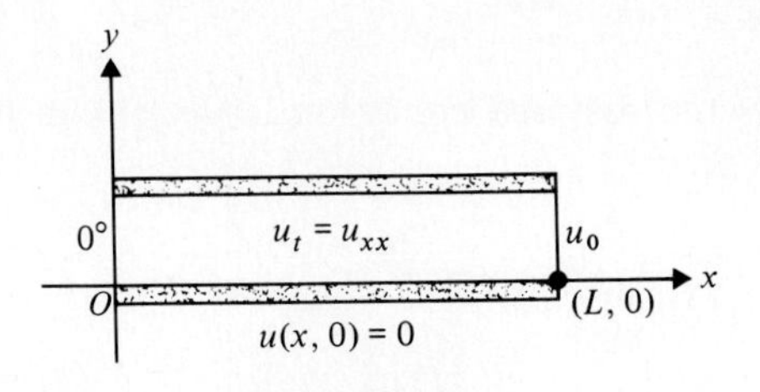

Figure 6.3–3

temperature u_0, a constant. The entire bar is initially at temperature zero. It is apparent from physical considerations that as $t \to \infty$, the temperature $u(x)$ for $0 \leq x \leq L$ will approach $u_0 x/L$. But this is just another way of saying that $u_0 x/L$ is the **steady-state solution**. This is the solution to the problem that can be stated as

$$u''(x) = 0, \qquad u(0) = 0, \qquad u(L) = u_0.$$

Thus the function $\phi(x)$ in (6.3–6) is the solution of the steady-state problem. When this solution is added to the **transient solution**, the result is the complete solution obtained in Exercise 8.

EXAMPLE 6.3–3 A cylindrical rod of length L is initially at temperature $f(x)$. Its left end is kept at temperature zero, while heat is transferred from the right end into the surroundings at temperature zero (Fig. 6.3–4). Find $u(x, t)$.

Solution Here the problem to be solved can be expressed as follows:

$$\begin{aligned} &\text{P.D.E.:} && u_t = u_{xx}, \quad 0 < x < L, \quad t > 0; \\ &\text{B.C.:} && \left.\begin{aligned} &u(0, t) = 0, \\ &-u_x(L, t) = hu(L, t), \quad h > 0, \end{aligned}\right\} \quad t > 0; \\ &\text{I.C.:} && u(x, 0) = f(x), \quad 0 < x < L. \end{aligned}$$

Using separation of variables, we have

$$\frac{X''}{X} = \frac{\dot{T}}{T} = -\lambda^2;$$

hence

$$X(x) = c_1 \cos \lambda x + c_2 \sin \lambda x.$$

The first boundary condition implies that $c_1 = 0$, whereas the second results in

$$\tan \lambda L = -\lambda/h. \tag{6.3–7}$$

Thus a solution can be written as

$$u(x, t) = \sum_{n=1} c_n \exp(-\lambda_n^2 t) \sin \lambda_n x, \tag{6.3–8}$$

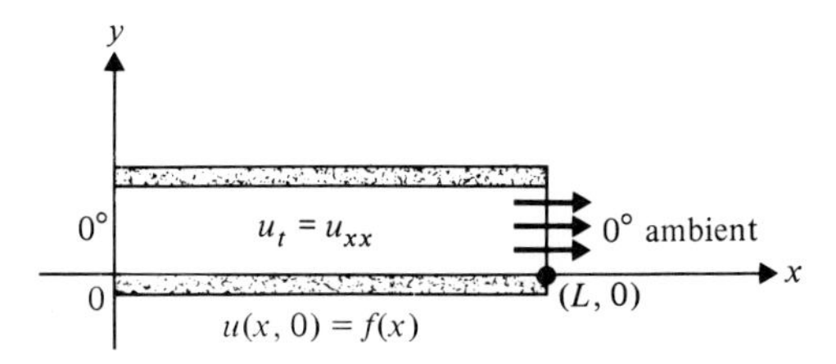

Figure 6.3–4

and if we apply the initial condition, this becomes

$$u(x, 0) = \sum_{n=1} c_n \sin \lambda_n x = f(x),$$

where the λ_n are solutions of Eq. (6.3–7).

Since the problem

$$X'' + \lambda^2 X = 0, \qquad X(0) = 0, \qquad hX(L) + X'(L) = 0, \tag{6.3–9}$$

is a regular Sturm–Liouville problem (Exercise 10), we know that to each eigenvalue λ_n there corresponds an eigenfunction $\sin \lambda_n x$. Moreover, the eigenfunctions form an orthogonal set on $[0, L]$ with weight function one. Finally, this orthogonal set is *complete* with respect to the class of piecewise smooth functions on $[0, L]$. Thus we *can* represent such a function $f(x)$ by a series of the functions $\sin \lambda_n x$. The coefficients c_n in Eq. (6.3–8) are given by (Exercise 11)

$$c_n = \frac{\int_0^L f(s) \sin \lambda_n s \, ds}{\int_0^L \sin^2 \lambda_n s \, ds}$$

$$= \frac{2(\lambda_n^2 + h^2)}{L(\lambda_n^2 + h^2) + h} \int_0^L f(s) \sin \lambda_n s \, ds. \quad \blacksquare \tag{6.3–10}$$

We conclude this section by mentioning that the one-dimensional Fermi* age equation for the **diffusion of neutrons** in a medium such as graphite is

$$\frac{\partial^2 q(x, \tau)}{\partial x^2} = \frac{\partial q(x, \tau)}{\partial \tau}.$$

Here q represents the number of neutrons that "slow down" (that is, fall below some given energy level) per second per unit volume. The **Fermi age**, τ, is a measure of the energy loss.

Key Words and Phrases

heat conduction
convection
Newton's law of cooling
ambient temperature
nonhomogeneous boundary condition
steady-state solution
transient solution
diffusion of neutrons
Fermi age

*Enrico Fermi (1901–1954), Italian nuclear physicist, who supervised the construction of the first nuclear reactor.

6.3 Exercises

• **1.** Show that a time scale change given by $\tau = kt$ will transform the diffusion equation (6.3–1) into

$$u_\tau = \nabla^2 u.$$

2. In Example 6.3–1, show that
(a) choosing $\lambda = 0$ leads to the trivial solution in $X(x)$;
(b) choosing $+\lambda^2$ for the separation constant leads to the trivial solution in $X(x)$.

3. Find the eigenvalues and eigenfunctions of the Sturm–Liouville problem

$$X'' + \lambda^2 X = 0, \qquad X'(0) = 0, \qquad X(L) = 0.$$

4. Explain why there is no a_0-term in the solution to Example 6.3–1.

5. Verify the solution to Example 6.3–1 completely.

6. Apply the method of separation of variables to the problem of Example 6.3–2, and show that the result is $u(x, t) = 0$. (*Hint*: Examine the problem in $T(t)$.)

7. Show that the solution of the problem in (6.3–6) is

$$\phi(x) = u_0 x/L.$$

8. Using the result in Exercise 7, solve the problem of Example 6.3–2.

9. In Example 6.3–3, show that:
(a) $\lambda = 0$ leads to the trivial solution in $X(x)$;
(b) the separation constant $+\lambda^2$ leads to the trivial solution in $X(x)$. (*Hint*: Show that $f(\lambda) = \tanh \lambda L + (\lambda/h)$ is zero for $\lambda = 0$ but nowhere else by examining $f'(\lambda)$.)

10. Verify that the problem in (6.3–9) is a regular Sturm–Liouville problem.

11. Obtain the coefficients c_n in Eq. (6.3–10). (*Hint*: After integrating, use Eq. (6.3–7) to find $\sin \lambda_n L$ and $\cos \lambda_n L$.)

•• **12.** Solve the following problem:

$$\begin{aligned} &\text{P.D.E.:} && u_t = u_{xx}, \quad 0 < x < L, \quad t > 0;\\ &\text{B.C.:} && \left.\begin{aligned} u_x(L, t) &= 0,\\ u(0, t) &= 0, \end{aligned}\right\} \quad t > 0;\\ &\text{I.C.:} && u(x, 0) = f(x), \quad 0 < x < L. \end{aligned}$$

13. Solve the problem of Exercise 12 given that

$$f(x) = \begin{cases} 0, & 0 \le x < \dfrac{L}{2},\\[2mm] L - x, & \dfrac{L}{2} \le x \le L. \end{cases}$$

14. A thin cylindrical bar of length L has both ends perfectly insulated and has an initial temperature distribution given by $f(x)$. (See Fig. 6.3–5.) Find the temperature at any point in the bar at any time t.

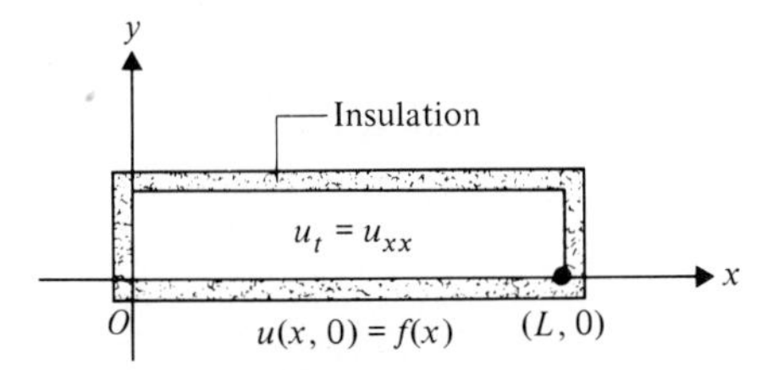

Figure 6.3-5

15. If heat is generated uniformly throughout a semi-infinite slab of width L at a constant rate C, then the one-dimensional heat equation has the form,

$$u_t = u_{xx} + C, \qquad C > 0.$$

Solve this equation under the assumptions that the edges $x = 0$ and $x = L$ of the slab are kept at temperature zero, the faces of the slab are insulated, and the initial temperature distribution is $f(x)$. (*Hint*: Make the following change of variable: $u(x, t) = U(x, t) + \phi(x)$.)

16. In Exercise 15, put $f(x) = 0$, and obtain the solution. (*Note*: This problem is a simplification of one that occurs in the manufacture of plywood, where heat is supplied by means of high-frequency heating.)

17. Solve the following problem:

$$\begin{aligned} &\text{P.D.E.:} \quad u_t = u_{xx}, \quad 0 < x < 2, \quad t > 0;\\ &\text{B.C.:} \quad \left.\begin{aligned} u_x(0, t) &= 0,\\ u(2, t) &= 0, \end{aligned}\right\} \quad t > 0;\\ &\text{I.C.:} \quad u(x, 0) = ax, \quad a > 0, \quad 0 < x < 2. \end{aligned}$$

18. Solve the following problem:

$$\begin{aligned} &\text{P.D.E.:} \quad u_t = u_{xx}, \quad 0 < x < 1, \quad t > 0;\\ &\text{B.C.:} \quad \left.\begin{aligned} u(0, t) &= 0,\\ u_x(1, t) &= 0, \end{aligned}\right\} \quad t > 0;\\ &\text{I.C.:} \quad u(x, 0) = u_0 x, \quad u_0 > 0, \quad 0 < x < 1. \end{aligned}$$

19. The faces $x = 0$, $x = a$, $y = 0$, and $y = b$ of a semi-infinite, solid, rectangular parallelepiped are kept at temperature zero, and there is an initial temperature distribution on the bottom face given by $f(x, y)$. Find the temperature in the interior at any time t. (*Hint*: Compare Example 6.1-1.)

20. A rod one unit in length is perfectly insulated along its length so that heat conduction can take place in only the x-direction. The left end is kept at temperature zero, and the right end is insulated. If the temperature is given initially by the function ax^2 (a is a positive constant), find the temperature in the rod at any time t.

21. **(a)** Solve the following problem:

$$\begin{aligned} &\text{P.D.E.:} \quad u_t = ku_{xx}, \quad 0 < x < \pi, \quad t > 0;\\ &\text{B.C.:} \quad u_x(0, t) = u_x(\pi, t) = 0, \quad t > 0;\\ &\text{I.C.:} \quad u(x, 0) = \sin^2 x, \quad 0 < x < \pi. \end{aligned}$$

(b) Verify the solution to part (a) completely.

22. A rod π units long is encased in an insulating blanket except for the two ends, which are kept at zero. If the initial temperature distribution is given by $x \sin x$, find the temperature at any point in the rod at any time t.

••• **23.** If $f(x)$ in Example 6.3–1 is defined as

$$f(x) = \begin{cases} 0, & 0 \le x < \dfrac{L}{2}, \\ L, & \dfrac{L}{2} \le x \le L, \end{cases}$$

obtain the result

$$\begin{aligned} u(x, t) = \frac{2L}{\pi}\Big[&\exp(-\pi^2 t/4L^2)(2 - \sqrt{2}) \cos(\pi x/2L) \\ &- \exp(-3\pi^2 t/4L^2)\left(\frac{2 + \sqrt{2}}{3}\right) \cos(3\pi x/2L) \\ &+ \exp(-5\pi^2 t/4L^2)\left(\frac{2 + \sqrt{2}}{5}\right) \cos 5\pi x/2L + \cdots\Big]. \end{aligned}$$

24. A rod of length L is insulated *at both ends* only, and the initial temperature distribution is given by $f(x)$. If there is a linear surface heat transfer between the rod and its surroundings at temperature zero, then the equation

$$v_t(x, t) = k v_{xx}(x, t) - h v(x, t),$$

where h is a positive constant, applies (Fig. 6.3–6). Using the substitution

$$v(x, t) = \exp(-ht) u(x, t),$$

reduce this problem to one that has been previously solved. (Compare Exercise 14.)

25. Solve Exercise 24, given that the ends of the rod are kept at temperature zero, rather than being insulated.

26. Consider the following problem.

P.D.E.: $u_t = k u_{xx}, \quad 0 < x < L, \quad t > 0;$

B.C.: $\left.\begin{aligned} u(0, t) &= T_1, \\ u(L, t) &= T_2, \end{aligned}\right\} \quad t > 0;$

I.C.: $u(x, 0) = f(x), \quad 0 < x < L,$

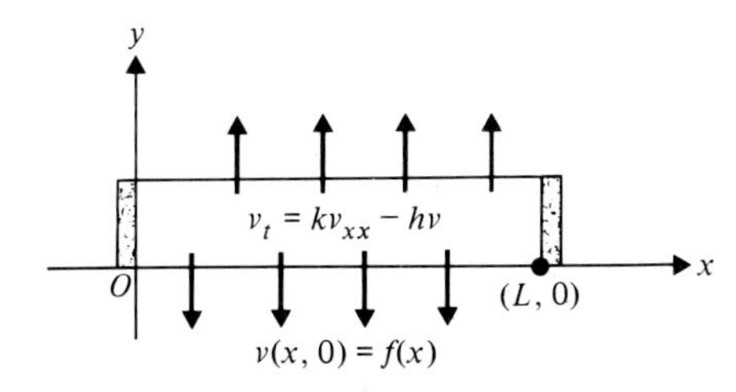

Figure 6.3–6

where T_1 and T_2 are constants. Solve this problem by making the change of variable

$$u(x, t) = v(x, t) + T_1 + \frac{x}{L}(T_2 - T_1).$$

27. Verify that the solution in Eq. (6.3–8) satisfies the boundary conditions in Example 6.3–3.

28. When an external force $\mathbf{F}$ acts on a diffusing particle, the diffusion equation has the more general form

$$\frac{\partial f_r}{\partial t} = \text{div}\,[D \text{ grad } f_r - (\mathbf{F}/M\beta)\, f_r].$$

When f_r is the density of a diffusing substance, the above equation, or its simpler version when $\mathbf{F} = \mathbf{0}$, is the diffusion equation. When f_r is the distribution function for a particle undergoing Brownian motion, the equation is called a *Fokker–Planck equation*. The solution for $\mathbf{F} = \mathbf{0}$ and for the particle starting at $\mathbf{r} = \mathbf{r}_0$ when $t = 0$

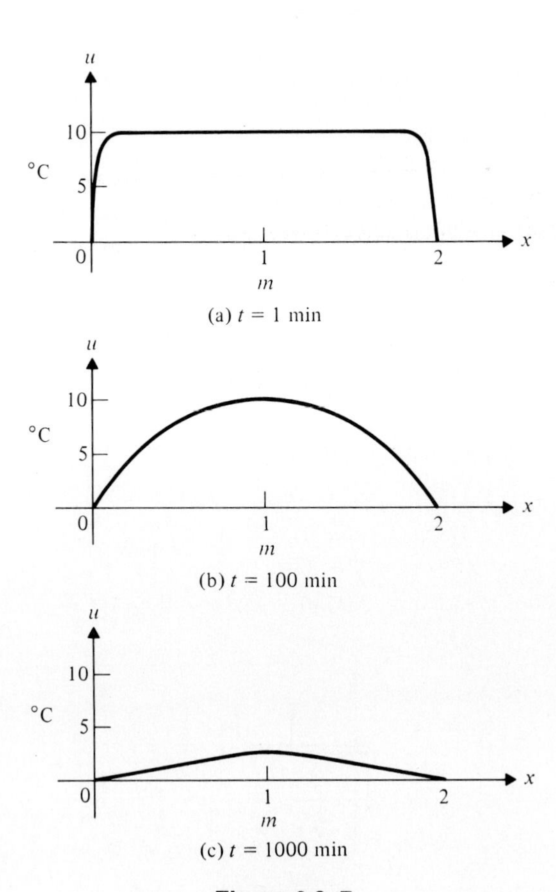

Figure 6.3–7

is

$$f_r(\mathbf{r}, t) = (4\pi DT)^{-3/2} \exp[-|\mathbf{r} - \mathbf{r}_0|^2/(4Dt)],$$

where $D = kT/M\beta$ is the diffusion constant, M is a measure of the mobility, k is the Boltzmann constant, T is temperature, t is time, and β is the thermal expansion coefficient. Verify the above solution.

29. **(a)** Suppose that the two ends of a uniform rod of mild steel ($k = 7.45 \times 10^{-4}$ m^2/min) 2 m long is held at 0°C. Suppose further that the lateral sides are insulated so that we are dealing with the one-dimensional diffusion equation. If the temperature in the rod is uniformly 10°C at $t = 0$, obtain the temperatures

$$u(x, t) = \frac{20}{\pi} \sum_{n=1} \frac{\sin[(2n-1)\pi x/2] \exp[-(2n-1)^2\pi^2 kt/4]}{2n-1}.$$

*__(b)__ Obtain the graphs shown in Fig. 6.3-7.

6.4 FOURIER AND LAPLACE TRANSFORM METHODS

Whenever a boundary-value problem is to be solved over an infinite or a semi-infinite domain, transform methods provide a good approach. In Sections 5.2, 5.3, and 6.1 we presented some examples using the Fourier transform. We give further examples in this section, using both the Fourier transform and the Laplace transform.

EXAMPLE 6.4-1 Find the bounded, harmonic function $v(x, y)$ in the semi-infinite strip $0 < x < c$, $y > 0$, that satisfies the following conditions:

$$\text{(a)}\quad v(0, y) = 0, \qquad \text{(b)}\quad v_y(x, 0) = 0, \qquad \text{(c)}\quad v_x(c, y) = f(y).$$

Give a physical interpretation of this problem.

Solution Laplace's equation

$$v_{xx} + v_{yy} = 0, \quad 0 < x < c, \qquad y > 0,$$

is to be solved. We use the Fourier *cosine* transform and transform the variable y because of boundary condition (b). Thus (refer to Eq. (5.2.9) in Section 5.2) we have

$$\frac{d^2\bar{v}(x, \alpha)}{dx^2} - \alpha^2\bar{v}(x, \alpha) = 0$$

$$\text{(a}'\text{)}\quad \bar{v}(0, \alpha) = 0, \qquad \text{(c}'\text{)}\quad \frac{d\bar{v}(c, \alpha)}{dx} = \bar{f}(\alpha),$$

where

$$\bar{f}(\alpha) = \int_0^\infty f(y) \cos \alpha y \, dy.$$

The general solution of the second-order ordinary differential equation is (Exercise 1)

$$\bar{v}(x, \alpha) = c_1(\alpha) \cosh(\alpha x) + c_2(\alpha) \sinh(\alpha x),$$

and $c_1(\alpha) = 0$ by condition (a′), whereas

$$c_2(\alpha) = \frac{\bar{f}(\alpha)}{\alpha \cosh(\alpha c)}$$

by condition (c′). Hence

$$\bar{v}(x, \alpha) = \frac{\bar{f}(\alpha) \sinh(\alpha x)}{\alpha \cosh(\alpha c)},$$

and the inverse cosine transform is given by

$$\begin{aligned} v(x, y) &= \frac{2}{\pi} \int_0^\infty \frac{\bar{f}(\alpha) \sinh(\alpha x)}{\alpha \cosh(\alpha c)} \cos(\alpha y) \, d\alpha \\ &= \frac{2}{\pi} \int_0^\infty \frac{\sinh(\alpha x) \cos(\alpha y)}{\alpha \cosh(\alpha c)} \, d\alpha \int_0^\infty f(s) \cos(\alpha s) \, ds. \qquad \textbf{(6.4–1)} \end{aligned}$$

Referring to Fig. 6.4–1, we can interpret the problem as finding the bounded steady-state temperatures in a semi-infinite slab of width c, given that the left edge is kept at temperature zero, the bottom edge is perfectly insulated, and the flux across the right edge is a given function $f(y)$. ■

In the next example we use the Fourier sine transform.

EXAMPLE 6.4–2 Solve the following problem (Fig. 6.4–2):

$$\begin{aligned} &\text{P.D.E.:} \quad u_t = u_{xx}, \quad x > 0, \quad t > 0; \\ &\text{B.C.:} \quad u(0, t) = u_0, \quad t > 0; \\ &\text{I.C.:} \quad u(x, 0) = 0, \quad x > 0, \end{aligned}$$

where $u_0 > 0$.

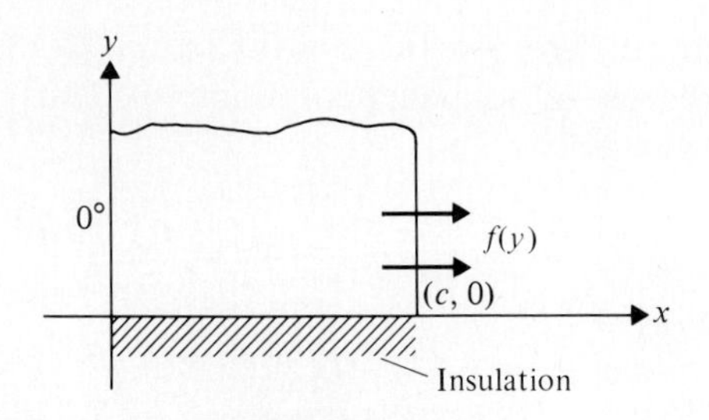

Figure 6.4–1

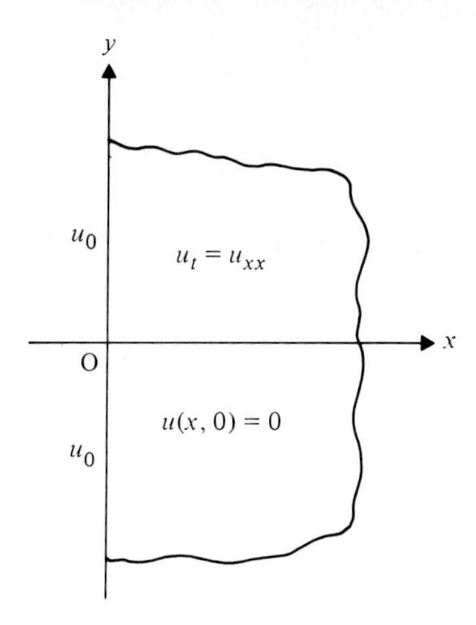

Figure 6.4-2

Solution It is appropriate to use the Fourier sine transform to transform u_{xx} because $0 < x < \infty$ *and* $u(0, t)$ is known. Thus (refer to Eq. 5.2-7)

$$\frac{d\bar{u}(\alpha, t)}{dt} = \alpha u_0 - \alpha^2 \bar{u}(\alpha, t),$$

and this first-order, nonhomogeneous, ordinary differential equation has the solution

$$\bar{u}(\alpha, t) = \frac{u}{\alpha}[1 - \exp(-\alpha^2 t)],$$

using the initial condition $\bar{u}(\alpha, 0) = 0$ (Exercise 2). The inverse Fourier sine transform of $\bar{u}(\alpha, t)$ is

$$u(x, t) = \frac{2u_0}{\pi} \int_0^\infty [1 - \exp(-\alpha^2 t)] \sin \alpha x \frac{d\alpha}{\alpha}.$$

But from Exercise 14 in Section 5.1,

$$\int_0^\infty \frac{\sin \alpha x \, d\alpha}{\alpha} = \frac{\pi}{2} \qquad \text{if } x > 0.$$

Hence

$$u(x, t) = u_0 - \frac{2u_0}{\pi} \int_0^\infty \exp(-\alpha^2 t) \sin \alpha x \frac{d\alpha}{\alpha},$$

and using the result

$$\frac{\sin \alpha x}{\alpha} = \int_0^x \cos \alpha s \, ds,$$

we have

$$u(x, t) = u_0 - \frac{2u_0}{\pi}\int_0^\infty \exp(-\alpha^2 t)\, d\alpha \int_0^x \cos \alpha s\, ds$$

$$= u_0 - \frac{2u_0}{\pi}\int_0^x ds \int_0^\infty \exp(-\alpha^2 t) \cos \alpha s\, d\alpha.$$

The α integral appears in tables of definite integrals; hence

$$u(x, t) = u_0 - \frac{u_0}{\sqrt{\pi}}\int_0^x \frac{\exp(-s^2/4t)\, ds}{\sqrt{t}}.$$

Now the substitution $v^2 = s^2/4t$ transforms the last result into (Exercise 3)

$$u(x, t) = u_0 - \frac{2u_0}{\sqrt{\pi}}\int_0^{x/2\sqrt{t}} \exp(-v^2)\, dv \qquad \textbf{(6.4–2)}$$

$$= u_0 - u_0 \operatorname{erf}(x/2\sqrt{t})$$

$$= u_0[1 - \operatorname{erf}(x/2\sqrt{t})]$$

$$= u_0 \operatorname{erfc}\left(\frac{x}{2\sqrt{t}}\right),$$

using the definition of erfc, the **complementary error function**:

$$\operatorname{erfc} x = 1 - \operatorname{erf} x = \frac{2}{\sqrt{\pi}}\int_x^\infty e^{-s^2}\, ds,$$

where the **error function**, erf x, is defined by

$$\operatorname{erf} x = \frac{2}{\sqrt{\pi}}\int_0^x e^{-s^2}\, ds. \qquad \blacksquare$$

An interesting observation can be drawn from the form of the solution in Eq. (6.4–2). A change in the boundary temperature u_0 is *instantly* transmitted to any other point on the x-axis. This implies that heat is propagated with an **infinite velocity**, which at first seems paradoxical. Recall that in Section 3.4, where we derived the diffusion equation, we used an *equilibrium* concept. Accordingly, $u(x, t)$ has meaning only if we assume that the system (rod, plate, etc.) is essentially in equilibrium for all t. On the other hand, conduction of heat is a result of randomly moving molecules transferring their kinetic energy as a result of collisions. Since the time scale of the latter motion is so different from that needed to reach an equilibrium temperature, we can assume that changes in temperature are propagated with infinite speed.

EXAMPLE 6.4–2(a) We solve the problem of Example 6.4–2 again, this time using the **Laplace transform**. Recall the definition* of the Laplace

*In the most general case, s may be a complex number with some restriction placed on its real part.

transform of $u(t)$,

$$U(s) = \int_0^\infty e^{-st}\, u(t)\, dt, \qquad s > 0.$$

Hence, using the fact that $u(x, 0) = 0$,

$$\int_0^\infty e^{-st} \frac{\partial u(x, t)}{\partial t} dt = e^{-st}\, u(x, t) \Big|_0^\infty + s \int_0^\infty e^{-st}\, u(x, t)\, dt$$

$$= sU(x, s),$$

so that the partial differential equation $u_t = u_{xx}$ becomes

$$\frac{d^2 U(x, s)}{dx^2} - sU(x, s) = 0$$

with

$$U(0, s) = \int_0^\infty e^{-st}\, u_0\, dt = \frac{u_0}{s}.$$

Now the solution of this second-order, homogeneous, ordinary differential equation is

$$U(x, s) = c_1(s)\, e^{\sqrt{s}x} + c_2(s)\, e^{-\sqrt{s}x}.$$

We take $c_1(s)$ to be zero so that $U(x, s)$ will remain bounded for $x > 0$, and applying the condition $U(0, s) = u_0/s$, we have (Exercise 4)

$$U(x, s) = \frac{u_0}{s}\, e^{-\sqrt{s}x}. \tag{6.4-3}$$

Using the table of Laplace transforms in Table II of the Appendix, we find

$$u(x, t) = u_0 \operatorname{erfc}\,(x/2\sqrt{t}\,)$$

as before. ■

The last example illustrates the effect produced by a nonhomogeneous boundary condition. Such a condition resulted in a *nonhomogeneous* differential equation in the transformed variable. In Section 6.3 we presented another method for dealing with a nonhomogeneous boundary condition. (See Example 6.3–2.) The following example shows how easily the Laplace transform can handle *some* nonhomogeneous boundary conditions.

EXAMPLE 6.4–3 Solve the following problem using the Laplace transform (Fig. 6.4–3).

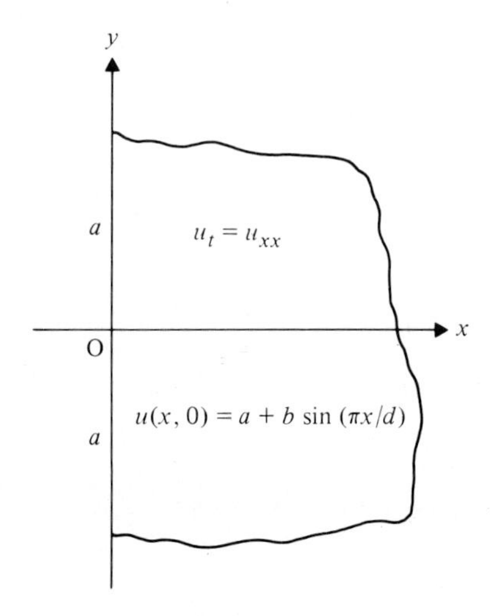

Figure 6.4-3

$$\begin{aligned}
&\text{P.D.E.:} && u_t = u_{xx}, \quad 0 < x < d, \quad t > 0; \\
&\text{B.C.:} && \left.\begin{aligned} u(0, t) &= a, \\ u(d, t) &= a, \end{aligned}\right\} \quad t > 0; \\
&\text{I.C.:} && u(x, 0) = a + b \sin\left(\frac{\pi}{d}x\right), \quad 0 < x < d,
\end{aligned}$$

where a and b are constants.

Solution Transforming the equation and the boundary conditions results in the problem

$$\frac{d^2U(x, s)}{dx^2} - sU(x, s) = -a - b \sin\left(\frac{\pi}{d}x\right),$$

$$U(0, s) = \frac{a}{s}, \qquad U(d, s) = \frac{a}{s}.$$

This is an ordinary differential equation with complementary solution (see Section 1.2)

$$U_c(x, s) = c_1(s)\, e^{\sqrt{s}x} + c_2(s)\, e^{-\sqrt{s}x}.$$

A particular integral,

$$U_p(x, s) = \frac{a}{s} + \frac{bd^2}{d^2s + \pi^2} \sin\left(\frac{\pi}{d}x\right),$$

can be found by the method of undetermined coefficients (see Section 1.2). The sum of $U_c(x, s)$ and $U_p(x, s)$ is the general solution, and evaluating $c_1(s)$ and

$c_2(s)$ with the aid of the conditions $U(0, s) = U(d, s) = a/s$, we have

$$U(x, s) = \frac{a}{s} + \frac{bd^2}{d^2s + \pi^2}\sin\left(\frac{\pi}{d}x\right).$$

In obtaining the inverse of this we note that $\sin(\pi x/d)$ can be treated as a constant. Thus

$$u(x, t) = a + b\sin\left(\frac{\pi}{d}x\right)\exp(-\pi^2 t/d^2). \qquad \textbf{(6.4–4)}$$

The details are left for the exercises. (See Exercise 5.) ■

EXAMPLE 6.4–4 Solve the following problem using the Laplace transform.

$$\begin{aligned}
&\text{P.D.E.:} && u_{tt} = u_{xx}, \quad 0 < x < c, \quad t > 0;\\
&\text{B.C.:} && \left.\begin{aligned} u(0, t) &= 0,\\ u(c, t) &= 0, \end{aligned}\right\} \quad t > 0;\\
&\text{I.C.:} && \left.\begin{aligned} u(x, 0) &= b\sin\left(\frac{\pi}{c}x\right),\\ u_t(x, 0) &= -b\sin\left(\frac{\pi}{c}x\right), \end{aligned}\right\} \quad 0 < x < c.
\end{aligned}$$

Solution Transforming the equation and the boundary conditions yields

$$\frac{d^2U(x, s)}{dx^2} = s^2U(x, s) - bs\sin\left(\frac{\pi}{c}x\right) + b\sin\left(\frac{\pi}{c}x\right),$$

$$U(0, s) = U(c, s) = 0,$$

which has the solution (Exercise 7),

$$U(x, s) = \frac{c^2b(s - 1)}{c^2s^2 + \pi^2}\sin\left(\frac{\pi}{c}x\right).$$

Hence

$$u(x, t) = b\sin\left(\frac{\pi}{c}x\right)\left[\cos\left(\frac{\pi}{c}t\right) - \frac{c}{\pi}\sin\left(\frac{\pi}{c}t\right)\right]. \quad ■ \qquad \textbf{(6.4–5)}$$

In the next example we *simplify* the conditions in Example 6.4–3 and show that, in spite of this, the problem is seemingly intractable by the method of Laplace transformation.

EXAMPLE 6.4–5 Solve the following problem using the Laplace transform (Fig. 6.4–4).

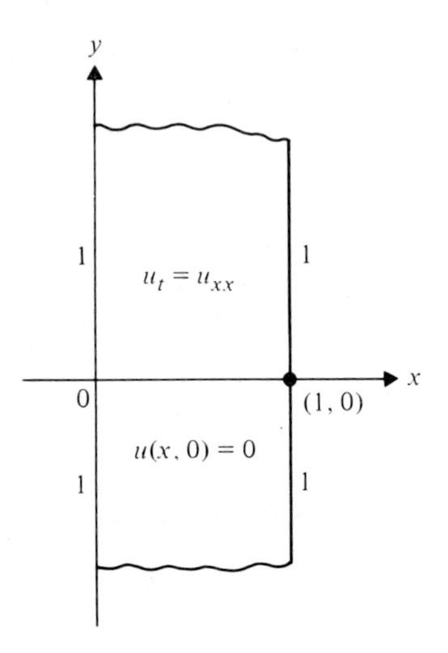

Figure 6.4-4

$$
\begin{array}{lll}
\text{P.D.E.:} & u_t = u_{xx}, \quad 0 < x < 1, & t > 0; \\
\text{B.C.:} & \left.\begin{array}{l} u(0, t) = 1, \\ u(1, t) = 1, \end{array}\right\} & t > 0; \\
\text{I.C.:} & u(x, 0) = 0, \quad 0 < x < 1. &
\end{array}
$$

The transformed problem is

$$
\begin{cases}
\dfrac{d^2U(x, s)}{dx^2} = sU(x, s), \\
U(0, s) = \dfrac{1}{s}, \qquad U(1, s) = \dfrac{1}{s},
\end{cases}
$$

which looks innocent enough. Its solution, however, is found to be (Exercise 9)

$$
\begin{aligned}
U(x, s) &= \frac{1}{s}\cosh \sqrt{s}x + \frac{(1 - \cosh \sqrt{s}\,)\sinh \sqrt{s}x}{s \sinh \sqrt{s}} \\
&= \frac{\sinh \sqrt{s}x + \sinh \sqrt{s}(1 - x)}{s \sinh \sqrt{s}}. \qquad \blacksquare
\end{aligned}
$$

Even with reasonably extensive tables of Laplace transforms at our disposal, we could not expect to find this particular function. What is needed here is a *general* method for obtaining the Laplace inverse of a function such as the method we used in Section 5.2 for obtaining the Fourier inverse transform. Finding the inverse Laplace transform involves *complex variables* and, in particular, *contour integration* and *residue theory*. For simpler func-

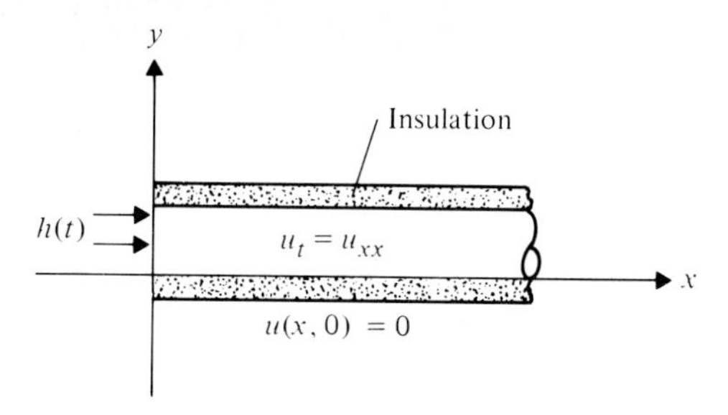

Figure 6.4-5

tions the methods of *partial fraction decomposition* and *convolution* may be helpful.*

We close this section with another example that involves a nonhomogeneous boundary condition. Consider a semi-infinite bar initially at temperature zero. If heat is supplied at the end $x = 0$ according to the function $h(t)$, then we have the problem stated in the following example.

EXAMPLE 6.4–6 Solve the following problem (Fig. 6.4–5):

$$\begin{aligned} \text{P.D.E.:}\quad & u_t = u_{xx}, \quad x > 0, \quad t > 0; \\ \text{B.C.:}\quad & u_x(0, t) = h(t), \quad t > 0; \\ \text{I.C.:}\quad & u(x, 0) = 0, \quad x > 0. \end{aligned}$$

Solution We may use the *Fourier cosine transform* to transform x because $0 < x < \infty$ and $u_x(0, t)$ is given. Then

$$\bar{u}(\alpha, t) = \int_0^\infty u(x, t) \cos \alpha x \, dx,$$

and u_{xx} is transformed into $-h(t) - \alpha^2\bar{u}$ (see Eq. 5.2–9). Thus the transformed problem becomes

$$\frac{d\bar{u}(\alpha, t)}{dt} + \alpha^2\bar{u}(\alpha, t) = -h(t),$$

which is a first-order linear differential equation. Multiplying by the **integrating factor**† $\exp(\alpha^2 t)$ produces

$$\bar{u}(\alpha, t) = -\exp(-\alpha^2 t)\left[\int_0^t \exp(\alpha^2 s)\, h(s)\, ds + C\right],$$

*See Chapters 3 and 10 of the author's *Advanced Engineering Mathematics* (Reading, Mass.: Addison-Wesley, 1982.)

†See Section 1.3 of the author's *Advanced Engineering Mathematics* (Reading, Mass.: Addison-Wesley, 1982).

and the condition $\bar{u}(\alpha, 0) = 0$ shows that $C = 0$. Hence (Exercise 11)

$$\bar{u}(\alpha, t) = -\exp(-\alpha^2 t) \int_0^t \exp(\alpha^2 s)\, h(s)\, ds$$
$$= -\int_0^t \exp[\alpha^2(s - t)]h(s)\, ds, \quad \textbf{(6.4-6)}$$

and

$$u(x, t) = -\frac{2}{\pi}\int_0^\infty \cos \alpha x\, d\alpha \int_0^t \exp[\alpha^2(s - t)]h(s)\, ds$$
$$= -\frac{2}{\pi}\int_0^t h(s)\, ds \int_0^\infty \exp[\alpha^2(s - t)] \cos \alpha x\, d\alpha$$
$$= -\frac{1}{\sqrt{\pi}}\int_0^t \frac{h(s) \exp[-x^2/4(t - s)]\, ds}{\sqrt{t - s}}, \quad \textbf{(6.4-7)}$$

assuming that the order of integration may be interchanged. ■

Up to this point, most of the solutions that we have obtained have been **formal solutions** in the sense that they have been obtained computationally. Since some of these formal solutions are phrased in terms of infinite series or improper integrals, it remains to verify their *validity*. This topic is discussed in the next section.

Key Words and Phrases

complementary error function (erfc x)
error function (erf x)
Laplace transform
integrating factor
formal solution

6.4 Exercises

• **1.** Solve the following two-point boundary-value problem (See Example 6.4-1.):

$$\frac{d^2\bar{v}(x, \alpha)}{dx^2} - \alpha^2\bar{v}(x, \alpha) = 0,$$

$$\bar{v}(0, \alpha) = 0, \qquad \frac{d\bar{v}(c, \alpha)}{dx} = \bar{f}(\alpha).$$

2. Solve the following problem:

$$\frac{d\bar{u}(\alpha, t)}{dt} + \alpha^2\bar{u}(\alpha, t) = \alpha u_0, \qquad \bar{u}(\alpha, 0) = 0.$$

(*Hint*: Use the fact that the given equation is linear and hence multiplying by the integrating factor exp $(\alpha^2 t)$ leads to a form that can be easily integrated. Alternatively, the method of undetermined coefficients may be used to obtain a particular solution.)

3. Obtain Eq. (6.4–2) from the previous step in the text.

4. Verify that Eq. (6.4–3) in the text is the correct solution of the given problem.

5. Given that

$$\frac{d^2U(x, s)}{dx^2} - sU(x, s) = -a - b \sin\left(\frac{\pi}{d}x\right),$$

$$U(0, s) = U(d, s) = \frac{a}{s}.$$

(a) Obtain the complementary solution $U_c(x, s)$.
(b) Obtain a particular integral by the method of undetermined coefficients.
(c) Obtain the complete solution

$$U(x, s) = \frac{a}{s} + \frac{bd^2}{d^2s + \pi^2} \sin\left(\frac{\pi}{d}x\right).$$

(d) Find the inverse Laplace transform of the function $U(x, s)$ that was found in part (c).

6. Verify completely the solution in Eq. (6.4–4).

7. Solve the following two-point boundary-value problem:

$$\frac{d^2U(x, s)}{dx^2} - s^2U(x, s) = b(1 - s) \sin\left(\frac{\pi}{c}x\right),$$

$$U(0, s) = U(c, s) = 0.$$

(Compare Example 6.4–4.)

8. Obtain the inverse Laplace transform of the function $U(x, s)$ of Example 6.4–4.

9. Solve the two-point boundary-value problem in Example 6.4–5 for $U(x, s)$.

10. In the problem of Example 6.4–5, make the substitution

$$u(x, t) = U(x, t) + \phi(x).$$

Then solve the problem by the method of separation of variables.

11. Obtain Eq. (6.4–6) from the given first-order linear differential equation.

12. Verify that the result in Eq. (6.4–7) follows from Eq. (6.4–6).

•• **13.** Solve Example 6.4–1 given that $f(y) = e^{-y}$.

14. **(a)** Solve the following problem using a Fourier transform (Fig. 6.4–6):

$$\begin{aligned} &\text{P.D.E.:} \quad u_t = u_{xx}, \quad x > 0, \quad t > 0;\\ &\text{B.C.:} \quad u_x(0, t) = u_0, \quad t > 0;\\ &\text{I.C.:} \quad u(x, 0) = 0, \quad x > 0. \end{aligned}$$

(b) Interpret this problem physically.

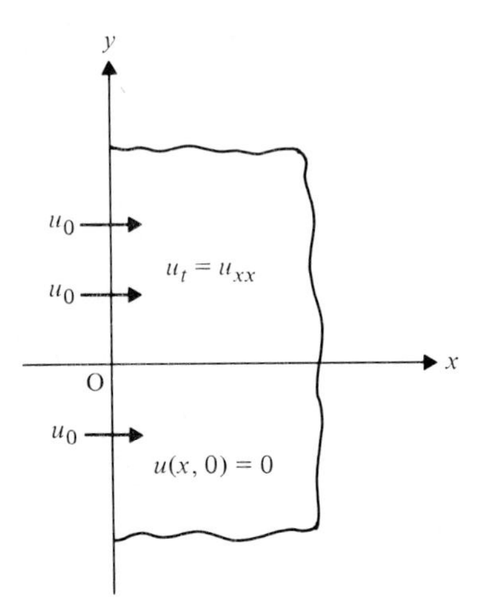

Figure 6.4–6

15. Solve Exercise 14 using the Laplace transform.

16. **(a)** Obtain the bounded harmonic function $v(x, y)$ over the semi-infinite strip $0 < x < c$, $y > 0$, that satisfies the following conditions:

(i) $v(0, y) = 0$ (ii) $v_y(x, 0) = 0$ (iii) $v_x(c, y) = f(y)$.

(b) Interpret this problem physically.

17. **(a)** Obtain the bounded harmonic function $v(x, y)$ over the semi-infinite strip $0 < y < b$, $x > 0$, that satisfies the following conditions:

(i) $v_y(x, 0) = 0$ (ii) $v_x(0, y) = 0$ (iii) $v(x, b) = f(x)$.

(b) Interpret this problem physically.

18. **(a)** Graph $f(x) = \exp(-x^2)$.
(b) Graph $f(x) = \operatorname{erf} x$. (*Hint*: Values of erf x can be found in tables.)
(c) Graph $f(x) = \operatorname{erfc} x$.
(d) Graph $u(x, t) = u_0 \operatorname{erfc}(x/2\sqrt{t})$ for various fixed values of t; take $u_0 = 1$ for convenience.

19. Solve the following problem:

$$\begin{aligned} &\text{P.D.E.:} \quad u_t = u_{xx}, \quad x > 0, \quad t > 0; \\ &\text{B.C.:} \quad u(0, t) = h(t), \quad t > 0; \\ &\text{I.C.:} \quad u(x, 0) = 0, \quad x > 0. \end{aligned}$$

(*Hint*: Pattern your solution after Example 6.4–6.)

20. In the result of Exercise 19, replace $h(t)$ by u_0, a constant, and then compare your answer with that for Example 6.4–2.

21. Show that

$$\int_0^\infty \frac{\exp(-ax)}{\sqrt{x}}\,dx = \sqrt{\pi/a}, \qquad a > 0.$$

*22. **(a)** Use the result in (6.4–2) to compute each of the following:

$$u(1, 1/4), \qquad u(1, 1), \qquad u(1, 4), \qquad u(1, 9).$$

(b) Graph the results obtained in part (a).

23. Show that the derivatives of the error function erf x are given by the following.

(a) $\dfrac{d}{dx}\operatorname{erf} x = \dfrac{2}{\sqrt{\pi}}\exp(-x^2)$

(b) $\dfrac{d^2}{dx^2}\operatorname{erf} x = -\dfrac{4}{\sqrt{\pi}}x\exp(-x^2)$

••• 24. Solve the following problem by using the Laplace transform:

P.D.E.: $u_{tt} = u_{xx} + \sin\dfrac{\pi x}{c}\sin\omega t, \quad 0 < x < c, \quad t > 0;$

B.C.: $u(0, t) = 0, \quad u(c, t) = 0, \quad t > 0;$

I.C.: $u(x, 0) = 0, \quad u_t(x, 0) = 0, \quad 0 < x < c.$

25. Explain the difficulties encountered when attempting to solve the problem of Example 6.4–6 by using the Laplace transform.

26. Verify the solution to Example 6.4–6 completely.

27. Using the value $u(1, 1)$ in Exercise 22, find the *time* required for the temperature to reach one-half this value at $x = 1$.

28. Read the article "Random Walks and Their Applications," by George H. Weiss, *Amer. Scientist 71*, No. 1, pp. 65–70. Comment on how the article ties in with the solution of Example 6.4–2.

29. Show that the solution to the problem in Example 6.4–2 can be written as $u(x, t) = \beta u_0$, where $0 \leq \beta \leq 1$, thus showing the maximum and minimum values of the result.

30. Verify the solution to Exercise 14(a) completely.

6.5 VERIFICATION OF SOLUTIONS

In all but a few cases the solutions of our boundary-value problems in the previous sections were obtained by using separation of variables (the Fourier method) or by using an integral transform method. The steps leading to a solution were often numerous and somewhat involved. As a result we were usually content with the solution if it satisfied the given boundary and initial conditions and if it seemed reasonable from physical considerations. Occasionally, there was an exercise in which a calculation and graph had to be made, and these helped to give a feeling of confidence in the solution.

In this section we discuss how a specific solution can be completely verified. To do this, we must show that the solution satisfies the given partial differential equation identically. This is not a trivial matter in case the solution is in the form of an infinite series or an improper integral. The differentiation of a series term by term and the interchange of differentiation and integration require that certain conditions be satisfied. It would be advisable at this point to review Sections 1.6, 2.5, and 4.5.

We give an example of the verification of a solution using one of the first solutions we obtained.

EXAMPLE 6.5–1 Given the problem of Example 3.4–1, namely,

$$\begin{aligned} &\text{P.D.E.:} \quad && u_t = ku_{xx}, \quad 0 < x < L, \quad t > 0; \\ &\text{B.C.:} \quad && u(0, t) = u(L, t) = 0, \quad t > 0; \\ &\text{I.C.:} \quad && u(x, 0) = f(x), \quad 0 < x < L, \end{aligned}$$

verify the solution

$$u(x, t) = \sum_{n=1} b_n \sin \frac{n\pi}{L} x \exp\left(-\frac{kn^2\pi^2}{L^2} t\right), \tag{6.5–1}$$

where

$$b_n = \frac{2}{L} \int_0^L f(s) \sin \frac{n\pi}{L} s \, ds.$$

Solution It is a simple matter to verify the boundary conditions. We need only observe that the b_n are bounded (see Theorem 2.5–1) and so is the function $\exp(-kn^2\pi^2 t/L^2)$. The initial condition is satisfied, since we assume that $f(x)$ is piecewise smooth and has a valid Fourier sine series representation.

Now consider the partial derivative

$$u_t = -\frac{k\pi^2}{L^2} \sum_{n=1} n^2 b_n \sin \frac{n\pi}{L} x \exp\left(-\frac{kn^2\pi^2}{L^2} t\right).$$

We have

$$|u_t| \leq Mk \sum_{n=1} n^2 \exp\left(-\frac{kn^2\pi^2}{L^2} t_0\right),$$

where M is the maximum value of $\pi^2 b_n/L^2$ and for all t_0 such that $0 < t_0 < \infty$. By the ratio test the last series of *positive constants* is convergent; hence the series defining u_t is uniformly convergent by the Weierstrass M-test. In a similar manner (Exercise 3), u_x and u_{xx} are defined by uniformly convergent series for all given values of the variables. Accordingly, term by term differentiation of (6.5–1) is valid by Theorem 1.6–2, and we have

$$u_{xx} = -\frac{\pi^2}{L^2} \sum_{n=1} b_n n^2 \sin \frac{n\pi}{L} x \exp\left(-\frac{kn^2\pi^2}{L^2} t\right)$$

so that $u_t = ku_{xx}$ is verified. ■

In the next example we consider a solution that is expressed in the form of an improper integral.

EXAMPLE 6.5–2 Given the problem of Example 5.2–1, namely,

$$\text{P.D.E.:} \quad u_t = u_{xx}, \qquad -\infty < x < \infty, \qquad t > 0;$$
$$\text{I.C.:} \quad u(x, 0) = f(x), \qquad -\infty < x < \infty,$$

verify the solution

$$u(x, t) = \frac{1}{2\sqrt{\pi t}} \int_{-\infty}^{\infty} f(s) \exp\left[-(s - x)^2/4t\right] ds. \tag{6.5–2}$$

Solution Clearly, the form of Eq. (6.5–2) is not the best one for verifying the initial condition. Referring to Section 5.2, we see that an equivalent form can be written as

$$u(x, t) = \frac{1}{\pi} \int_{-\infty}^{\infty} \int_{0}^{\infty} f(s) \cos \alpha(s - x) \exp(-\alpha^2 t)\, d\alpha\, ds. \tag{6.5–3}$$

Using this version of the solution, we have $u(x, 0) = \lim_{t \to 0^+} u(x, t)$, and, if we assume for the moment that we can interchange limit and integration, the result is

$$u(x, 0) = \frac{1}{\pi} \int_{-\infty}^{\infty} \int_{0}^{\infty} f(s) \cos \alpha(s - x)\, d\alpha\, ds, \tag{6.5–4}$$

which agrees with Eq. (5.1–9), the Fourier integral representation of the function $f(x)$. Eq. (6.5–4) represents $f(x)$ provided that this function is piecewise smooth and absolutely integrable on the interval $(-\infty, \infty)$. At points of discontinuity the representation gives the mean of the left- and right-hand limits of $f(x)$ at that point.

Since $f(x)$ is absolutely integrable, there is a positive constant M such that

$$|f(x)| < M, \qquad -\infty < x < \infty.$$

Thus for $t > 0$ and using Eq. (6.5–2),

$$|u(x, t)| < \frac{M}{2\sqrt{\pi t}} \int_{-\infty}^{\infty} \exp\left[-(s - x)^2/4t\right] ds$$
$$= \frac{M}{\sqrt{\pi}} \int_{-\infty}^{\infty} \exp(-\xi^2)\, d\xi = M,$$

where we have used the substitution

$$\xi = \frac{s - x}{2\sqrt{t}}. \tag{6.5–5}$$

We have shown that the improper integral in the solution (6.5–2) is bounded.

Before we continue, we state a theorem analogous to the one that describes the Weierstrass M-test for series (see Section 1.6).

THEOREM 6.5–1*

Let $\phi(x, s)$ be continuous in the region

$$R: \quad -\infty < x < \infty, \qquad -\infty < s < \infty.$$

Then if a function $M(s)$ has the properties

$$|\phi(x, s)| \le M(s) \qquad \text{in } R$$

and

$$\int_{-\infty}^{\infty} M(s)\, ds$$

converges, the integral

$$\Phi(x) = \int_{-\infty}^{\infty} \phi(x, s)\, ds$$

converges absolutely and uniformly in the interval $-\infty < x < \infty$.

Proof *By the hypotheses*

$$\left| \int_{b}^{\infty} \phi(x, s)\, ds \right| \le \int_{b}^{\infty} |\phi(x, s)|\, ds \le \int_{b}^{\infty} M(s)\, ds < \epsilon,$$

when b is sufficiently large, say, $b > B$. Note that B is independent of x. □

Accordingly, the improper integral in Eq. (6.5–2) converges absolutely and uniformly for $t > 0$ and all x satisfying $-\infty < x < \infty$. Moreover, the integrals obtained from Eq. (6.5–2) by differentiating partially under the integral sign with respect to x and t are uniformly convergent in the neighborhood of the point (x, t). Hence it follows that $u(x, t)$ satisfies the diffusion equation for $t > 0$ and $-\infty < x < \infty$ as required.

There is one more point that requires investigation. We would expect that a bona fide solution of the problem would have the property that a small change in the initial condition would result in a small change in the solution. Another way of saying this is that $u(x, t)$ should be continuous for $t \ge 0$. We have already determined that $u(x, t)$ is continuous for $t > 0$; hence we need only investigate what happens at a point $u(x_0, 0)$, where $-\infty < x_0 < \infty$. We make the change of variable shown in Eq. (6.5–5). Then

$$u(x, t) = \frac{1}{\sqrt{\pi}} \int_{-\infty}^{\infty} f(x + 2\xi\sqrt{t}) \exp(-\xi^2)\, d\xi,$$

*The theorem is credited to Charles J. de la Vallée Poussin (1866–1962), a French mathematician.

and

$$\lim_{t \to 0^+} u(x_0, t) = f(x_0),$$

where we have interchanged integration and the limit operation by virtue of the uniform convergence of the integral. ■

The analysis in the last example leads to the following definition.

DEFINITION 6.5–1
A boundary-value problem is said to be **well-posed** *if the following three requirements are met:*

1. *A solution of the problem exists.*
2. *The solution is unique.*
3. *The solution depends continuously on the data (initial and boundary conditions).*

A discussion of the conditions under which a given problem is well-posed would lead us too far afield. We will, however, comment briefly on the three requirements in Definition 6.5–1.

We have already seen that a solution to the Neumann problem can be found only to within an additive constant. Hence such a problem is not well-posed because it does not have a unique solution. Hadamard* has given a number of examples of problems in which the solution does not depend continuously on the data. One of these can be found in Exercise 21. With respect to existence, an applications-oriented person might be tempted to say, "I have found a solution by some method and have verified it; hence I am satisfied." The flaw in this logic is that one could spend an inordinate amount of effort trying to obtain a solution when, in fact, a solution did not exist. Hence we present a definition and an existence theorem at this point.

DEFINITION 6.5–2
The **Cauchy problem** *for a linear second-order partial differential equation*

$$Au_{xx} + Bu_{xy} + Cu_{yy} + Du_x + Eu_y + Fu = G, \qquad \textbf{(6.5–6)}$$

where the coefficients $A, B, \ldots, G$ are functions of x and y, can be stated as follows. Let R be a region of the xy-plane in which $A, B, \ldots, G$ are continuous. Let C_0 be a smooth arc† in R defined by the parametric

*Jacques Hadamard (1865–1963), a French mathematician.

†See Section 5.2 of the author's *Advanced Engineering Mathematics* (Reading, Mass.: Addison-Wesley, 1982).

equations

$$x = x(s), \qquad y = y(s), \qquad a < s < b.$$

Determine a solution $u = f(x, y)$ *of Eq. (6.5–6) that also satisfies*

$$u[x(s), y(s)] = \phi(s) \qquad \text{and} \qquad \frac{\partial u[x(s), y(s)]}{\partial n} = \psi(s)$$

in some neighborhood of the curve C_0. *The functions* $x(s)$, $y(s)$, $\phi(s)$, *and* $\psi(s)$ *are called the* ***Cauchy*** *or* ***initial data****, and* C_0 *is called the* ***initial curve****.*

Stated more succinctly, the Cauchy problem is the following: Given a smooth arc C_0 in the space where (x, y, u) are rectangular cartesian coordinates, does there exist a unique solution of Eq. (6.5–6) for which u_x and u_y satisfy prescribed values on C_0? Conditions for which the Cauchy problem can be solved were first given by Cauchy and then in more complete form by Kovalevski* in 1875. The following theorem is a special case of the more general one called the **Cauchy–Kovalevski Theorem.**

THEOREM 6.5–2

Let the coefficients $A, B, \ldots, G$ *of Eq. (6.5–6) be analytic in a region* R *of the xy-plane containing the origin with* $C(x, y) \neq 0$ *in* R. *Furthermore, let* $\phi(x)$ *and* $\psi(x)$ *be arbitrary analytic functions on the segment of the x-axis in* R. *Then there exists a neighborhood* R_0 *of the origin and a unique analytic solution* $u = f(x, y)$ *of Eq. (6.5–6) in* R_0 *such that*

$$f(x, 0) = \phi(x) \qquad \text{and} \qquad \frac{\partial f(x, 0)}{\partial y} = \psi(x)$$

on the segment of the x-axis contained in R_0.

We observe that Theorem 6.5–2 is a *local* theorem, that is, existence and uniqueness are guaranteed only in the neighborhood of a point (in this case the origin). The requirement that the coefficient C in Eq. (6.5–6) be nonzero in R ensures that the initial curve (in this case a portion of the x-axis) is not a characteristic curve (see Section 3.5). If Eq. (6.5–6) is elliptic in R, this does not present a problem, since elliptic equations have no real characteristics. In the case of hyperbolic and parabolic equations, unique solutions are guaranteed only in the neighborhoods of points of the initial curve at which the tangent does not coincide with the direction of a characteristic.

An excellent discussion of the Cauchy problem may be found in Rene Dennemeyer's *Introduction to Partial Differential Equations and Boundary Value Problems* (New York: McGraw-Hill, 1968), Chapter 2.

---

*Sonya Kovalevski (1850–1891) was a native of Russia, received her mathematical education in Germany, and taught in Sweden.

Our final example concerns the verification of a solution to Laplace's equation.

EXAMPLE 6.5–3 Given the problem of Example 6.1–3, namely,

$$\begin{aligned}
&\text{P.D.E.:} && u_{xx} + u_{yy} = 0, \quad 0 < x < a, \quad 0 < y < b;\\
&\text{B.C.:} && \left.\begin{aligned} u_x(0, y) &= 0,\\ u_x(a, y) &= 0,\end{aligned}\right\} \quad 0 < y < b,\\
& && \left.\begin{aligned} u(x, b) &= 0,\\ u(x, 0) &= \sin(\pi x/a),\end{aligned}\right\} \quad 0 < x < a,
\end{aligned}$$

verify the solution

$$u(x, y) = \frac{2}{\pi b}(b - y) + \frac{4}{\pi}\sum_{n=1} \frac{\cos(2n\pi x/a)\sinh[2n\pi(b - y)/a]}{(1 - 4n^2)\sinh(2n\pi b/a)}. \tag{6.5–7}$$

Solution To simplify the notation, we put $\omega_n = 2n\pi/a$. Then

$$u_x = \frac{4}{\pi}\sum_{n=1} \frac{\omega_n \sin \omega_n x \sinh \omega_n(b - y)}{(4n^2 - 1)\sinh \omega_n b}, \tag{6.5–8}$$

and consider the term

$$\frac{\sinh \omega_n(b - y)}{\sinh \omega_n b}, \tag{6.5–9}$$

which varies between zero and one inclusive for all n. Further analysis shows that (Exercise 5)

$$\sinh \omega_n(b - y) < \frac{1}{2}\exp[\omega_n(b - y)],$$

whereas (Exercise 6)

$$\sinh \omega_n b \geq \frac{1}{2}[1 - \exp(-2b)]\exp(\omega_n b).$$

Hence

$$\frac{\sinh \omega_n(b - y)}{\sinh \omega_n b} \leq \frac{\exp(-\omega_n y)}{1 - \exp(-2b)}.$$

Thus for constant y_0 such that $0 < y_0 < b$ the series of positive constants

$$\sum_{n=1} \frac{n \exp(-\omega_n y_0)}{1 - \exp(-2b)}$$

converges (uniformly) by the ratio test (Exercise 7). Hence by Abel's test (see below) the series (6.5–8) is uniformly convergent for all x in $(0, a)$ and all y in

$(0, b)$. This establishes not only the validity of the series expression for u_x, but also, by Theorem 1.6–2, the uniform convergence of $u(x, y)$ in Eq. (6.5–7) for y in $(0, b)$.

> **Abel's Test.*** If $\sum \phi_n(x)$ converges uniformly for $a \leq x \leq b$, if for every fixed value of x the sequence $\{\psi_n(x)\}$ is real and monotone (that is, either nonincreasing or nondecreasing), and if for every n and every x in $[a, b]$ the functions $|\psi_n(x)| < M$ for some number M, then the series $\sum \phi_n(x)\psi_n(x)$ is uniformly convergent for x in $[a, b]$.

In a similar manner it can be shown that the series representing u_{xx} converges uniformly in x for each y in $(0, b)$ (Exercise 8). Moreover, u_{yy} converges uniformly in y for each x in $(0, a)$, since u_{yy} also contains the term (6.5–9).

It remains to examine the partial derivative with respect to y. We have

$$u_y = -\frac{2}{\pi b} + \frac{4}{\pi} \sum_{n=1} \frac{\omega_n \cos \omega_n x \cosh \omega_n(b - y)}{(4n^2 - 1) \sinh \omega_n b}, \tag{6.5–10}$$

with ω_n as before. Now we have (Exercise 9)

$$\frac{\cosh \omega_n(b - y)}{\sinh \omega_n b} \leq \frac{2 \exp(-\omega_n y)}{1 - \exp(-2b)},$$

and by the same argument used previously the series in Eq. (6.5–10) is also uniformly convergent for $0 < x < a$ and $0 < y < b$. Thus the solution (6.5–7) may be differentiated term by term, and we can verify that it is a harmonic function in the open rectangular region. ■

In this section we have given an indication of what is involved in verifying the solution to a boundary-value problem. In Chapter 7 we will continue to obtain formal solutions to our problems so that we do not lose sight of our main objective.

We also discussed briefly the topics of existence of solutions and their continuous dependence on the initial data. The remaining topic of uniqueness can be found in a number of texts that are listed at the end of this chapter.

Key Words and Phrases

well-posed problem
Cauchy problem
Cauchy data
initial curve
Cauchy–Kovalevski theorem
Abel's test

*After Niels Henrik Abel (1802–1829), a Norwegian mathematician.

6.5 Exercises

• **1.** Explain why it is not sufficient to say that the boundary conditions in Example 6.5–1 are satisfied simply because $\sin(n\pi x/L)$ is zero when $x = 0$ and $x = L$.

2. Apply the ratio test to the series of positive constants

$$\sum_{n=1} n^2 \exp\left(-\frac{kn^2\pi^2}{L^2}t_0\right)$$

to show that the series converges for all t_0 such that $0 < t_0 < \infty$.

3. **(a)** Using an argument similar to the one in the text in Example 6.5–1, show that u_x is defined by a uniformly convergent series.
(b) Repeat part (a) for u_{xx}.

4. In Example 6.5–2, verify that the substitution (6.5–5) produces the result shown.

5. Show that

$$\sinh \omega_n(b - y) < \frac{1}{2}\exp[\omega_n(b - y)]$$

in Example 6.5–3.

6. Verify that

$$\sinh \omega_n b \geq \frac{1}{2}[1 - \exp(-2b)]\exp(\omega_n b)$$

in Example 6.5–3.

7. Apply the ratio test to the series

$$\sum_{n=1} \frac{n\exp(-\omega_n y_0)}{1 - \exp(-2b)},$$

where y_0 is in the interval $(0, b)$ and $\omega_n = 2n\pi/a$ to show that the series converges.

8. Use an argument similar to the one in the text to show that the series representing u_{xx} converges uniformly in x for each y in $(0, b)$ in Example 6.5–3.

9. Verify that

$$\frac{\cosh \omega_n(b - y)}{\sinh \omega_n b} \leq \frac{2\exp(-\omega_n y)}{1 - \exp(-2b)}.$$

10. Show that the solution in Eq. (6.5–7) represents a harmonic function in the open region $0 < x < a$, $0 < y < b$.

•• **11.** Given the problem of Example 3.2–2, namely,

$$\begin{aligned}
\text{P.D.E.:}\quad & u_{xx} + u_{yy} = 0, \quad 0 < x < \pi, \quad 0 < y < b;\\
\text{B.C.:}\quad & u(0, y) = u(\pi, y) = 0, \quad 0 < y < b,\\
& u(x, b) = 0, \quad u(x, 0) = f(x), \quad 0 < x < \pi,
\end{aligned}$$

verify the solution

$$u(x, y) = \frac{2}{\pi}\sum_{n=1} \frac{\sin nx \sinh n(b - y)}{\sinh nb}\int_0^\pi f(s)\sin ns\,ds.$$

Assume that $f(x)$ is *continuous* and that $f'(x)$ is piecewise continuous in $(0, \pi)$ and that $f(0) = f(\pi) = 0$.

12. If in the problem of Exercise 11 we let $f(x) = 100$, $0 < x < \pi$, the solution as given in Eq. (3.2–15) is

$$u(x, y) = \frac{400}{\pi} \sum_{m=1} \frac{\sin (2m - 1)x \sinh (2m - 1)(b - y)}{(2m - 1) \sinh (2m - 1)b}.$$

Verify this solution to the extent possible.

13. **(a)** Given the problem of Example 3.4–1, namely,

$$\begin{aligned} \text{P.D.E.:} \quad & u_t = ku_{xx}, \quad 0 < x < L, \quad t > 0, \quad k > 0; \\ \text{B.C.:} \quad & u(0, t) = u(L, t) = 0, \quad t > 0; \\ \text{I.C.:} \quad & u(x, 0) = f(x), \quad 0 < x < L, \end{aligned}$$

verify the solution

$$u(x, t) = \frac{2}{L} \sum_{n=1} \sin \frac{n\pi}{L} x \exp(-kn^2\pi^2 t/L^2) \int_0^L f(s) \sin \frac{n\pi}{L} s \, ds.$$

(b) Under what conditions is the problem well-posed?

14. The case of an infinite string acted upon by an external force and subject to initial conditions leads to the following problem.

$$\begin{aligned} \text{P.D.E.:} \quad & u_{tt} = a^2 u_{xx} + \phi(x, t), \quad -\infty < x < \infty, \quad t > 0; \\ \text{I.C.:} \quad & u(x, 0) = f(x), \quad u_t(x, 0) = g(x), \quad -\infty < x < \infty. \end{aligned}$$

The solution of this problem is given by

$$u(x, t) = \frac{1}{2}[f(x + at) + f(x - at)] + \frac{1}{2a} \int_{x-at}^{x+at} g(\xi)\, d\xi$$

$$+ \frac{1}{2a} \int_0^t d\tau \int_{x-a(t-\tau)}^{x+a(t-\tau)} \phi(\xi, \tau)\, d\xi.$$

Verify this solution.

15. At points in the xy-plane such that $(x, y) \neq (x_0, y_0)$ the function

$$u(x, y) = \log \frac{1}{\sqrt{(x - x_0)^2 + (y - y_0)^2}}$$

is called the *fundamental solution* of Laplace's equation in the plane. Verify this solution.

••• **16.** Investigate the uniform convergence of the improper integral

$$\int_0^\infty e^{-xt}\, dt, \qquad x > 0.$$

17. Show that the improper integral

$$\int_0^\infty e^{-t} \cos xt \, dt$$

converges for all x.

18. The *error function* erf x is defined as follows:

$$\operatorname{erf} x = \frac{2}{\sqrt{\pi}} \int_0^x e^{-\xi^2}\, d\xi.$$

Prove the following properties of this function.

(a) $\operatorname{erf} \infty = 1$

(b) $|\operatorname{erf} x| \leq 1$

(c) $\operatorname{erf}(-x) = -\operatorname{erf} x$

(d) $\operatorname{erf} x = \dfrac{2}{\sqrt{\pi}} \displaystyle\sum_{n=0} \frac{(-1)^n x^{2n+1}}{(2n+1)n!}$

(*Hint*: Expand the integrand in the definition as a Maclaurin series.)

19. Prove that if a function is absolutely integrable on an interval, then it is bounded on the interval.

20. Consider the function

$$y(\alpha, t) = \alpha e^{-\alpha^2 t}, \qquad t > 0.$$

(a) Show that $y(\alpha, t)$ is absolutely integrable for $0 < \alpha < \infty$.

(b) Show that $y(\alpha, t)$ is bounded on $0 < \alpha < \infty$.

(c) Graph $y(\alpha, t)$ for a constant value of t, say, $t = 1$.

21. The following example of an ill-posed problem is due to Hadamard. The problem

$$\text{P.D.E.:} \quad u_{xx} + u_{yy} = 0, \qquad -\infty < x < \infty, \qquad y > 0;$$

$$\text{B.C.:} \quad \left.\begin{aligned} u(x, 0) &= 0, \\ u_y(x, 0) &= \frac{1}{n}\sin nx, \end{aligned}\right\} \quad -\infty < x < \infty,$$

has solution $u \equiv 0$ when $u_y(x, 0) = 0$, but for positive values of n the solution is

$$u(x, y) = \frac{1}{n^2} \sin nx \sinh ny.$$

(a) Verify the above solutions.

(b) Explain why the problem is not well-posed.

22. Consider the following problem.

$$\text{P.D.E.:} \quad u_t = u_{xx}, \qquad -\infty < x < \infty, \qquad t > 0;$$

$$\text{I.C.:} \quad u(x, 0) = \begin{cases} u_0, & |x| < 1 \\ 0, & |x| > 1, \end{cases}$$

where u_0 is a constant.

(a) Show that the solution can be written as

$$\begin{aligned} u(x, t) &= \frac{u_0}{\sqrt{\pi}} \int_{-(1+x)/2\sqrt{t}}^{(1-x)/2\sqrt{t}} \exp(-\xi^2)\, d\xi \\ &= \frac{u_0}{2}\left[\operatorname{erf} \frac{1-x}{2\sqrt{t}} + \operatorname{erf} \frac{1+x}{2\sqrt{t}}\right]. \end{aligned}$$

(b) Verify the solution.

(c) Compute $u(1, t)$.

(d) Determine whether or not the problem is well-posed.

REFERENCES

Carrier, G. F., and C. E. Pearson, *Partial Differential Equations: Theory and Practice*. New York: Academic Press, 1976.

Churchill, R. V., and J. W. Brown, *Fourier Series and Boundary Value Problems*, 3d ed. New York: McGraw-Hill, 1978.

Copson, E. T., *Partial Differential Equations*. Cambridge: Cambridge University Press, 1975.

Dennemeyer, Rene, *Introduction to Partial Differential Equations and Boundary Value Problems*. New York: McGraw-Hill, 1968.

Friedman, Avner, *Partial Differential Equations of Parabolic Type*. Englewood Cliffs, N. J.: Prentice-Hall, 1964.

Sagan, Hans, *Boundary and Eigenvalue Problems in Mathematical Physics*. New York: John Wiley, 1961.

Tychonov, A. N., and A. A. Samarski, *Partial Differential Equations of Mathematical Physics*, Vol. I. San Francisco: Holden-Day, 1964.

Weinberger, H. F., *A First Course in Partial Differential Equations with Complex Variables and Transform Methods*. Waltham, Mass.: Blaisdell, 1965.

Williams, W. E., *Partial Differential Equations*. Oxford: Clarendon Press, 1980.

Zachmanoglou, E. C., and D. W. Thoe, *Introduction to Partial Differential Equations with Applications*. Baltimore: Williams and Wilkins, 1976.

7 Boundary-Value Problems in Other Coordinate Systems

7.1 POLAR COORDINATES

Whenever a region has circular symmetry, there is usually an advantage in using **polar coordinates**. The relationship between rectangular and polar coordinates is shown in Fig. 7.1-1 and the equations accompanying the figure.

From these relations we can compute the partial derivatives:

$$\frac{\partial x}{\partial \rho} = \cos \phi, \qquad \frac{\partial y}{\partial \rho} = \sin \phi,$$

$$\frac{\partial x}{\partial \phi} = -\rho \sin \phi, \qquad \frac{\partial y}{\partial \phi} = \rho \cos \phi.$$

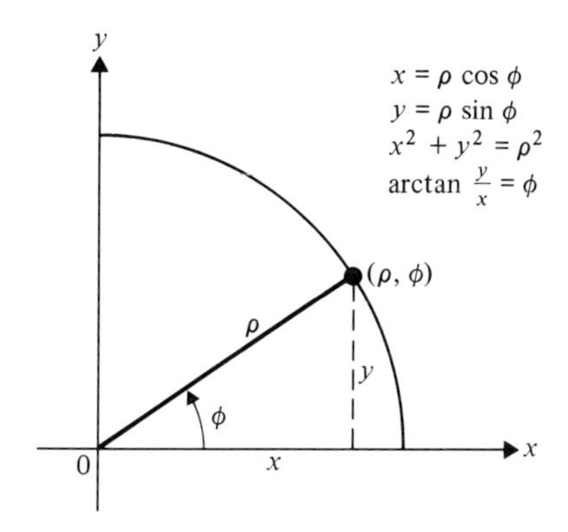

Figure 7.1-1
Polar Coordinates.

If $u(x, y)$ is a twice-differentiable function of x and y, then by the chain rule,

$$\frac{\partial u}{\partial \rho} = \frac{\partial u}{\partial x}\frac{\partial x}{\partial \rho} + \frac{\partial u}{\partial y}\frac{\partial y}{\partial \rho} = \frac{\partial u}{\partial x}\cos\phi + \frac{\partial u}{\partial y}\sin\phi. \tag{7.1-1}$$

Hence

$$\frac{\partial^2 u}{\partial \rho^2} = \frac{\partial}{\partial \rho}\left(\frac{\partial u}{\partial x}\right)\cos\phi + \frac{\partial u}{\partial x}\frac{\partial}{\partial \rho}(\cos\phi) + \frac{\partial}{\partial \rho}\left(\frac{\partial u}{\partial y}\right)\sin\phi + \frac{\partial u}{\partial y}\frac{\partial}{\partial \rho}(\sin\phi)$$

$$= \frac{\partial}{\partial \rho}\left(\frac{\partial u}{\partial x}\right)\cos\phi + \frac{\partial}{\partial \rho}\left(\frac{\partial u}{\partial y}\right)\sin\phi,$$

since the other two terms are zero.

In order to evaluate a term like $\partial/\partial\rho\,(\partial u/\partial x)$ we observe that Eq. (7.1-1) applies not only to $u(x, y)$ but to *any* differentiable function of x and y. We can, in fact, write Eq. (7.1-1) symbolically as

$$\frac{\partial(\ )}{\partial \rho} = \frac{\partial(\ )}{\partial x}\cos\phi + \frac{\partial(\ )}{\partial y}\sin\phi.$$

Thus

$$\frac{\partial u_x}{\partial \rho} = u_{xx}\cos\phi + u_{xy}\sin\phi,$$

and

$$\frac{\partial u_y}{\partial \rho} = u_{yx}\cos\phi + u_{yy}\sin\phi.$$

Therefore

$$\frac{\partial^2 u}{\partial \rho^2} = u_{xx}\cos^2\phi + 2u_{xy}\sin\phi\cos\phi + u_{yy}\sin^2\phi,$$

assuming that $u_{xy} = u_{yx}$, which is the case for the functions with which we will be dealing.

In a similar manner we find

$$\frac{1}{\rho^2}\frac{\partial^2 u}{\partial \phi^2} = u_{xx}\sin^2\phi - 2u_{xy}\sin\phi\cos\phi + u_{yy}\cos^2\phi$$

$$- \frac{1}{\rho}u_x\cos\phi - \frac{1}{\rho}u_y\sin\phi,$$

and since

$$\frac{1}{\rho}\frac{\partial u}{\partial \rho} = \frac{1}{\rho}u_x\cos\phi + \frac{1}{\rho}u_y\sin\phi,$$

we have

$$\frac{\partial^2 u}{\partial \rho^2} + \frac{1}{\rho^2}\frac{\partial^2 u}{\partial \phi^2} + \frac{1}{\rho}\frac{\partial u}{\partial \rho} = \frac{\partial^2 u}{\partial x^2} + \frac{\partial^2 u}{\partial y^2}. \tag{7.1-2}$$

In Eq. (7.1–2) we have the Laplacian in two dimensions in both polar and rectangular coordinates. If u is *independent of* ϕ, then we have the simpler result

$$\frac{\partial^2 u}{\partial x^2} + \frac{\partial^2 u}{\partial y^2} = \frac{\partial^2 u}{\partial \rho^2} + \frac{1}{\rho}\frac{\partial u}{\partial \rho} = \frac{1}{\rho}\frac{\partial}{\partial \rho}\left(\rho\frac{\partial u}{\partial \rho}\right). \qquad \textbf{(7.1–3)}$$

Observe that Eqs. (7.1–2) and (7.1–3) must be dimensionally correct. For example,* if $[u] = T$ and $[x] = [y] = L$, then $[u_{xx}] = [u_{yy}] = T/L^2$. Hence each term in the polar coordinate versions of the Laplacian must also have the same dimension. Recalling that angles are expressed in dimensionless radians will now help us to remember Eq. (7.1–2).

EXAMPLE 7.1–1 Find the steady-state, bounded temperatures in a circular metallic disk of radius c if the temperatures on the edge are given by $f(\phi)$. See Fig. 7.1–2.

Solution We first observe that $f(\phi)$ is assumed to be a piecewise smooth function. We also note that $u(\rho, -\pi) = u(\rho, \pi)$ and $u_\phi(\rho, -\pi) = u_\phi(\rho, \pi)$, these restrictions being necessary so that the temperatures will be *uniquely* determined. Then we have the following problem:

P.D.E.: $\dfrac{1}{\rho}\dfrac{\partial}{\partial \rho}(\rho u_\rho) + \dfrac{1}{\rho^2}u_{\phi\phi} = 0, \qquad 0 < \rho < c, \qquad -\pi < \phi \le \pi;$

B.C.: $u(c, \phi) = f(\phi), \qquad -\pi < \phi \le \pi.$

It appears that we do not have enough boundary conditions to solve the problem, but the fact that the temperatures must be *bounded* will provide additional information. We use the method of separation of variables and assume

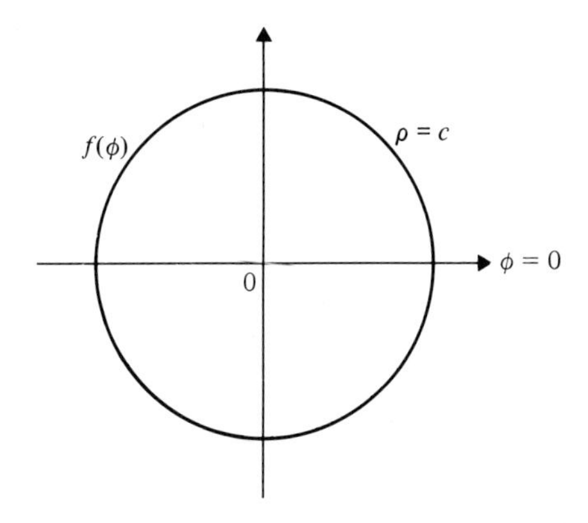

Figure 7.1–2
Temperatures in a disk.

*The square bracket is read "has the dimension of." Standard usage gives T (temperature), t (time), L (length), m (mass), etc.

we can write

$$u(\rho, \phi) = R(\rho)\, \Phi(\phi).$$

Then

$$\frac{\Phi}{\rho} \frac{d}{d\rho}\left(\rho \frac{dR}{d\rho}\right) + \frac{R}{\rho^2} \frac{d^2\Phi}{d\phi^2} = 0,$$

or dividing by $R\Phi/\rho^2$, we have

$$\frac{\rho}{R} \frac{d}{d\rho}\left(\rho \frac{dR}{d\rho}\right) = -\frac{1}{\Phi} \frac{d^2\Phi}{d\phi^2} = n^2, \qquad n = 0, 1, 2, \ldots.$$

Choosing the separation constant as we have will insure that Φ [and $u(\rho, \phi)$] will be periodic of period 2π in ϕ. This choice is dictated by the physical nature of the problem.

Thus the equations

$$\frac{d^2\Phi}{d\phi^2} + n^2\Phi = 0, \qquad n = 0, 1, 2, \ldots,$$

have solutions (Exercise 1)

$$\Phi_n(\phi) = a_n \cos n\phi + b_n \sin n\phi.$$

The second ordinary differential equation can be written as

$$\rho^2 \frac{d^2R_n}{d\rho^2} + \rho \frac{dR_n}{d\rho} - n^2R_n = 0. \qquad \textbf{(7.1-4)}$$

We first take the case $n = 0$ and solve the resulting equation by reduction of order* to obtain (Exercise 2)

$$R_0(\rho) = c_1 \log \rho + c_2.$$

In order that this solution remain bounded in the neighborhood of $\rho = 0$ we must take $c_1 = 0$. Hence the solution corresponding to $n = 0$ is a constant that we may take to be unity.

Equations (7.1-4) are Cauchy-Euler equations (see Section 1.3) whose solutions are (Exercise 3)

$$R_n(\rho) = A_n\rho^n + B_n\rho^{-n},$$
$$n = 1, 2, \ldots,$$

and we must have $B_n = 0$ so that $R_n(\rho)$ [and $u(\rho, \phi)$] will remain bounded at $\rho = \epsilon > 0$, that is, in the neighborhood of $\rho = 0$. Without loss of generality we can also take $A_n = 1$. Then

$$u_n(\rho, \phi) = (a_n \cos n\phi + b_n \sin n\phi)\rho^n$$

*Let $v_n = dR_n/d\rho$, divide the result by ρ, and obtain a first-order differential equation that can be integrated, since it is the derivative of a product.

are bounded functions that satisfy the given partial differential equation. To satisfy the boundary condition, we take

$$u(\rho, \phi) = \frac{1}{2}a_0 + \sum_{n=1} (a_n \cos n\phi + b_n \sin n\phi)\rho^n \qquad \textbf{(7.1-5)}$$

and leave it as an exercise (Exercise 4) to show that the solution to the problem is given by Eq. (7.1-5) with the constants a_n and b_n defined as follows:

$$\begin{aligned} a_n &= \frac{1}{\pi c^n}\int_{-\pi}^{\pi} f(s) \cos ns\, ds, \qquad n = 0, 1, 2, \ldots, \\ b_n &= \frac{1}{\pi c^n}\int_{-\pi}^{\pi} f(s) \sin ns\, ds, \qquad n = 1, 2, \ldots. \end{aligned} \qquad \textbf{(7.1-6)}$$

Since the function $f(\phi)$ is being represented by a Fourier series, this function must satisfy the conditions necessary for such representation (see Section 4.5). ■

We observe that $\rho = 0$ is a regular singular point of the Cauchy–Euler equations (7.1–4). Hence a natural question that might be asked is, "What is the value of the solution in Example 7.1–1 when $\rho = 0$ and is this result reasonable?" From Eq. (7.1–5) we have

$$\lim_{\rho \to 0} u(\rho, \phi) = \frac{1}{2}a_0,$$

since the coefficients a_n and b_n are bounded. But we also have from Eq. (7.1–6)

$$\frac{1}{2}a_0 = \frac{1}{2\pi}\int_{-\pi}^{\pi} f(s)\, ds,$$

which is the *mean value* of the function f on the boundary $\rho = c$. In other words, the value of the harmonic function at the center of the circle is the average or mean of its values on the circumference. We can confirm this result in another way and, at the same time, obtain a form more suitable for verification of the solution.

If we substitute the coefficients a_n and b_n as given by Eqs. (7.1–6) into Eq. (7.1–5), we have (Exercise 5)

$$u(\rho, \phi) = \frac{1}{\pi}\int_{-\pi}^{\pi} f(s)\left[\frac{1}{2} + \sum_{n=1} \left(\frac{\rho}{c}\right)^n \cos n(\phi - s)\right] ds. \qquad \textbf{(7.1-7)}$$

Now let $(\rho/c) = r$, $\phi - s = \theta$, and consider

$$\begin{aligned} \sum_{n=1} r^n \cos n\theta &= \frac{1}{2}\sum_{n=1} [(re^{i\theta})^n + (re^{-i\theta})^n] \\ &= \frac{1}{2}\left[\frac{1}{1 - re^{i\theta}} + \frac{1}{1 - re^{-i\theta}} - 2\right], \end{aligned}$$

provided that $|r \exp(\pm i\theta)| < 1$, which is the case here. With some algebraic manipulation (Exercise 6) the last expression can be written as

$$\frac{1}{2}\,\frac{2r\cos\theta - 2r^2}{1 - 2r\cos\theta + r^2}$$

so that

$$u(\rho, \phi) = \frac{1}{2\pi}\int_{-\pi}^{\pi} \frac{(c^2 - \rho^2)f(s)}{c^2 - 2\rho c\cos(\phi - s) + \rho^2}\,ds, \qquad \rho < c. \tag{7.1-8}$$

This form of the solution is called **Poisson's integral**. The Poisson integral is the solution of the Dirichlet problem over the interior of a circular region. (See also Exercise 24.) Although the Poisson integral gives the solution in closed form, there may be formidable difficulties in evaluating it except in special cases (Exercise 12).

The study of harmonic functions in circular regions can produce other useful properties. We start with a version of the divergence theorem known as Green's theorem,* namely,

$$\iint_S \left(f\frac{\partial g}{\partial n} - g\frac{\partial f}{\partial n}\right) dS = \iiint_V (f\nabla^2 g - g\nabla^2 f)\,dV,$$

which holds for twice-differentiable functions f and g in a region V and on its boundary S. If f is harmonic in V and $g = 1$ there and on S, then

$$\iint_S \frac{\partial f}{\partial n}\,dS = 0. \tag{7.1-9}$$

In the plane, Eq. (7.1-9) reduces to

$$\int_C \frac{\partial f}{\partial n}\,ds = 0. \tag{7.1-10}$$

Hence a *necessary* condition for the existence of a solution to the Neumann problem

P.D.E.: $\frac{1}{\rho}(\rho u_\rho)_\rho + \frac{1}{\rho^2}u_{\phi\phi} = 0, \quad 0 < \rho < c, \quad -\pi < \phi \le \pi;$

B.C.: $u_\rho(c, \phi) = f(\phi), \quad -\pi < \phi \le \pi,$

is that

$$\int_{-\pi}^{\pi} f(\phi)\,d\phi = 0,$$

that is, that the mean value of the normal derivative on the boundary be zero.

*After George Green (1793–1841), an English mathematician.

If C is a circle of radius r with center at (x_0, y_0), then Eq. (7.1–10) can be written (Fig. 7.1–3)

$$\int_C \frac{\partial}{\partial r}[f(x_0 + r\cos\theta, y_0 + r\sin\theta)]\, r\, d\theta = 0,$$

or

$$\frac{\partial}{\partial r}\int_{-\pi}^{\pi} f(x_0 + r\cos\theta, y_0 + r\sin\theta)\, d\theta = 0.$$

Thus the integral is independent of r and has the same value for *any* r, in particular for $r = 0$. For this value of r the integral has the value $2\pi f(x_0, y_0)$; hence it follows that

$$f(x_0, y_0) = \frac{1}{2\pi}\int_{-\pi}^{\pi} f(x_0 + r\cos\theta, y_0 + r\sin\theta)\, d\theta.$$

We have proved the following theorem.

THEOREM 7.1–1
Let R be a circular region of radius r centered at (x_0, y_0) in which f is harmonic. Then f has the ***mean-value property***

$$f(x_0, y_0) = \frac{1}{2\pi}\int_{-\pi}^{\pi} f(x_0 + r\cos\theta, y_0 + r\sin\theta)\, d\theta. \tag{7.1–11}$$

The theorem can be readily extended to higher dimensions.

In the next example we consider a problem in three dimensions, that is, we use **cylindrical coordinates.**

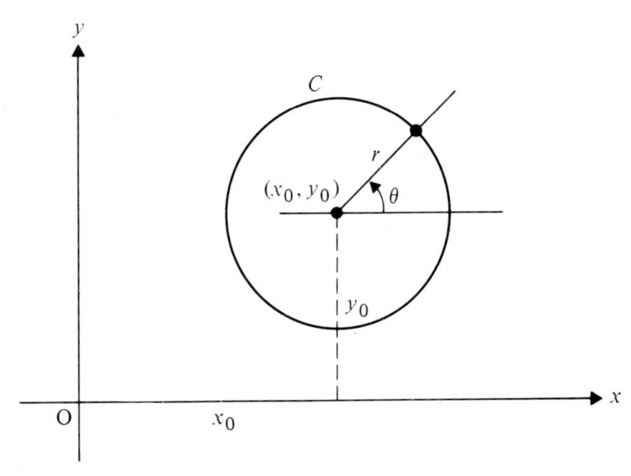

Figure 7.1–3

EXAMPLE 7.1–2 Solve the following boundary-value problem.

P.D.E.: $\nabla^2 u = 0$ in cylindrical coordinates, $b < \rho < c$, $-\pi < \phi \le \pi$, $-\infty < z < \infty$;

B.C.: $u(b, \phi, z) = f(\phi)$, $-\pi < \phi \le \pi$, $-\infty < z < \infty$,
$u(c, \phi, z) = 0$, $-\pi < \phi \le \pi$, $-\infty < z < \infty$.

Solution The boundary conditions indicate that u is independent of z; consequently, this three-dimensional problem reduces to a two-dimensional one (Fig. 7.1–4). Using separation of variables, we obtain the following ordinary differential equations and boundary conditions (compare Example 7.1–1):

$$\Phi'' + n^2\Phi = 0, \quad \Phi(-\pi) = \Phi(\pi), \quad \Phi'(-\pi) = \Phi'(\pi), \quad n = 0, 1, 2, \ldots;$$
$$\rho^2 R_n'' + \rho R_n' - n^2 R_n = 0, \quad R_n(c) = 0.$$

The solutions of these equations can be written as (Exercises 9 and 10)

$$\Phi(n\phi) = A_n \cos n\phi + B_n \sin n\phi$$

and

$$R_n(\rho) = (c/\rho)^n - (\rho/c)^n, \quad n = 1, 2, \ldots,$$

respectively. The case $n = 0$ must be taken separately (Why?), and results in $\Phi =$ constant and $R = \log(\rho/c)$.

Forming a linear combination of the products, we have

$$u(\rho, \phi) = A_0 \log(\rho/c) + \sum_{n=1} [(c/\rho)^n - (\rho/c)^n](A_n \cos n\phi + B_n \sin n\phi).$$

The nonhomogeneous boundary condition now produces

$$u(b, \phi) = f(\phi) = A_0 \log(b/c) + \sum_{n=1} [(c/b)^n - (b/c)^n](A_n \cos n\phi + B_n \sin n\phi).$$

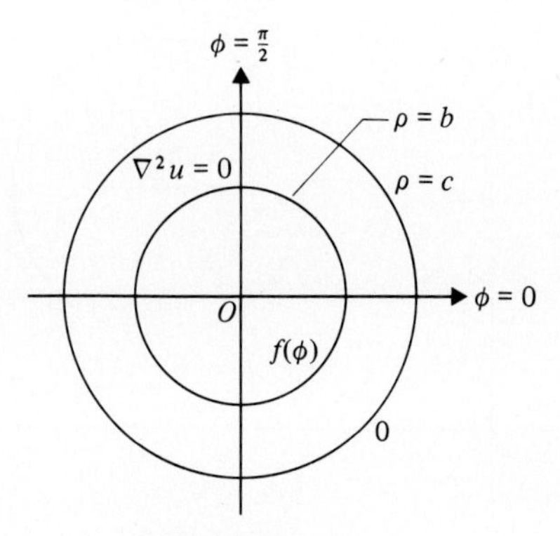

Figure 7.1–4

This is a Fourier series representation of $f(\phi)$; hence, making the substitutions,

$$\frac{1}{2}a_0 = A_0 \log(b/c), \qquad a_n = [(c/b)^n - (b/c)^n]A_n,$$
$$b_n = [(c/b)^n - (b/c)^n]B_n,$$

we have the final result,

$$u(\rho, \phi) = \frac{\log(\rho/c)}{2\log(b/c)}a_0 + \sum_{n=1}^{\infty} \frac{(c/\rho)^n - (\rho/c)^n}{(c/b)^n - (b/c)^n}(a_n \cos n\phi + b_n \sin n\phi), \tag{7.1-12}$$

where a_0, a_n, and b_n are the Fourier coefficients in the expansion of $f(\phi)$. ■

Before we can consider the solution of the wave and diffusion equations in polar and cylindrical coordinates, we will need to have a more complete knowledge of Bessel functions. Accordingly, we take up this topic in the next section.

Key Words and Phrases

polar coordinates
Poisson's integral
mean-value property
cylindrical coordinates
periodic boundary conditions

7.1 Exercises

• **1.** **(a)** In Example 7.1–1, explain in physical terms why $u(\rho, \phi)$ must be periodic of period 2π in ϕ.
(b) Explain why the case $n = 0$ was included in Example 7.1–1.

2. Solve the equation

$$\rho^2 \frac{d^2R}{d\rho^2} + \rho \frac{dR}{d\rho} = 0$$

by reduction of order. (Compare Example 7.1–1.)

3. Solve the Cauchy–Euler equations.

$$\rho^2 \frac{d^2R_n}{d\rho^2} + \rho \frac{dR_n}{d\rho} - n^2R_n = 0, \qquad n = 1, 2, \ldots.$$

(Compare Example 7.1–1.)

4. Supply the details in Example 7.1–1 to show that the solution is given by Eq. (7.1–5) with the constants defined as in Eq. (7.1–6).

5. Obtain Eq. (7.1–7).

6. Show that

$$\frac{1}{1 - re^{i\theta}} + \frac{1}{1 - re^{-i\theta}} - 2 = \frac{2r\cos\theta - 2r^2}{1 - 2r\cos\theta + r^2}.$$

7. Give a physical interpretation of the problem in Example 7.1–2.

8. Explain why the boundary conditions imply that u is independent of z in Example 7.1–2. Is this consistent with the physical situation?

9. In Example 7.1–2, show that the solutions of

$$\Phi'' + n^2\Phi = 0, \qquad \Phi(-\pi) = \Phi(\pi),$$
$$\Phi'(-\pi) = \Phi'(\pi), \qquad n = 0, 1, 2, \ldots,$$

are

$$\Phi(n\phi) = A_n \cos n\phi + B_n \sin n\phi.$$

[*Note*: The boundary conditions here are called **periodic boundary conditions** (see Section 2.7).]

10. In Example 7.1–2, show that the solutions of

$$\rho^2 R_n'' + \rho R_n' - n^2 R_n = 0, \qquad R_n(c) = 0,$$

are

$$R_n(\rho) = (c/\rho)^n - (\rho/c)^n, \qquad n = 1, 2, \ldots,$$

and

$$R_0(\rho) = \log(\rho/c).$$

•• 11. In Example 7.1–1, obtain the solution if $f(\phi) = u_0$, a constant. Does your result agree with the physical facts? Explain.

12. **(a)** In Example 7.1–1, obtain the solution if $f(\phi)$ is given by

$$f(\phi) = \begin{cases} 0, & \text{for } -\pi < \phi < 0, \\ 100, & \text{for } 0 < \phi < \pi. \end{cases}$$

(b) Compute $u(c, 0)$, $u(c, \pi/2)$, $u(c, \pi)$, $u(c, -\pi)$, and $u(0, 0)$. (*Hint*: Use the Poisson integral form.)

13. Change Example 7.1–2 so that the outer surface is insulated, and solve the resulting problem. What restrictions must be placed on $f(\phi)$?

14. Find the harmonic function in the region $1 < \rho < c$, $0 < \phi < \pi$, if $u = u_0$ when $\rho = c$ and all other boundaries are kept at zero (Fig. 7.1–5).

15. Find the steady-state temperatures in the quadrant $1 < \rho < c$, $0 < \phi < \pi/2$, if the boundaries $\phi = 0$ and $\phi = \pi/2$ are kept at zero, the boundary $\rho = c$ is insulated, and the remaining boundary is kept at temperature u_0, a constant (Fig. 7.1–6).

16. Find the steady-state temperatures in the region $1 < \rho < c$, $0 < \phi < \pi/2$, if the temperatures of the boundaries $\rho = 1$ and $\rho = c$ are kept at zero and $f(\phi)$, respectively, and the remaining two boundaries are insulated (Fig. 7.1–7).

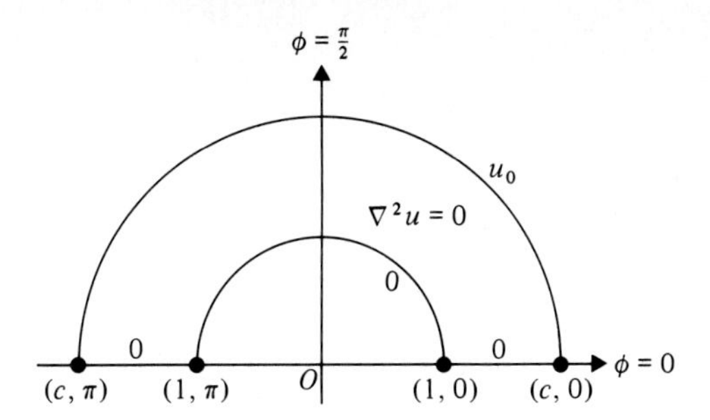

Figure 7.1-5

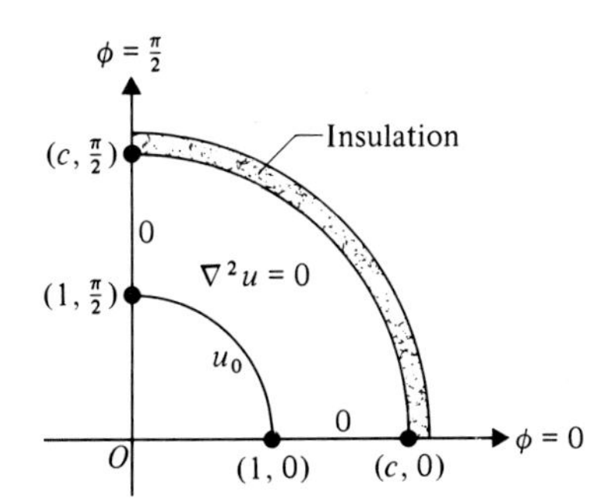

Figure 7.1-6

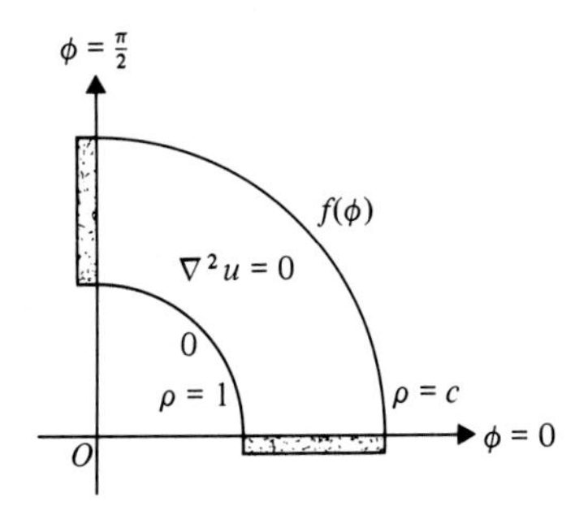

Figure 7.1-7

17. Find the bounded, steady-state temperatures in the *unbounded* region $\rho > c$ if $u(c, \phi) = f(\phi), \ -\pi < \phi \leq \pi$.

18. **(a)** The faces of a thin annular region (Fig. 7.1-8) are insulated, the inner edge $\rho = a$ is kept at 100°, and the outer edge $\rho = b$ is kept at 0°. Find the temperatures in the region.

*__(b)__ Compute the temperatures for $\rho = a = 1, 1.5, 2, 2.5, 3, 3.5, 4 = b$ in part (a).

(c) Graph the results in part (b).

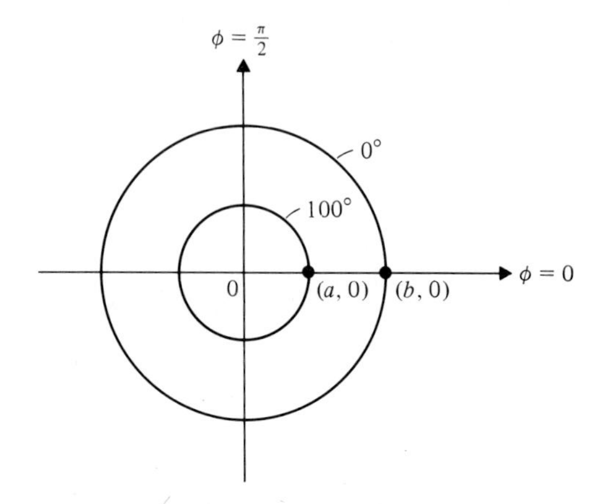

Figure 7.1-8

19. Solve the following problem:

$$\text{P.D.E.:} \quad \nabla^2 u = 0 \quad \text{in polar coordinates,}$$
$$0 < \rho < c, \qquad 0 \le \phi < 2\pi;$$

$$\text{B.C.:} \quad u(c, \phi) = 100, \qquad 0 < \phi < \frac{\pi}{4},$$

$$u(c, \phi) = 0, \qquad \frac{\pi}{4} < \phi < 2\pi.$$

20. Find the steady-state temperatures in a 45° sector of the unit circle if the temperatures of the line segments are held at zero and that of the curved portion is given by $f(\phi)$.

••• **21.** Verify the result obtained in Example 7.1-1.

22. **(a)** Solve the following problem relating to the static, transverse displacement of a membrane:

$$\text{D.E.:} \quad \frac{d}{d\rho}\left(\rho \frac{dz}{d\rho}\right) = 0, \qquad 1 < \rho < \rho_0;$$

$$\text{B.C.:} \quad z(1) = 0, \qquad z(\rho_0) = z_0.$$

(b) Interpret the problem in part (a) in physical terms.

(c) Solve the problem in part (a) given that the inner circle has radius 10 cm, the outer circle has radius 20 cm, and the outer rim is displaced 2 cm initially.

***(d)** Graph the solution in part (c).

23. In Example 7.1-2, examine the solution as $b \to 0$. Next, try separation of variables with $b = 0$, and carry the work as far as possible.

24. Given the following problem:

$$\text{P.D.E.:} \quad \frac{1}{\rho}(\rho u_\rho)_\rho + \frac{1}{\rho^2} u_{\phi\phi} = 0, \qquad 0 < \rho < 1, \qquad -\pi < \phi \le \pi;$$

$$\text{B.C.:} \quad u(1, \phi) = 100, \qquad 0 < \phi < \pi,$$
$$u(1, \phi) = 0, \qquad \pi < \phi < 2\pi.$$

Obtain the solution

$$u(\rho, \phi) = \frac{100}{\pi} \arctan \left[\frac{1 - \rho^2}{1 + \rho^2} \cot (\pi - \phi)\right].$$

25. The solution in Eq. (7.1–8) is for the internal Dirichlet problem. Derive the following solution for the *external* Dirichlet problem:

$$u(\rho, \phi) = \frac{1}{\pi} \int_{-\infty}^{\infty} f(s) \left[\frac{1}{2} + \sum_{n=1} \left(\frac{c}{\rho}\right)^n \cos n(\phi - s)\right] ds, \qquad \rho > c,$$

$$= \frac{1}{2\pi} \int_{-\infty}^{\infty} \frac{(\rho^2 - c^2) f(s)}{c^2 - 2\rho c \cos (\phi - s) + \rho^2}\, ds, \qquad \rho > c.$$

26. Change the region in Exercise 24 from $0 < \rho < 1$ to $\rho > 1$, everything else remaining the same, and then obtain the bounded solution.

27. **(a)** Verify the solution of Exercise 24.
(b) Is it reasonable to ask what the value of $u(1, \pi)$ is? Explain.

7.2 CYLINDRICAL COORDINATES; BESSEL FUNCTIONS

Laplace's equation in cylindrical coordinates (ρ, ϕ, z) has the following form (Exercise 1):

$$\nabla^2 u = \frac{1}{\rho} \frac{\partial}{\partial \rho} \left(\rho \frac{\partial u}{\partial \rho}\right) + \frac{1}{\rho^2} \frac{\partial^2 u}{\partial \phi^2} + \frac{\partial^2 u}{\partial z^2} = 0. \qquad \textbf{(7.2–1)}$$

We apply the method of separation of variables by assuming that u is a product of functions of ρ, ϕ, and z, that is,

$$u = R(\rho)\Phi(\phi)Z(z),$$

and substitute the appropriate derivatives into the partial differential equation. Then

$$\frac{\Phi Z}{\rho} \frac{d}{d\rho} \left(\rho \frac{dR}{d\rho}\right) + \frac{RZ}{\rho^2} \frac{d^2\Phi}{d\phi^2} + R\Phi \frac{d^2 Z}{dz^2} = 0,$$

and dividing by $R\Phi Z/\rho^2$ yields

$$\frac{\rho}{R} \frac{d}{d\rho} \left(\rho \frac{dR}{d\rho}\right) + \frac{\rho^2}{Z} \frac{d^2 Z}{dz^2} = -\frac{1}{\Phi} \frac{d^2\Phi}{d\phi^2}.$$

Since the left-hand member of the last equation is independent of ϕ, the equation can be satisfied only if both members are equal to a constant. Hence

$$-\frac{1}{\Phi} \frac{d^2\Phi}{d\phi^2} = n^2, \qquad n = 0, 1, 2, \ldots,$$

and a second separation yields

$$\frac{1}{\rho R}\frac{d}{d\rho}\left(\rho\frac{dR}{d\rho}\right) - \frac{n^2}{\rho^2} = -\frac{1}{Z}\frac{d^2Z}{dz^2} = -\lambda^2.$$

We have called the first *separation constant* n^2 because this will force Φ (and u) to be periodic of period 2π in ϕ. This is the desired situation in many applied problems. We have called the second separation constant λ^2 because we do *not* want Z (and u) to be periodic in z.

The values of the separation constants n and λ are actually dictated by the nature of the boundary conditions that u must satisfy. We have chosen those values that conform to the boundary conditions most often found in problems dealing with this general topic at this level. We have also kept in mind that our task is made easier if we obtain Sturm–Liouville problems after separating variables.

Thus by separation of variables we have reduced Laplace's equation to the following three ordinary, linear, homogeneous differential equations:

$$\frac{d^2Z}{dz^2} - \lambda^2 Z = 0, \tag{7.2-2}$$

$$\frac{d^2\Phi}{d\phi^2} + n^2\Phi = 0, \qquad n = 0, 1, 2, \ldots, \tag{7.2-3}$$

$$\frac{d^2R}{d\rho^2} + \frac{1}{\rho}\frac{dR}{d\rho} + \left(\lambda^2 - \frac{n^2}{\rho^2}\right)R = 0. \tag{7.2-4}$$

The solutions to the first two are straightforward and are given by

$$Z(\lambda z) = Ae^{\lambda z} + Be^{-\lambda z} \tag{7.2-5}$$

and

$$\Phi(n\phi) = C\cos n\phi + D\sin n\phi, \tag{7.2-6}$$

respectively. Equation (7.2–4) is a Bessel equation (see Example 1.5–5) with linearly independent solutions $J_n(\lambda\rho)$ and $Y_n(\lambda\rho)$. The first of these is called the **Bessel function of the first kind of order** n, and the second is the **Bessel function of the second kind of order** n.* Hence the general solution of Eq. (7.2–4) can be written as

$$R_n(\lambda\rho) = EJ_n(\lambda\rho) + FY_n(\lambda\rho). \tag{7.2-7}$$

We remark that the solutions to Laplace's equation in cylindrical coordinates are *products* of the functions shown in Eqs. (7.2–5), (7.2–6), and (7.2–7). A function u that satisfies $\nabla^2 u = 0$ is called a *harmonic function*; hence the products referred to are sometimes called **cylindrical harmonics**. Since $J_n(\lambda\rho)$ is defined at $\rho = 0$ while $Y_n(\lambda\rho)$ is not (as we shall see later), we

*Bessel functions of the second kind will be discussed later in this section.

choose the arbitrary constant F to be zero if u must be bounded at the origin. We would also choose $A = 0$ if it is required that $\lim_{z\to\infty} |u|$ exist and λ is nonnegative. Moreover, one of the boundary conditions may require that either C or D be zero. Thus in practice the cylindrical harmonics are not as formidable as they may seem.

In Section 7.4 we will look at some applications that involve Bessel functions, but in this section we examine these functions in more detail. We have solved the equivalent of Eq. (7.2–4) (see Exercise 2) and obtained the solution

$$J_n(x) = \sum_{m=0} \frac{(-1)^m}{m!(m+n)!}\left(\frac{x}{2}\right)^{2m+n}, \qquad n = 0, 1, 2, \ldots, \tag{7.2–8}$$

in Example 1.5–5.

We next examine $J_0(x)$ and $J_1(x)$ in some detail. From Eq. (7.2–8) we have

$$J_0(x) = 1 - \frac{x^2}{2^2} + \frac{x^4}{2^2 \cdot 4^2} - \frac{x^6}{2^2 \cdot 4^2 \cdot 6^2} + - \cdots,$$

$$J_1(x) = \frac{x}{2} - \frac{x^3}{2^2 \cdot 4} + \frac{x^5}{2^2 \cdot 4^2 \cdot 6} - \frac{x^7}{2^2 \cdot 4^2 \cdot 6^2 \cdot 8} + - \cdots.$$

Comparing these functions with the Maclaurin's series for $\cos x$ and $\sin x$,

$$\cos x = 1 - \frac{x^2}{2!} + \frac{x^4}{4!} - \frac{x^6}{6!} + - \cdots,$$

$$\sin x = x - \frac{x^3}{3!} + \frac{x^5}{5!} - \frac{x^7}{7!} + - \cdots,$$

we note a similarity between $J_0(x)$ and $\cos x$ and also between $J_1(x)$ and $\sin x$. For example (Exercise 4),

$$\begin{aligned}
J_0(0) &= 1, & \cos 0 &= 1;\\
J_0(-x) &= J_0(x), & \cos(-x) &= \cos x;\\
J_0'(0) &= 0, & \frac{d}{dx}(\cos x)\Big|_{x=0} &= 0;\\
J_1(0) &= 0, & \sin 0 &= 0;\\
J_1(-x) &= -J_1(x), & \sin(-x) &= -\sin x;\\
J_0'(x) &= -J_1(x), & \frac{d}{dx}(\cos x) &= -\sin x.
\end{aligned}$$

These similarities can be seen in the graphs of $J_0(x)$ and $J_1(x)$ shown in Fig. 7.2–1. Note that the Bessel functions of the first kind are almost periodic with a variable period nearly equal to 2π. In fact, the period approaches 2π as $x \to \infty$. Note also that the maximum amplitudes are decreasing with increasing x.

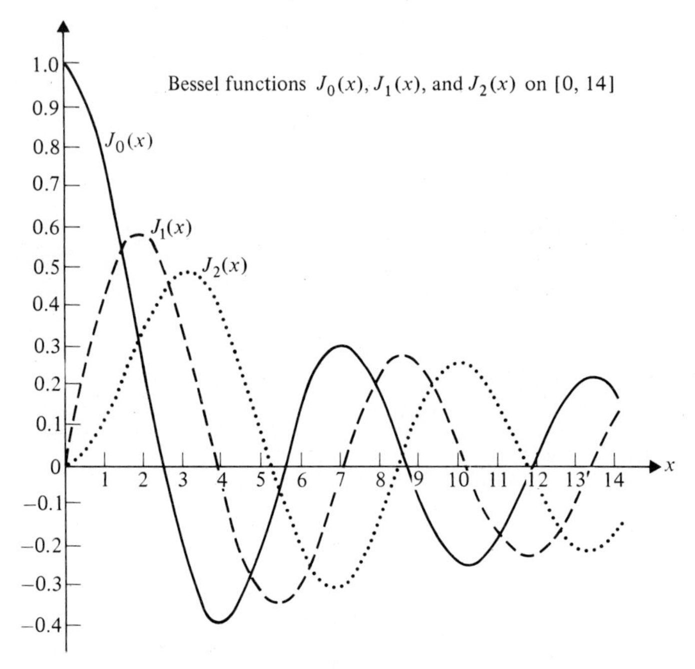

Figure 7.2-1

In solving boundary-value problems the zeros of $J_n(x)$, that is, the roots of $J_n(x) = 0$, are of importance. The approximate values of the first few zeros of $J_0(x)$ and $J_1(x)$ are listed in the table below.

	1st	*2nd*	*3rd*	*4th*	*5th*	*6th*
$J_0(x)$	2.405	5.520	8.654	11.792	14.931	18.071
$J_1(x)$	0.0	3.832	7.016	10.173	13.323	16.471

Another useful relation, namely,

$$\frac{d}{dx}[x^n J_n(x)] = x^n J_{n-1}(x), \qquad n = 1, 2, \ldots, \tag{7.2-9}$$

can be easily obtained (Exercise 4) from Eq. (7.2-8). In differential form, Eq. (7.2-9) becomes

$$d[x^n J_n(x)] = x^n J_{n-1}(x)\, dx,$$

and integrating from 0 to c ($c > 0$), we have

$$x^n J_n(x) \Big|_0^c = \int_0^c x^n J_{n-1}(x)\, dx$$

or

$$\int_0^c x^n J_{n-1}(x)\, dx = c^n J_n(c).$$

When $n = 1$, this reduces to

$$\int_0^c xJ_0(x)\,dx = cJ_1(c). \tag{7.2–10}$$

a result that will be useful later.

Orthogonality of Bessel Functions

Bessel functions of the first kind satisfy an **orthogonality relation** under certain conditions, and we look at this property next.

Bessel's differential equation of order n can be written as

$$x^2 \frac{d^2u}{dx^2} + x\frac{du}{dx} + (\lambda^2x^2 - n^2)u = 0. \tag{7.2–11}$$

A particular solution of this equation is $u = J_n(\lambda x)$, the Bessel function of the first kind of order n. Similarly, $v = J_n(\mu x)$ is a particular solution of

$$x^2 \frac{d^2v}{dx^2} + x\frac{dv}{dx} + (\mu^2x^2 - n^2)v = 0. \tag{7.2–12}$$

Now multiply Eq. (7.2–11) by v/x and Eq. (7.2–12) by u/x, and subtract to obtain

$$vx\frac{d^2u}{dx^2} + v\frac{du}{dx} + (\lambda^2x^2 - n^2)\frac{uv}{x} - ux\frac{d^2v}{dx^2} - u\frac{dv}{dx} - (\mu^2x^2 - n^2)\frac{uv}{x} = 0.$$

This can be written as

$$\begin{aligned}(\lambda^2 - \mu^2)xuv &= ux\frac{d^2v}{dx^2} - vx\frac{d^2u}{dx^2} + u\frac{dv}{dx} - v\frac{du}{dx}\\ &= \frac{d}{dx}\left[x\left(u\frac{dv}{dx} - v\frac{du}{dx}\right)\right].\end{aligned}$$

Hence for some $c > 0$,

$$(\lambda^2 - \mu^2)\int_0^c xuv\,dx = \int_0^c d\left[x\left(u\frac{dv}{dx} - v\frac{du}{dx}\right)\right],$$

or, if we replace u and v by their values $J_n(\lambda x)$ and $J_n(\mu x)$, respectively,

$$\begin{aligned}(\lambda^2 - \mu^2)\int_0^c xJ_n(\lambda x)J_n(\mu x)\,dx &= x[\mu J_n(\lambda x)J_n'(\mu x) - \lambda J_n(\mu x)J_n'(\lambda x)]\Big|_0^c\\ &= c[\mu J_n(\lambda c)J_n'(\mu c) - \lambda J_n(\mu c)J_n'(\lambda c)].\end{aligned}$$

Thus we have

$$\int_0^c xJ_n(\lambda x)J_n(\mu x)\,dx = \frac{c}{\lambda^2 - \mu^2}[\mu J_n(\lambda c)J_n'(\mu c) - \lambda J_n(\mu c)J_n'(\lambda c)].$$

From this it follows that

$$\int_0^c xJ_n(\lambda x)J_n(\mu x)\,dx = 0,$$

provided that $\lambda \neq \mu$ *and*

$$\mu J_n(\lambda c)J_n'(\mu c) - \lambda J_n(\mu c)J_n'(\lambda c) = 0. \tag{7.2–13}$$

Now Eq. (7.2–13) holds if λc and μc are different roots of

1. $J_n(x) = 0$ because then $J_n(\lambda c) = 0$ and $J_n(\mu c) = 0$;
2. $J_n'(x) = 0$ because then $J_n'(\mu c) = 0$ and $J_n'(\lambda c) = 0$;
3. $hJ_n(x) + xJ_n'(x) = 0$, where $h > 0$.

To see this last condition, note that if λc and μc are different roots of $hJ_n(x) + xJ_n'(x) = 0$, then it follows that

$$hJ_n(\lambda c) + \lambda cJ_n'(\lambda c) = 0 \qquad \text{and} \qquad hJ_n(\mu c) + \mu cJ_n'(\mu c) = 0.$$

Multiplying the first of these by $\mu J_n'(\mu c)$ and the second by $\lambda J_n'(\lambda c)$ results in

$$h\mu J_n'(\mu c)J_n(\lambda c) + c\lambda\mu J_n'(\lambda c)J_n'(\mu c) = 0,$$
$$h\lambda J_n'(\lambda c)J_n(\mu c) + c\lambda\mu J_n'(\lambda c)J_n'(\mu c) = 0.$$

Subtracting, we have

$$h[\mu J_n(\lambda c)J_n'(\mu c) - \lambda J_n(\mu c)J_n'(\lambda c)] = 0$$

or

$$\mu J_n(\lambda c)J_n'(\mu c) - \lambda J_n(\mu c)J_n'(\lambda c) = 0,$$

which is identical to Eq. (7.2–13).

Finally, observe that if $h = 0$ in condition 3 above, then condition 3 is equivalent to condition 2. Hence we may take $h \geq 0$ in condition 3.

In the foregoing we have shown that the Bessel functions $J_n(\lambda x)$ and $J_n(\mu x)$ are orthogonal on the interval $0 < x < c$ *with weight function* x provided that $\lambda \neq \mu$ and one of conditions 1, 2, or 3 above holds.

Fourier-Bessel Series

We note that although Bessel's differential equation does not meet the conditions required of a *regular* Sturm–Liouville system (Section 2.2), it does fall into a special class of *singular* systems discussed in Section 2.7. We have, in fact, proved a particular case of Theorem 2.7–2 by showing that the eigenfunctions of the singular Sturm–Liouville system are orthogonal.

It can also be shown* that the normalized eigenfunctions that we will obtain in this section form a complete set (see Section 2.5) with respect to the

*See Andrew Gray and G. B. Mathews, *A Treatise on Bessel Functions and Their Application to Physics*, 2nd ed. (Philadelphia: Dover, 1966), pp. 94 ff.

class of piecewise smooth functions on the interval $(0, c)$. Thus we will be able to represent piecewise smooth functions by a series of Bessel functions of the first kind, called a **Fourier–Bessel series.**

Let $c\lambda_j$, $j = 1, 2, 3, \ldots$, be the positive roots (zeros) of $J_n(\lambda c) = 0$. These roots can be found tabulated in various mathematical tables for computational purposes. For example, if $n = 0$, then, approximately, $\lambda_1 c = 2.405$, $\lambda_2 c = 5.520$, $\lambda_3 c = 8.654$, etc.

Now consider the representation on $(0, c)$ of the function f:

$$f(x) = A_1 J_n(\lambda_1 x) + A_2 J_n(\lambda_2 x) + A_3 J_n(\lambda_3 x) + \cdots. \qquad \textbf{(7.2–14)}$$

If we wish to find the value of A_2, say, then we multiply each term by $xJ_n(\lambda_2 x)$ and integrate with respect to x from 0 to c. Then we have

$$\begin{aligned}\int_0^c xf(x)J_n(\lambda_2 x)\,dx &= A_1 \int_0^c xJ_n(\lambda_1 x)J_n(\lambda_2 x)\,dx \\ &+ A_2 \int_0^c xJ_n(\lambda_2 x)J_n(\lambda_2 x)\,dx \\ &+ A_3 \int_0^c xJ_n(\lambda_3 x)J_n(\lambda_2 x)\,dx + \cdots.\end{aligned}$$

Because of the orthogonality of the Bessel functions, all integrals on the right with the exception of the second one are zero. Thus we have

$$\int_0^c xf(x)J_n(\lambda_2 x)\,dx = A_2 \int_0^c xJ_n^2(\lambda_2 x)\,dx,$$

from which we obtain

$$A_2 = \frac{\displaystyle\int_0^c xf(x)J_n(\lambda_2 x)\,dx}{\displaystyle\int_0^c xJ_n^2(\lambda_2 x)\,dx}.$$

Since the same procedure can be used for *any* coefficient, we have, in general,

$$A_j = \frac{\displaystyle\int_0^c xf(x)J_n(\lambda_j x)\,dx}{\displaystyle\int_0^c xJ_n^2(\lambda_j x)\,dx}, \qquad j = 1, 2, 3, \ldots. \qquad \textbf{(7.2–15)}$$

Next we evaluate the denominator in the last expression. To do this, we go back to Bessel's differential equation of order n,

$$xu'' + u' + \left(\lambda_j^2 x - \frac{n^2}{x}\right)u = 0.$$

A particular solution of this equation is $u = J_n(\lambda_j x)$. Multiplying by the *integrating factor* $2u'x$ yields

$$2x^2u'u'' + 2(u')^2x + \left(\lambda_j^2 x - \frac{n^2}{x}\right)2xu'u = 0$$

or

$$2xu'(xu'' + u') + (\lambda_j^2 x^2 - n^2)2uu' = 0. \tag{7.2–16}$$

Using the fact that

$$\frac{d}{dx}(xu')^2 = 2xu'(xu'' + u')$$

and

$$\frac{d}{dx}(u^2) = 2uu',$$

we can write the differential equation (7.2–16) as

$$\frac{d}{dx}(xu')^2 + (\lambda_j^2 x^2 - n^2)\frac{d}{dx}(u^2) = 0.$$

Hence integrating from 0 to c, we have

$$\int_0^c d(xu')^2 + \int_0^c (\lambda_j^2 x^2 - n^2)\, d(u^2) = 0.$$

Using integration by parts as shown for the second integral gives us

$$w = \lambda_j^2 x^2 - n^2 \qquad dv = d(u^2)$$

$$dw = 2\lambda_j^2 x\, dx \qquad v = u^2$$

$$(xu')^2\Big|_0^c + u^2(\lambda_j^2 x^2 - n^2)\Big|_0^c - 2\lambda_j^2 \int_0^c xu^2\, dx = 0.$$

But $u = J_n(\lambda_j x)$, $u' = \lambda_j J_n'(\lambda_j x)$, so the last expression becomes

$$\lambda_j^2 c^2 [J_n'(\lambda_j c)]^2 + J_n^2(\lambda_j c)(\lambda_j^2 c^2 - n^2) + n^2 J_n^2(0) = 2\lambda_j^2 \int_0^c xJ_n^2(\lambda_j x)\, dx. \tag{7.2–17}$$

Since $J_n(\lambda_j c) = 0$ and $J_n(0) = 0$ for $n = 1, 2, 3, \ldots$, Eq. (7.2–17) reduces to

$$\int_0^c xJ_n^2(\lambda_j x)\, dx = \frac{c^2}{2}[J_n'(\lambda_j c)]^2. \tag{7.2–18}$$

Hence the coefficients in Eq. (7.2–15) become

$$A_j = \frac{2}{c^2[J_n'(\lambda_j c)]^2} \int_0^c xf(x)J_n(\lambda_j x)\, dx, \qquad j = 1, 2, 3, \ldots. \tag{7.2–19}$$

Noting that x in the above is a dummy variable in a definite integral and can thus be replaced by any other letter, we can write the Fourier–Bessel series in Eq. (7.2–14) as

$$f(x) = \frac{2}{c^2} \sum_{j=1} \frac{J_n(\lambda_j x)}{[J_n'(\lambda_j c)]^2} \int_0^c sf(s)J_n(\lambda_j s)\, ds.$$

The equality in this representation of $f(x)$ is not to be taken literally. As in the case of Fourier series, this series will converge to the *mean value* of the function at points where $f(x)$ has a jump discontinuity. At points where f is continuous the series will converge to the functional values at those points.

Equation (7.2–18) expresses the **square of the norm** of the eigenfunction $J_n(\lambda_j x)$. Thus we have for the **norm** (indicated by the double bars)

$$\|J_n(\lambda_j x)\| = \frac{c}{\sqrt{2}} J_n'(\lambda_j c).$$

The set

$$\left\{ \frac{\sqrt{2}}{c} \frac{J_n(\lambda_1 x)}{J_n'(\lambda_1 c)}, \frac{\sqrt{2}}{c} \frac{J_n(\lambda_2 x)}{J_n'(\lambda_2 c)}, \ldots \right\}$$

is an orthonormal set on the interval $(0, c)$ with weight function x *if* the λ_j are such that $J_n(\lambda_j c) = 0$.

If the λ_j are such that $hJ_n(\lambda_j c) + \lambda_j c J_n'(\lambda_j c) = 0$, then solving for $J_n'(\lambda_j c)$, we have

$$J_n'(\lambda_j c) = -\frac{h}{\lambda_j c} J_n(\lambda_j c). \tag{7.2–20}$$

Substituting this value into Eq. (7.2–17) produces

$$\int_0^c x J_n^2(\lambda_j x)\, dx = \frac{\lambda_j^2 c^2 - n^2 + h^2}{2\lambda_j^2} J_n^2(\lambda_j c).$$

Thus in this case we use, in place of Eq. (7.2–19), the following formula for the coefficients in the Fourier–Bessel series:

$$A_j = \frac{2\lambda_j^2}{(\lambda_j^2 c^2 - n^2 + h^2) J_n^2(\lambda_j c)} \int_0^c x f(x) J_n(\lambda_j x)\, dx, \qquad j = 1, 2, 3, \ldots . \tag{7.2–21}$$

The formula (7.2–21) is not valid for $j = 1$ in the single case when $h = 0$ and $n = 0$. Then Eq. (7.2–20) becomes

$$J_0'(\lambda_j c) = 0,$$

that is, $\lambda_j c$ is a zero of $J_0'(x) = 0$. The *first* zero of $J_0'(x)$ is at $x = 0$; hence $\lambda_1 = 0$, the first zero of $J_0'(\lambda c)$. But then $J(0) = 1$, and the coefficient of the integral in Eq. (7.2–21) can be evaluated by L'Hôpital's rule. Hence

$$A_1 = \frac{2}{c^2} \int_0^c x f(x)\, dx. \tag{7.2–22}$$

We summarize the various cases in the following list. It can be seen from this list that the *definition* of the λ_j is an important part of the Fourier–Bessel representation of a function. Accordingly, this information about the λ_j should be given for every Fourier–Bessel series.

Fourier–Bessel series: $f(x) \sim \sum_{j=1} A_j J_n(\lambda_j x)$

Source of λ_j: $J_n(\lambda_j c) = 0, \quad n = 0, 1, 2, \ldots$

Coefficients: $$A_j = \frac{2}{c^2[J_n'(\lambda_j c)]^2} \int_0^c xf(x)J_n(\lambda_j x)\,dx, \quad j = 1, 2, 3, \ldots$$

Source of λ_j: $hJ_n(\lambda_j c) + \lambda_j c J_n'(\lambda_j c) = 0, \quad h \geq 0, \quad n = 1, 2, \ldots$

Coefficients: $$A_j = \frac{2\lambda_j^2}{(\lambda_j^2 c^2 - n^2 + h^2)J_n^2(\lambda_j c)} \int_0^c xf(x)J_n(\lambda_j x)\,dx, \quad j = 1, 2, 3, \ldots$$

Source of λ_j: $hJ_n(\lambda_j c) + \lambda_j c J_n'(\lambda_j c) = 0, \quad h > 0, \quad n = 0, 1, 2, \ldots$

Coefficients: $$A_j = \frac{2\lambda_j^2}{(\lambda_j^2 c^2 - n^2 + h^2)J_n^2(\lambda_j c)} \int_0^c xf(x)J_n(\lambda_j x)\,dx, \quad j = 1, 2, 3, \ldots$$

Source of λ_j: $J_0'(\lambda_j c) = 0$

Coefficients: $$A_1 = \frac{2}{c^2} \int_0^c xf(x)\,dx$$

$$A_j = \frac{2}{c^2 J_0^2(\lambda_j c)} \int_0^c xf(x)J_0(\lambda_j c)\,dx, \quad j = 2, 3, 4, \ldots$$

EXAMPLE 7.2–1 Evaluate

$$\int_0^1 x^3 J_0(x)\,dx.$$

Solution We use integration by parts with $u = x^2$ and $dv = xJ_0(x)\,dx$. Thus

$$\int_0^1 x^3 J_0(x)\,dx = x^3 J_1(x)\Big|_0^1 - 2\int_0^1 x^2 J_1(x)\,dx,$$

after observing that

$$\int xJ_0(x)\,dx = \int^x sJ_0(s)\,ds = xJ_1(x),$$

using Eq. (7.2–10). Now, applying Eq. (7.2–9), we have

$$\int x^2 J_1(x)\,dx = x^2 J_2(x).$$

Hence

$$\int_0^1 x^3 J_0(x)\,dx = x^3 J_1(x) - 2x^2 J_2(x)\Big|_0^1$$
$$= J_1(1) - 2J_2(1),$$

a result that can be reduced. From Exercise 4(e) we have

$$J_n'(x) + \frac{n}{x} J_n(x) = J_{n-1}(x),$$

while from Exercise 17(a) we obtain

$$J_n'(x) - \frac{n}{x} J_n(x) = -J_{n+1}(x).$$

Subtracting this last equation from the previous one produces

$$\frac{2n}{x} J_n(x) = J_{n-1}(x) + J_{n+1}(x). \qquad \textbf{(7.2–23)}$$

Equation (7.2–23) with $n = 2$ and $x = 1$ can now be used to write

$$J_2(1) = 2J_1(1) - J_0(1)$$

so that

$$\int_0^1 x^3 J_0(x)\,dx = 2J_0(1) - 3J_1(1) \doteq 0.210. \quad \blacksquare$$

EXAMPLE 7.2–2 Obtain the Fourier–Bessel series representation of $f(x) = 1$ in terms of $J_0(\lambda_j x)$ on the interval $0 < x < 1$, where the λ_j are such that $J_0(\lambda_j) = 0$, $j = 1, 2, \ldots$.

Solution From Eq. (7.2–19) we have for $n = 0$, $c = 1$,

$$A_j = \frac{2}{J_1^2(\lambda_j)} \int_0^1 x J_0(\lambda_j x)\,dx, \qquad j = 1, 2, \ldots,$$

and by making the substitution $s = \lambda_j x$, this becomes

$$A_j = \frac{2}{[\lambda_j J_1(\lambda_j)]^2} \int_0^{\lambda_j} s J_0(s)\,ds = \frac{2}{J_1(\lambda_j)},$$

using Eq. (7.2–10). Hence we have

$$f(x) = 1 \sim 2 \sum_{j=1}^{\infty} \frac{J_0(\lambda_j x)}{\lambda_j J_1(\lambda_j)}. \quad \blacksquare \qquad \textbf{(7.2–24)}$$

We leave it for an exercise to show that (7.2–24) is valid for $x = 0$ (see Exercise 27). Observe, however, that $f(1) = 0$ and that for x near zero the series converges very slowly. In Fig. 7.2–2 we show the result of adding the first ten terms of the representation in (7.2–24).

Bessel Functions of the Second Kind

We conclude this section with a brief reference to the Bessel function of the *second* kind of order n. It is a second linearly independent solution of Bessel's differential equation, which can be obtained by the method of variation of parameters.* The computation will be omitted, and we give details only of $Y_0(x)$ and graphs of this function as well as $Y_1(x)$ and $Y_2(x)$ in Fig. 7.2–3. The Bessel function of the second kind of order zero is given by †

$$Y_0(x) = \frac{2}{\pi}\left[J_0(x)\left(\log\frac{x}{2} + \gamma\right)\right] + \frac{2}{\pi}\left(\frac{x^2}{4} - \frac{3x^4}{128} + \frac{11x^6}{13{,}824} + - \cdots\right), \tag{7.2–25}$$

where the term γ is called Euler's constant. It is an irrational number defined by

$$\gamma = \lim_{n\to\infty}\left(\sum_{k=1}^{n}\frac{1}{k} - \log n\right) \doteq 0.577215.$$

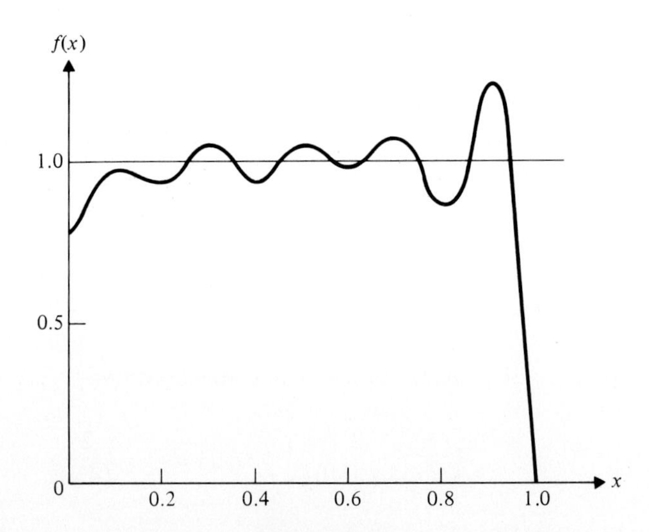

Figure 7.2–2
A representation by a Fourier-Bessel series.

---

*See Edgar A. Kraut, *Fundamentals of Mathematical Physics* (New York: McGraw-Hill, 1967), pp. 264 ff.

†Some authors call this the *Neumann function* of order zero, while others call it Weber's Bessel function of the second kind of order zero.

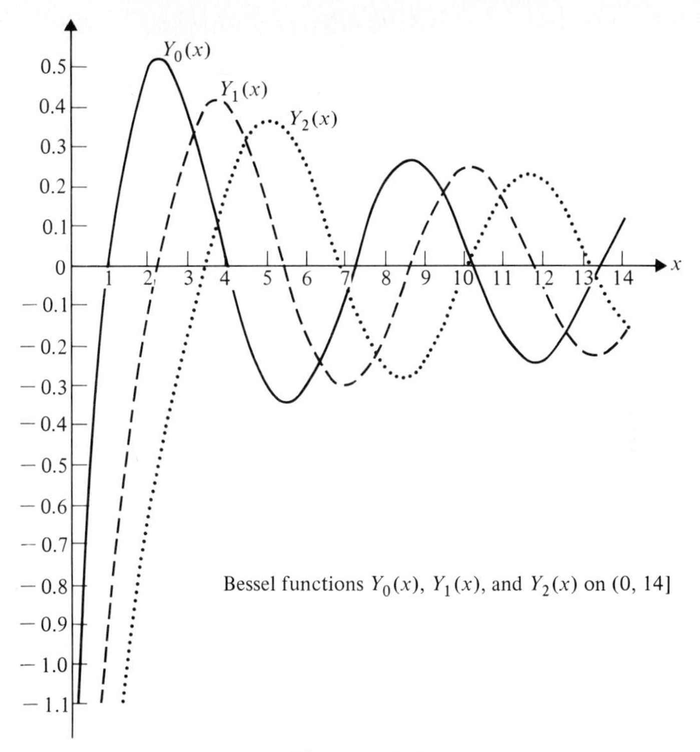

Figure 7.2-3

Of importance in our future work is the fact that *all* Bessel functions of the second kind contain the term log $(x/2)$. Consequently, $Y_n(0)$ is undefined, as Fig. 7.2-3 indicates. In view of the fact that the applications involving Bessel functions in this text are such that $x > 0$, we will not be using Bessel functions of the second kind. It should be noted, however, that the latter are required in certain instances—for example, in solving problems involving electromagnetic waves in coaxial cables.

In connection with Bessel functions we should also mention the **Hankel transform.*** This integral transform pair is defined as follows:

$$
\begin{aligned}
F_n(\lambda) &= \mathcal{H}_n\,\{f(\rho)\} = \int_0^\infty f(\rho)J_n(\lambda\rho)\,\rho\,d\rho, \\
f(\rho) &= \mathcal{H}_n^{-1}\,\{F_n(\lambda)\} = \int_0^\infty F_n(\lambda)J_n(\lambda\rho)\,\lambda\,d\lambda.
\end{aligned}
\tag{7.2-26}
$$

This transform arises naturally in connection with the Laplacian in cylindrical coordinates. The Hankel transform may lead to significant simplification in problems involving Bessel's equation.

*After Hermann Hankel (1839–1873), a German mathematician.

Key Words and Phrases

Bessel function of the first and second kind of order n
cylindrical harmonics
orthogonality relation
Fourier–Bessel series
square of the norm
norm
Hankel transform

7.2 Exercises

• 1. Obtain Eq. (7.2–1) from Eq. (7.1–2).

2. By making the substitutions $y = R$ and $x = \lambda\rho$, show that Eq. (7.2–4) can be reduced to Eq. (1.5–6).

3. Use the ratio test to show that the series (7.2–8) representing Bessel's function of the first kind of order n converges for all x.

4. Use Eq. (7.2–8) to prove each of the following:

(a) $J_0'(0) = 0$

(b) $J_1(0) = 0$

(c) $J_1(-x) = -J_1(x)$

(d) $J_0'(x) = -J_1(x)$

(e) $xJ_n'(x) = -nJ_n(x) + xJ_{n-1}(x), \qquad n = 1, 2, \ldots$

(f) $\dfrac{d}{dx}[x^n J_n(x)] = x^n J_{n-1}(x), \qquad n = 1, 2, \ldots$

5. Verify that $x = 0$ is a regular singular point of Bessel's differential equation (7.2–11). (*Hint*: See Section 1.5.)

6. Obtain Eq. (7.2–23).

•• 7. Obtain the general solution of each of the following differential equations *by inspection*:

(a) $\dfrac{d}{dx}\left(x\dfrac{dy}{dx}\right) + xy = 0$

(b) $4xy'' + 4y' + y = 0$

(c) $\dfrac{d^2y}{dx^2} + ye^x = 0 \qquad$ (*Hint*: Let $u = e^x$.)

8. If $J_0(\lambda_j) = 0$, then obtain each of the following:

(a) $\displaystyle\int_0^1 J_1(\lambda_j s)\,ds = 1/\lambda_j$

(b) $\displaystyle\int_0^{\lambda_j} J_1(s)\,ds = 1$

(c) $\displaystyle\int_0^{\infty} J_1(\lambda_j s)\,ds = 0$

9. Obtain each of the following:

(a) $\int^x J_0(s)J_1(s)\,ds = -\frac{1}{2}[J_0(x)]^2$

(b) $\int^x s^2 J_0(s)J_1(s)\,ds = \frac{1}{2}x^2\,[J_1(x)]^2$

10. Expand each of the following functions in a Fourier–Bessel series of functions $J_0(\lambda_j x)$ on the interval $0 < x < c$, where $J_0(\lambda_j c) = 0$. (*Note*: The coefficients are given by Eq. (7.2–19), but the integral need not be evaluated in all cases.)

(a) $f(x) = 1$

(b) $f(x) = x^2$ (*Note*: Use the following reduction formula:

$$\int_0^x s^n J_0(s)\,ds = x^n J_1(x) + (n-1)x^{n-1}J_0(x) - (n-1)^2 \int_0^x s^{n-2}J_0(s)\,ds, \qquad n = 2, 3, \ldots .)$$

(c) $f(x) = \begin{cases} 0, & \text{for } 0 < x < 1, \\ 1/x, & \text{for } 1 \le x \le 2. \end{cases}$

(See Fig. 7.2–4.)

11. Obtain the Fourier–Bessel series representations of each of the following functions:

(a) $f(x) = x$ on $-1 < x < 1$ in terms of $J_1(\lambda_j x)$ where the λ_j are positive roots of $J_1(\lambda) = 0$.

(b) $f(x) = x^3$ on $0 \le x < 1$ in terms of $J_1(\lambda_j x)$ where the λ_j are positive roots of $J_1(\lambda) = 0$.

12. Bessel's differential equation is notorious for its many disguises. Show that each of the following is a Bessel's differential equation.

(a) $\frac{dy}{dx} + ay^2 + \frac{1}{x}y + \frac{1}{a} = 0$

(This is a *Riccati equation*, but the substitution

$$y = \frac{1}{az}\frac{dz}{dx}$$

transforms it into a Bessel equation.)

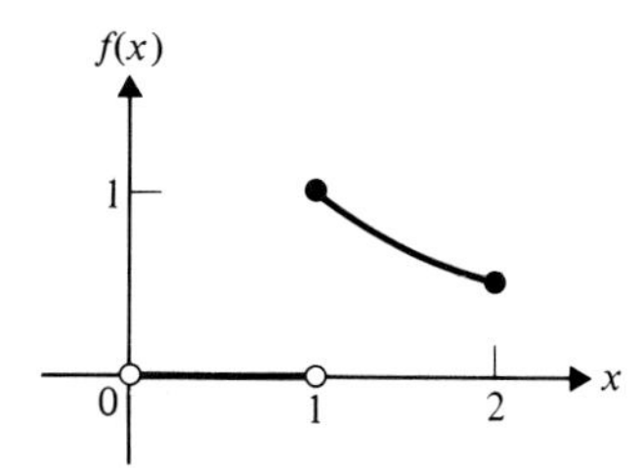

Figure 7.2–4

(b) $r^2 \dfrac{d^2R}{dr^2} + 2r\dfrac{dR}{dr} + [\lambda^2 r^2 - n(n+1)]R = 0$

(This equation arises when the *Helmholtz equation** in spherical coordinates is solved by separation of variables. Make the substitution

$$R(\lambda r) = \frac{Z(\lambda r)}{(\lambda r)^{1/2}}$$

to transform it into a Bessel equation of order $n + \frac{1}{2}$.)

(c) $\dfrac{d^2y}{dx^2} + \dfrac{1}{x}\dfrac{dy}{dx} + \dfrac{n}{k}y = 0$

(This is *Fourier's equation*, but the substitution $x\sqrt{(n/k)} = z$ will transform it into a Bessel equation.)

13. In Bessel's differential equation of order 1/2, make the substitution $y = u/\sqrt{x}$ to obtain

$$\frac{d^2u}{dx^2} + u = 0.$$

Solve this equation to obtain

$$y = c_1\frac{\sin x}{\sqrt{x}} + c_2\frac{\cos x}{\sqrt{x}},$$

and explain the qualitative nature of $J_{1/2}(x)$, that is, the decaying peak amplitude of this function.

14. **(a)** Expand the function of Exercise 10(c) in a Fourier–Bessel series in terms of $J_1(\lambda_j x)$ where λ_j satisfies $J_1(2\lambda_j) = 0$.

(b) To what value will the series converge when $x = 1$? Explain.

15. **(a)** Prove that

$$J_{1/2}(x) = \sqrt{2/\pi x}\,\sin x.$$

(*Hint*: Use the Maclaurin's series for $\sin x$.)

(b) Prove that

$$J_{-1/2}(x) = \sqrt{2/\pi x}\,\cos x.$$

16. Evaluate

$$\int^x s^n J_{n-1}(s)\,ds.$$

(*Hint*: Use Exercise 4(f).)

17. **(a)** Prove that

$$\frac{d}{dx}[x^{-n}J_n(x)] = -x^{-n}J_{n+1}(x).$$

*After Hermann von Helmholtz (1821–1894), a German military surgeon who turned to mathematical physics in 1871.

(b) Evaluate

$$\int^{x} \frac{J_{n+1}(s)}{s^n}\, ds.$$

18. **(a)** The equation

$$y'' + \frac{1}{x}y' - y = 0$$

is called the *modified Bessel equation of order zero*. Show that a solution is

$$J_0(ix) = 1 + \frac{x^2}{2^2} + \frac{x^4}{2^2 \cdot 4^2} + \frac{x^6}{2^2 \cdot 4^2 \cdot 6^2} + \cdots .$$

We also write $I_0(x) = J_0(ix)$ where $I_0(x)$ is called the *modified Bessel function of the first kind of order zero*. (See Fig. 7.2–5.)

(b) Find the interval of convergence of $I_0(x)$.

19. Prove that

$$Y_0'(x) = -Y_1(x).$$

20. **(a)** By dividing Bessel's differential equation (7.2–11) by x, show that it has the form of the Sturm–Liouville equation (2.2–6).

(b) Using the notation of Section 2.2, show that $r(a) = 0$ and $a_1 = a_2 = 0$.

(c) Show that the second boundary condition of (2.2–7) is satisfied by the Bessel function of the first kind.

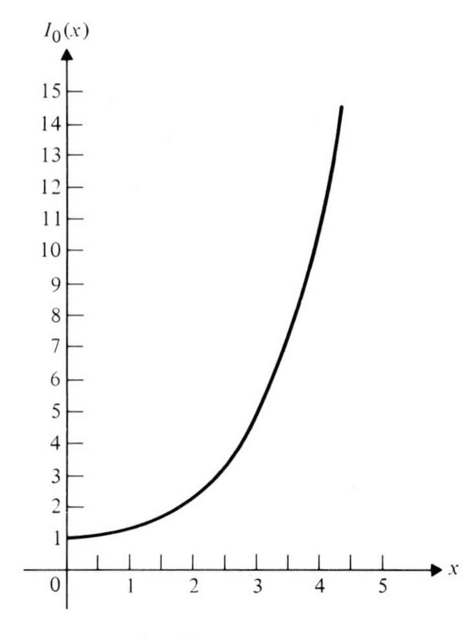

Figure 7.2–5

21. If λ_j is such that $J_0(\lambda_j) = 0$ for $j = 1, 2, \ldots$, obtain each of the following series representations:

(a) $$x^2 \sim 2 \sum_{j=1}^{\infty} \frac{\lambda_j^2 - 4}{\lambda_j^3 J_1(\lambda_j)} J_0(\lambda_j x), \qquad 0 < x < 1$$

(b) $$J_0(kx) \sim 2J_0(k) \sum_{j=1}^{\infty} \frac{\lambda_j J_0(\lambda_j x)}{(\lambda_j^2 - k^2) J_1(\lambda_j)}, \qquad 0 < x < 1,$$

where $k \neq \lambda_j$, $j = 1, 2, \ldots$.

22. If λ_j is a positive root of $J_1(x) = 0$, obtain each of the following series representations:

(a) $$x^2 \sim \frac{1}{2} + 4 \sum \frac{J_0(\lambda_j x)}{\lambda_j^2 J_0(\lambda_j)}, \qquad 0 < x < 1$$

(b) $$(1 - x^2)^2 \sim \frac{1}{2} - 64 \sum \frac{J_0(\lambda_j x)}{\lambda_j^4 J_0(\lambda_j)}, \qquad 0 < x < 1$$

23. **(a)** Show that

$$f(x) = 1 - x^2 \sim 8 \sum_{j=1}^{\infty} \frac{J_0(\lambda_j x)}{\lambda_j^3 J_1(\lambda_j)}, \qquad 0 \le x \le 1,$$

where the λ_j are the positive roots of $J_0(x) = 0$. (*Hint*: Use Eq. (7.2–24) and Exercise 21(a).)

(b) Explain why the representation in part (a) is valid for $-1 \le x \le 1$.

***(c)** Compute the first five coefficients in the representation in part (a) to three decimals.

***(d)** Compute $f(0)$ in the representation in part (a) to three decimals.

24. Obtain the recursion formula shown in Exercise 10(b).

25. **(a)** Obtain the Fourier–Bessel series representation of x in terms of $J_1(\lambda_j x)$ on the interval $0 < x < 1$, where λ_j are such that $J_1'(\lambda_j) = 0$, $j = 1, 2, \ldots$.

(b) Show that the series obtained in part (a) is valid for $-1 < x < 1$.

***(c)** Compute the first five terms of the series of part (a) to three decimals at $x = 1/2$. How does the sum of these terms compare with the true value at $x = 1/2$?

26. Show that

$$J_{n-2}(x) = \frac{2(n-1)}{x} J_{n-1}(x) - J_n(x),$$

for $n = 2, 3, 4, \ldots$.

27. Consider the series

$$2 \sum_{j=1}^{\infty} [\lambda_j J_1(\lambda_j)]^{-1}$$

obtained from (7.2–24) when $x = 0$.

(a) Show that the series is an *alternating* series.

(b) Using the Leibniz criteria for convergence of alternating series in Section 1.4, prove that the series converges.

*(c) Obtain a graphical interpretation of the slow convergence of the series by computing the partial sums S_N for $N = 1, 2, \ldots, 10$ and graphing these values.

(d) Show that the fortieth term in the series is approximately -0.224.

28. Obtain each of the following:

(a) $$\int_0^1 x^2 J_0(\lambda x)\,dx = \frac{1}{\lambda}J_1(\lambda) + \frac{1}{\lambda^2}J_0(\lambda) - \frac{1}{\lambda^3}\int_0^\lambda J_0(s)\,ds$$

(b) $$\int_0^1 x^3 J_0(\lambda x)\,dx = \frac{\lambda^2 - 4}{\lambda^3}J_1(\lambda) + \frac{2}{\lambda^2}J_0(\lambda)$$

(c) $$\int_0^1 x(1 - x^2)J_0(\lambda x)\,dx = \frac{4}{\lambda^3}J_1(\lambda) - \frac{2}{\lambda^2}J_0(\lambda)$$

29. Bessel's function of the first kind of integral order can be written in the form of an integral as follows:

$$J_n(x) = \frac{1}{\pi}\int_0^\pi \cos(n\theta - x\sin\theta)\,d\theta.$$

(a) Show that for $n = 0$ the above becomes

$$J_0(x) = \frac{2}{\pi}\int_0^{\pi/2} \cos(x\sin\theta)\,d\theta.$$

(b) Verify that the integral in part (a) satisfies Bessel's differential equation.

(c) Obtain the equivalent integral form

$$J_0(x) = \frac{2}{\pi}\int_0^1 \frac{\cos xt}{\sqrt{1 - t^2}}\,dt.$$

••• **30.** Bessel's differential equation of *nonintegral* order $\nu\ (\geq 0)$ is

$$x^2y'' + xy' + (x^2 - \nu^2)y = 0, \qquad \textbf{(7.2-27)}$$

where ν is a real number. One solution is the Bessel function of the first kind of order ν given by

$$J_\nu(x) = \sum_{m=0} \frac{(-1)^m}{m!\,\Gamma(\nu + m + 1)}\left(\frac{x}{2}\right)^{2m+\nu},$$

where the *gamma function* (Fig. 7.2-6) is defined by

$$\Gamma(\nu) = \int_0^\infty t^{\nu-1}\exp(-t)\,dt, \qquad \nu > 0.$$

(a) Show that

$$\Gamma(\nu + 1) = \nu\Gamma(\nu), \qquad \nu > 0.$$

(b) Obtain the relation

$$\Gamma(n + 1) = n!, \qquad n = 0, 1, 2, \ldots.$$

(c) Show that

$$\Gamma(1/2) = \sqrt{\pi}.$$

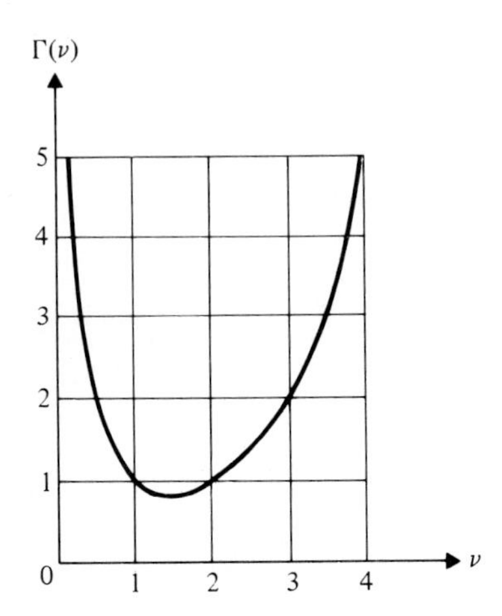

Figure 7.2-6
The gamma function.

31. Obtain the following approximation for small values of x ($x \to 0$):

$$J_1(x) \doteq \frac{x}{2}[\Gamma(2)]^{-1} = \frac{x}{2}$$

32. Obtain the following approximation for large values of x ($x \to \infty$):

$$J_1(x) \doteq \frac{2}{\sqrt{\pi x}} \cos\left(x - \frac{3\pi}{4}\right)$$

33. **(a)** Prove that between consecutive positive zeros of $J_n(x)$ there is at least one zero of $J_{n+1}(x)$.

(b) Prove that between consecutive positive zeros of $J_n(x)$ there is at least one zero of $J_{n-1}(x)$.

(c) Prove that the zeros of $J_n(x)$ and $J_{n+1}(x)$ separate each other.

7.3 SPHERICAL COORDINATES; LEGENDRE POLYNOMIALS

Spherical coordinates (r, ϕ, θ) may be defined as shown in Fig. 7.3-1. We have $r \geq 0$, $0 \leq \phi < 2\pi$, and $0 \leq \theta \leq \pi$. A word of caution is in order here. Some authors use ρ instead of r, some interchange ϕ and θ, and some follow both practices. It is essential, therefore, to note a particular author's definition.

In *cylindrical coordinates* (ρ, ϕ, z) the Laplacian is

$$\nabla^2 u = \frac{\partial^2 u}{\partial \rho^2} + \frac{1}{\rho}\frac{\partial u}{\partial \rho} + \frac{1}{\rho^2}\frac{\partial^2 u}{\partial \phi^2} + \frac{\partial^2 u}{\partial z^2}. \tag{7.3-1}$$

The relation between rectangular and spherical coordinates is given by

$$x = r \sin \theta \cos \phi, \qquad y = r \sin \theta \sin \phi, \qquad z = r \cos \theta,$$

and we could obtain the Laplacian in spherical coordinates from these. It is a bit simpler and more instructive, however, to begin with Eq. (7.3–1) and evaluate the four terms of that sum in spherical coordinates.

If we hold ϕ fixed, then u is a function of ρ and z, and by virtue of

$$z = r \cos \theta, \qquad \rho = r \sin \theta,$$

u is a function of r and θ. Thus we have by the chain rule

$$\frac{\partial u}{\partial \rho} = \frac{\partial u}{\partial r}\frac{\partial r}{\partial \rho} + \frac{\partial u}{\partial \theta}\frac{\partial \theta}{\partial \rho}$$

$$= \frac{\rho}{r}\frac{\partial u}{\partial r} + \frac{z}{r^2}\frac{\partial u}{\partial \theta}. \qquad \textbf{(7.3–2)}$$

In obtaining Eq. (7.3–2) we have used the relations $z^2 + \rho^2 = r^2$ and $\rho/z = \tan \theta$ to find $\partial r/\partial \rho$ and $\partial \theta/\partial \rho$. Note that z is held constant when differentiating r and θ partially with respect to ρ.

Then

$$\frac{\partial^2 u}{\partial \rho^2} = \frac{\partial}{\partial \rho}\left(\frac{\rho}{r}\frac{\partial u}{\partial r} + \frac{z}{r^2}\frac{\partial u}{\partial \theta}\right),$$

and we use Eq. (7.3–2), since it serves as a formula for differentiating *any* function of r and θ. Symbolically,

$$\frac{\partial(\ )}{\partial \rho} = \frac{\partial(\ )}{\partial r}\frac{\rho}{r} + \frac{\partial(\ )}{\partial \theta}\frac{z}{r^2}.$$

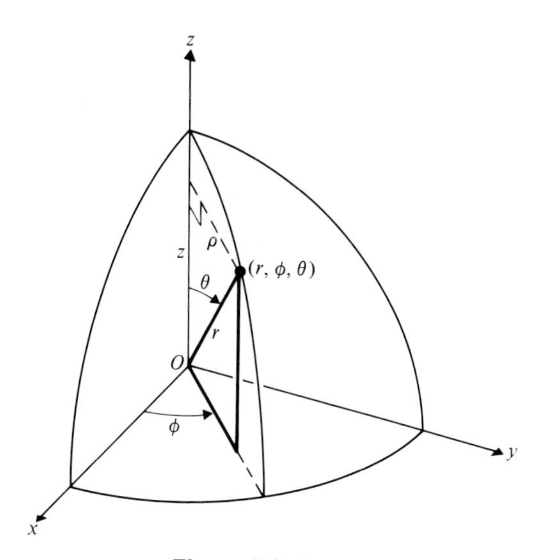

Figure 7.3–1
Spherical coordinates.

Hence

$$\frac{\partial^2 u}{\partial \rho^2} = \frac{\partial u_r}{\partial \rho}\frac{\rho}{r} + \frac{1}{r}u_r + u_r\rho\frac{\partial}{\partial \rho}\left(\frac{1}{r}\right) + \frac{\partial u_\theta}{\partial \rho}\frac{z}{r^2} + u_\theta z\frac{\partial}{\partial \rho}\left(\frac{1}{r^2}\right)$$

$$= \frac{\rho}{r}\left(\frac{\partial^2 u}{\partial r^2}\frac{\rho}{r} + \frac{\partial^2 u}{\partial \theta \partial r}\frac{z}{r^2}\right) + \frac{1}{r}u_r + u_r\rho\left(-\frac{1}{r^2}\frac{\rho}{r}\right)$$

$$+ \frac{z}{r^2}\left(\frac{\partial^2 u}{\partial r \partial \theta}\frac{\rho}{r} + \frac{\partial^2 u}{\partial \theta^2}\frac{z}{r^2}\right) + u_\theta z\left(\frac{2}{r^3}\frac{\rho}{r}\right)$$

$$= \frac{\rho^2}{r^2}\frac{\partial^2 u}{\partial r^2} + \frac{2\rho}{r^3}\frac{\partial^2 u}{\partial r \partial \theta} + \frac{1}{r}\frac{\partial u}{\partial r} - \frac{\rho^2}{r^3}\frac{\partial u}{\partial r} + \frac{z^2}{r^4}\frac{\partial^2 u}{\partial \theta^2} - \frac{2\rho z}{r^4}\frac{\partial u}{\partial \theta}.$$

In a similar manner we can calculate $\partial^2 u/\partial z^2$. We have

$$\frac{\partial u}{\partial z} = \frac{\partial u}{\partial r}\frac{\partial r}{\partial z} + \frac{\partial u}{\partial \theta}\frac{\partial \theta}{\partial z} = \frac{\partial u}{\partial r}\frac{z}{r} + \frac{\partial u}{\partial \theta}\left(-\frac{\rho}{r^2}\right)$$

$$\frac{\partial^2 u}{\partial z^2} = \frac{\partial}{\partial z}\left(\frac{z}{r}\frac{\partial u}{\partial r}\right) - \frac{\partial}{\partial z}\left(\frac{\rho}{r^2}\frac{\partial u}{\partial \theta}\right)$$

$$= \frac{1}{r}\frac{\partial u}{\partial r} + z\frac{\partial u}{\partial r}\left(-\frac{1}{r^2}\right)\frac{z}{r} + \frac{z}{r}\left(\frac{\partial^2 u}{\partial r^2}\frac{z}{r} - \frac{\rho}{r^2}\frac{\partial^2 u}{\partial r \partial \theta}\right)$$

$$- \rho\frac{\partial u}{\partial \theta}\left(-\frac{2}{r^3}\right)\frac{z}{r} - \frac{\rho}{r^2}\left[\frac{\partial^2 u}{\partial \theta \partial r}\frac{z}{r} - \frac{\partial^2 u}{\partial \theta^2}\left(-\frac{\rho}{r^2}\right)\right]$$

$$= \frac{1}{r}\frac{\partial u}{\partial r} - \frac{z^2}{r^3}\frac{\partial u}{\partial r} + \frac{z^2}{r^2}\frac{\partial^2 u}{\partial r^2} - \frac{2\rho z}{r^3}\frac{\partial^2 u}{\partial r \partial \theta} + \frac{2\rho z}{r^4}\frac{\partial u}{\partial \theta} + \frac{\rho^2}{r^4}\frac{\partial^2 u}{\partial \theta^2}$$

Hence

$$u_{zz} + u_{\rho\rho} = \frac{\partial u}{\partial r}\left(\frac{r^2 - z^2 + r^2 - \rho^2}{r^3}\right) + \frac{\partial^2 u}{\partial r^2}\left(\frac{z^2 + \rho^2}{r^2}\right) + \frac{\partial^2 u}{\partial \theta^2}\left(\frac{z^2 + \rho^2}{r^4}\right)$$

$$= \frac{1}{r}\frac{\partial u}{\partial r} + \frac{\partial^2 u}{\partial r^2} + \frac{1}{r^2}\frac{\partial^2 u}{\partial \theta^2}.$$

Finally, adding the equivalent of the terms

$$\frac{1}{\rho}\frac{\partial u}{\partial \rho} \quad \text{and} \quad \frac{1}{\rho^2}\frac{\partial^2 u}{\partial \phi^2},$$

we have

$$\nabla^2 u = \frac{\partial^2 u}{\partial r^2} + \frac{2}{r}\frac{\partial u}{\partial r} + \frac{1}{r^2 \sin^2\theta}\frac{\partial^2 u}{\partial \phi^2} + \frac{1}{r^2}\frac{\partial^2 u}{\partial \theta^2} + \frac{\cot\theta}{r^2}\frac{\partial u}{\partial \theta}, \tag{7.3–3}$$

which is the Laplacian in spherical coordinates. It corresponds to Eq. (7.3–1) in cylindrical coordinates and to

$$\nabla^2 u = \frac{\partial^2 u}{\partial x^2} + \frac{\partial^2 u}{\partial y^2} + \frac{\partial^2 u}{\partial z^2}$$

in rectangular coordinates.

We point out as in Section 7.2 that Eq. (7.3–3) must be dimensionally correct. Thus since $[u_{rr}] = T/L^2$ in a heat conduction problem, each of the addends must have the same dimension.

Solution of Laplace's Equation in Spherical Coordinates

Laplace's equation (also called the potential equation) in spherical coordinates can be written as

$$\nabla^2 u = \frac{\partial^2 u}{\partial r^2} + \frac{2}{r}\frac{\partial u}{\partial r} + \frac{1}{r^2 \sin^2\theta}\frac{\partial^2 u}{\partial \phi^2} + \frac{1}{r^2}\frac{\partial^2 u}{\partial \theta^2} + \frac{\cot\theta}{r^2}\frac{\partial u}{\partial \theta} = 0.$$

An equivalent form of this equation is (Exercise 1)

$$\frac{1}{r^2}\frac{\partial}{\partial r}\left(r^2 \frac{\partial u}{\partial r}\right) + \frac{1}{r^2 \sin\theta}\frac{\partial}{\partial \theta}\left(\sin\theta \frac{\partial u}{\partial \theta}\right) + \frac{1}{r^2 \sin^2\theta}\frac{\partial^2 u}{\partial \phi^2} = 0. \tag{7.3–4}$$

We seek a solution by the method of separation of variables. Assume that

$$u(r, \phi, \theta) = R(r)\Phi(\phi)\Theta(\theta)$$

and substitute into Eq. (7.3–4). Then

$$\frac{1}{r^2}\frac{d}{dr}\left(r^2\Phi\Theta \frac{dR}{dr}\right) + \frac{1}{r^2 \sin\theta}\frac{d}{d\theta}\left(R\Phi \sin\theta \frac{d\Theta}{d\theta}\right) + \frac{1}{r^2 \sin^2\theta} R\Theta \frac{d^2\Phi}{d\phi^2} = 0.$$

Next, divide each term by $R\Phi\Theta/r^2 \sin^2\theta$ to obtain

$$\frac{\sin^2\theta}{R}\frac{d}{dr}\left(r^2 \frac{dR}{dr}\right) + \frac{\sin\theta}{\Theta}\frac{d}{d\theta}\left(\sin\theta \frac{d\Theta}{d\theta}\right) = -\frac{1}{\Phi}\frac{d^2\Phi}{d\phi^2}.$$

Since the left-hand member is independent of ϕ, we have

$$-\frac{1}{\Phi}\frac{d^2\Phi}{d\phi^2} = m^2, \qquad m = 0, 1, 2, \ldots, \tag{7.3–5}$$

where the first separation constant m is chosen to be a nonnegative integer in order that the function Φ (and also u) be periodic of period 2π in ϕ. This is often necessary from physical considerations, as we shall see later.

Separating variables again produces

$$\frac{1}{R}\frac{d}{dr}\left(r^2 \frac{dR}{dr}\right) = -\left[\frac{1}{\Theta \sin\theta}\frac{d}{d\theta}\left(\sin\theta \frac{d\Theta}{d\theta}\right) - \frac{m^2}{\sin^2\theta}\right] = \lambda,$$

where we have called the second separation constant λ. At this point, nothing further is known about this quantity.

Thus we have reduced Laplace's equation to the following three second-order, linear, homogeneous ordinary differential equations:

$$\frac{d^2\Phi}{d\phi^2} + m^2\Phi = 0, \tag{7.3-6}$$

$$\frac{1}{\sin\theta}\frac{d}{d\theta}\left(\sin\theta\frac{d\Theta}{d\theta}\right) + \left(\lambda - \frac{m^2}{\sin^2\theta}\right)\Theta = 0, \tag{7.3-7}$$

$$\frac{d}{dr}\left(r^2\frac{dR}{dr}\right) - \lambda R = 0. \tag{7.3-8}$$

Note that the first and third equations each contain one of the two separation constants, while the second contains both constants. Products of solutions of these three equations are called **spherical harmonics**.

Equations (7.3–6) have constant coefficients, so their solution presents no difficulty. They are

$$\Phi(m\phi) = A_m \cos m\phi + B_m \sin m\phi, \qquad m = 0, 1, 2, \ldots, \tag{7.3-9}$$

where A_m and B_m are arbitrary constants that can be determined from given boundary conditions.

Next we look at Eq. (7.3–8), which can be written in the equivalent form

$$r^2\frac{d^2R}{dr^2} + 2r\frac{dR}{dr} - \lambda R = 0.$$

This is a Cauchy-Euler equation. It can be solved by making the substitution $R = r^k$, in which case the differential equation becomes

$$r^2k(k-1)r^{k-2} + 2rkr^{k-1} - \lambda r^k = 0$$

or

$$(k^2 + k - \lambda)r^k = 0.$$

Hence $R = r^k$ is a solution of Eq. (7.3–8) provided that $k^2 + k - \lambda = 0$. To find a second, linearly independent solution may not be so simple. If we choose $k = n$, then $\lambda = n(n+1)$, and if we further (wisely) choose $k = -(n+1)$, then *also* $\lambda = n(n+1)$. Thus with $\lambda = n(n+1)$, Eq. (7.3–8) has the two linearly independent solutions r^n and $r^{-(n+1)}$ (Exercise 2), so the general solution can be written as

$$R_n(r) = C_n r^n + D_n r^{-(n+1)}. \tag{7.3-10}$$

In order to solve Eq. (7.3–7) we make the following substitutions:

$$x = \cos\theta, \qquad \Theta(\theta) = y(x), \qquad \frac{d}{d\theta} = \frac{dx}{d\theta}\frac{d}{dx} = -\sin\theta\frac{d}{dx}.$$

Then

$$\frac{d}{d\theta}\left(\sin\theta\frac{d\Theta}{d\theta}\right) = -\sin\theta\frac{d}{dx}\left(\sin\theta\,\frac{dx}{d\theta}\frac{d\Theta}{dx}\right)$$

$$= \sin\theta\frac{d}{dx}\left(\sin^2\theta\,\frac{dy}{dx}\right)$$

$$= \sqrt{1-x^2}\,\frac{d}{dx}\left[(1-x^2)\frac{dy}{dx}\right].$$

With these substitutions, Eq. (7.3–7) becomes

$$\frac{d}{dx}\left[(1-x^2)\frac{dy}{dx}\right] + \left[n(n+1) - \frac{m^2}{1-x^2}\right]y = 0$$

or, in equivalent form,

$$(1-x^2)\frac{d^2y}{dx^2} - 2x\frac{dy}{dx} + \left[n(n+1) - \frac{m^2}{1-x^2}\right]y = 0. \qquad \textbf{(7.3–11)}$$

Equation (7.3–11) is **Legendre's associated differential equation.** Its general solution, found by a series method, is

$$y_{n,m}(x) = c_{n,m}P_n^m(x) + d_{n,m}Q_n^m(x),$$

where $P_n^m(x)$ and $Q_n^m(x)$ are called **associated Legendre functions of the first and second kind**, respectively. The use of subscripts and superscripts indicates that these functions depend on m and n as well as on the argument x. The arbitrary constants $c_{n,m}$ and $d_{n,m}$ depend on m and n as well.

If $m = 0$, then Eq. (7.3–11) becomes

$$(1-x^2)\frac{d^2y}{dx^2} - 2x\frac{dy}{dx} + n(n+1)y = 0, \qquad \textbf{(7.3–12)}$$

which is known as **Legendre's differential equation.** A particular solution is $y = P_n(x)$, the Legendre polynomial of degree n, $n = 0, 1, 2, \ldots$ (see Example 1.5–6). Note that n must be a nonnegative integer in order to obtain a solution of Eq. (7.3–12) that is *bounded* on $-1 \le x \le 1$.

A second linearly independent solution is $Q_n(x)$. Since $Q_n(x)$ has a singularity at $x = \pm 1$ (as we will show later), it can be used only when $x \ne 1$ $(\theta \ne 0)$ and when $x \ne -1$ $(\theta \ne \pi)$.

The case where u is independent of ϕ implies $m = 0$ (see Eq. (7.3–9)). In this case Laplace's equation in spherical coordinates (7.3–4) reduces to

$$\frac{1}{r^2}\frac{\partial}{\partial r}\left(r^2\frac{\partial u}{\partial r}\right) + \frac{1}{r^2\sin\theta}\frac{\partial}{\partial\theta}\left(\sin\theta\,\frac{\partial u}{\partial\theta}\right) = 0,$$

with solutions that are products of

$$R_n(r) = C_nr^n + D_nr^{-(n+1)}$$

and

$$\Theta_n(\theta) = E_n P_n(\cos\theta) + F_n Q_n(\cos\theta), \qquad n = 0, 1, 2, \ldots.$$

Admittedly, we have made a number of seemingly restrictive simplifying assumptions in order to get to this point in solving Laplace's equation in spherical coordinates. This was done not merely to simplify the mathematical aspects. We shall see in Section 7.4 that many of the applications will yield to our simplified approach. It should be kept in mind, however, that the *nature* of the separation constants m and λ depends on the boundary conditions in any given case.

Legendre Polynomials

In Example 1.5–6 we solved Legendre's differential equation (7.3–12) by the method of Frobenius and obtained the particular solutions $P_n(x)$, called **Legendre polynomials**. For reference we list the first few Legendre polynomials here:

$$P_0(x) = 1, \qquad P_1(x) = x, \qquad P_2(x) = \frac{1}{2}(3x^2 - 1),$$

$$P_3(x) = \frac{1}{2}(5x^3 - 3x), \qquad P_4(x) = \frac{1}{8}(35x^4 - 30x^2 + 3).$$

Graphs of $P_2(x)$, $P_3(x)$, and $P_{10}(x)$ are shown in Fig. 7.3–2.

Some properties of the Legendre polynomials that are useful in solving certain boundary-value problems are listed below. (See also Exercise 3.)

$$\begin{aligned}
&\text{(a)} \quad P_{2n+1}(0) = 0\\
&\text{(b)} \quad P_n(1) = 1\\
&\text{(c)} \quad P_n(-1) = (-1)^n\\
&\text{(d)} \quad P'_{n+1}(x) - xP'_n(x) = (n+1)P_n(x), \qquad n = 1, 2, \ldots\\
&\text{(e)} \quad xP'_n(x) - P'_{n-1}(x) = nP_n(x), \qquad n = 1, 2, \ldots\\
&\text{(f)} \quad P'_{n+1}(x) - P'_{n-1}(x) = (2n+1)P_n(x), \qquad n = 1, 2, \ldots
\end{aligned} \tag{7.3–13}$$

Note that property (f) is the sum of properties (d) and (e). We can prove property (d) from the **definition of the Legendre polynomials**,

$$P_n(x) = \frac{1}{2^n} \sum_{k=0}^{N} \frac{(-1)^k (2n-2k)!}{k!(n-2k)!(n-k)!} x^{n-2k}, \tag{7.3–14}$$

where $N = n/2$ if n is even and $N = (n - 1)/2$ if n is odd. We have

$$P_{n+1}(x) = \frac{1}{2^{n+1}} \sum_{k=0}^{N} \frac{(-1)^k(2n-2k+2)!}{k!(n-2k+1)!(n-k+1)!} x^{n-2k+1}$$

$$P'_{n+1}(x) = \frac{1}{2^{n+1}} \sum_{k=0}^{N} \frac{(-1)^k(2n-2k+2)!(n-2k+1)}{k!(n-2k+1)!(n-k+1)!} x^{n-2k}$$

$$P'_n(x) = \frac{1}{2^n} \sum_{k=0}^{N} \frac{(-1)^k(2n-2k)!(n-2k)}{k!(n-2k)!(n-k)!} x^{n-2k-1}$$

$$xP'_n(x) = \frac{1}{2^n} \sum_{k=0}^{N} \frac{(-1)^k(2n-2k)!(n-2k)}{k!(n-2k)!(n-k)!} x^{n-2k}$$

$$P'_{n+1}(x) - xP'_n(x) = \frac{1}{2^{n+1}} \sum_{k=0}^{N} \frac{(-1)^k(2n-2k+2)(2n-2k+1)(2n-2k)!}{k!(n-2k)!(n-k+1)(n-k)!} x^{n-2k}$$

$$- \frac{1}{2^n} \sum_{k=0}^{N} \frac{(-1)^k(2n-2k)!(n-2k)}{k!(n-2k)!(n-k)!} x^{n-2k}$$

$$P'_{n+1}(x) - xP'_n(x) = (2n-2k+1-n+2k) \frac{1}{2^n} \sum_{k=0}^{N} \frac{(-1)^k(2n-2k)!}{k!(n-2k)!(n-k)!} x^{n-2k}$$

$$= (n+1)P_n(x).$$

Property (e) may be shown in a similar manner (Exercise 3).

Orthogonality of Legendre Polynomials

Now we show under what conditions the Legendre polynomials are orthogonal. This property of orthogonality is essential in solving boundary-value problems as we will see in the following section.

We begin with the fact that the Legendre polynomials $P_n(x)$ satisfy Legendre's differential equation,

$$\frac{d}{dx} [(1 - x^2)P'_n(x)] + n(n + 1)P_n(x) = 0, \qquad n = 0, 1, 2, \ldots .$$

Multiplying this equation by $P_m(x)$ and integrating from -1 to 1 results in

$$\int_{-1}^{1} P_m(x) \frac{d}{dx} [(1 - x^2)P'_n(x)]\, dx + n(n + 1) \int_{-1}^{1} P_m(x)P_n(x)\, dx = 0. \tag{7.3-15}$$

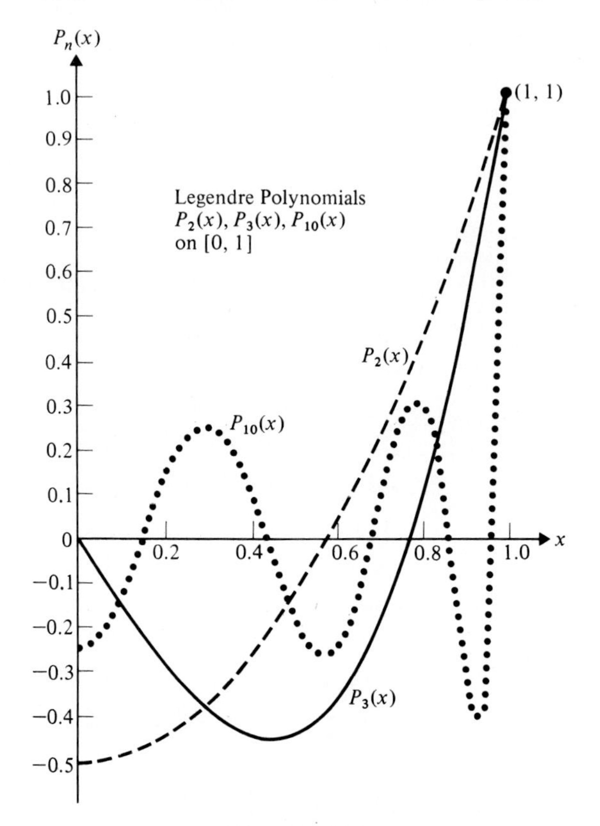

Figure 7.3-2
Legendre polynomials.

The first integral may be integrated by parts, putting

$$u = P_m(x), \qquad du = P_m'(x)\,dx,$$
$$dv = \frac{d}{dx}[(1 - x^2)P_n'(x)]\,dx, \qquad v = (1 - x^2)P_n'(x).$$

Then

$$\int_{-1}^{1} P_m(x)\frac{d}{dx}[(1 - x^2)P_n'(x)]\,dx$$
$$= P_m(x)P_n'(x)(1 - x^2)\Big|_{-1}^{1} - \int_{-1}^{1} (1 - x^2)P_n'(x)P_m'(x)\,dx.$$

The first term on the right vanishes at both limits by virtue of the term $(1 - x^2)$. Thus Eq. (7.3–15) reduces to

$$-\int_{-1}^{1} (1 - x^2)P_n'(x)P_m'(x)\,dx + n(n + 1)\int_{-1}^{1} P_m(x)P_n(x)\,dx = 0.$$

In this last equation, m and n have no special significance except that they are both nonnegative integers. Hence we may interchange m and n to obtain

$$-\int_{-1}^{1} (1 - x^2)P_m'(x)P_n'(x)\,dx + m(m + 1)\int_{-1}^{1} P_n(x)P_m(x)\,dx = 0.$$

Subtracting this equation from the previous one yields

$$(n - m)(n + m + 1)\int_{-1}^{1} P_m(x)P_n(x)\,dx = 0.$$

Now suppose that $n \neq m$. Then $n - m \neq 0$ and $n + m + 1 = 0$ is impossible. (Why?) Hence we are left with the result

$$\int_{-1}^{1} P_m(x)P_n(x)\,dx = 0, \qquad m \neq n. \tag{7.3–16}$$

This shows that the set

$$\{P_0(x), P_1(x), P_2(x), \ldots\}$$

is an orthogonal set on $[-1, 1]$ with weight function one. See also Exercise 24.

In applications the Legendre polynomials are often expressed in terms of θ. Let $x = \cos\theta$, $dx = -\sin\theta\,d\theta$, and change the limits accordingly. Then Eq. (7.3–16) becomes

$$\int_{\pi}^{0} P_m(\cos\theta)P_n(\cos\theta)(-\sin\theta d\theta) = 0, \qquad m \neq n$$

or

$$\int_{0}^{\pi} \sin\theta\, P_m(\cos\theta)P_n(\cos\theta)\,d\theta = 0, \qquad m \neq n.$$

Thus the set

$$\{P_0(\cos\theta), P_1(\cos\theta), P_2(\cos\theta), \ldots\}$$

is an orthogonal set on the interval $0 \leq \theta \leq \pi$ with weight function $\sin\theta$.

If in Eq. (7.3–16) we replace n by $2n$ and m by $2m$, then

$$\int_{-1}^{1} P_{2m}(x)P_{2n}(x)\,dx = 2\int_{0}^{1} P_{2m}(x)P_{2n}(x)\,dx = 0, \qquad n \neq m.$$

In other words, the Legendre polynomials of *even* degree are orthogonal on the interval $0 \leq x \leq 1$ with weight function one. Similarly, the Legendre polynomials of *odd* degree are orthogonal on the interval $0 \leq x \leq 1$ with weight function one (Exercise 4).

Legendre Series*

The orthogonality property of the Legendre polynomials makes it possible to represent certain functions f by a **Legendre series**, that is, a series of Legendre polynomials. This representation is possible because Legendre's differential

*Some authors call this a *Fourier–Legendre series*.

equation (7.3–12), together with an appropriate boundary condition, constitutes a singular Sturm–Liouville problem of the type discussed in Section 2.7 (see Exercise 23). Moreover, it can be shown* that the normalized Legendre polynomials that we will obtain in this section form a **complete orthonormal set** with respect to piecewise smooth functions on $(-1, 1)$.

For such a function we can write

$$f(x) = A_0P_0(x) + A_1P_1(x) + A_2P_2(x) + A_3P_3(x) + \cdots.$$

In order to find A_2, for example, we multiply the above by $P_2(x)$ and integrate from -1 to 1. Then

$$\begin{aligned}\int_{-1}^{1} f(x)P_2(x)\,dx = A_0 &\int_{-1}^{1} P_0(x)P_2(x)\,dx \\ + A_1 &\int_{-1}^{1} P_1(x)P_2(x)\,dx \\ + A_2 &\int_{-1}^{1} P_2(x)P_2(x)\,dx \\ + A_3 &\int_{-1}^{1} P_3(x)P_2(x)\,dx + \cdots.\end{aligned}$$

Because of the orthogonality property of the $P_n(x)$, each integral on the right with the exception of the third one is zero. Thus

$$\int_{-1}^{1} f(x)P_2(x)\,dx = A_2 \int_{-1}^{1} [P_2(x)]^2\,dx,$$

from which we get

$$A_2 = \frac{\int_{-1}^{1} f(x)P_2(x)\,dx}{\int_{-1}^{1} [P_2(x)]^2\,dx}.$$

Each coefficient A_n can be found in the same way so that, in general,

$$A_n = \frac{\int_{-1}^{1} f(x)P_n(x)\,dx}{\int_{-1}^{1} [P_n(x)]^2\,dx}, \qquad n = 0, 1, 2, \ldots. \tag{7.3–17}$$

Next we evaluate the denominator in Eq. (7.3–17). This quantity is referred to as the **square of the norm** and is denoted by $\|P_n\|^2$. In order to evaluate this term we first need a result known as **Rodrigues' formula.**†

---

*See Ruel V. Churchill and J. W. Brown, *Fourier Series and Boundary Value Problems*, 3rd ed., (New York: McGraw-Hill, 1978), pp. 234 ff.

†After Olinde Rodrigues (1794–1851), a French economist and mathematician.

We begin with the binomial expansion of $(x^2 - 1)^n$, which can be written as

$$(x^2 - 1)^n = \sum_{k=0}^{n} (-1)^k \frac{n!}{k!(n-k)!} x^{2n-2k}.$$

Differentiating this n times yields (Exercise 16)

$$\frac{d^n(x^2-1)^n}{dx^n} = \sum_{k=0}^{N} \frac{(-1)^k n!(2n-2k)!}{k!(n-k)!(n-2k)!} x^{n-2k}, \qquad \textbf{(7.3–18)}$$

where the last term is a constant. But $n - 2N = 0$ implies that $N = n/2$, whereas $n - 1 - 2N = 0$ implies that $N = (n-1)/2$. Then, since N is a nonnegative integer, it is defined in Eq. (7.3–18) as $n/2$ if n is even and $(n-1)/2$ if n is odd.

Recall the definition of $P_n(x)$ in summation form,

$$P_n(x) = \frac{1}{2^n} \sum_{k=0}^{N} \frac{(-1)^k(2n-2k)!}{k!(n-2k)!(n-k)!} x^{n-2k}, \qquad \textbf{(7.3–14)}$$

where N is defined as in Eq. (7.3–18). A comparison of Eqs. (7.3–14) and (7.3–18) shows that

$$P_n(x) = \frac{1}{2^n n!} \frac{d^n(x^2-1)^n}{dx^n}, \qquad n = 0, 1, 2, \ldots, \qquad \textbf{(7.3–19)}$$

which is Rodrigues' formula.

We can use Eq. (7.3–19) to evaluate $\|P_n\|^2$ as follows. We have

$$\int_{-1}^{1} [P_n(x)]^2\, dx = \int_{-1}^{1} P_n(x) \frac{1}{2^n n!} \frac{d^n}{dx^n}(x^2-1)^n\, dx,$$

and integrating by parts with

$$u = P_n(x) \qquad dv = \frac{d^n}{dx^n}(x^2-1)^n\, dx$$

$$du = P_n'(x)\, dx \qquad v = \frac{d^{n-1}}{dx^{n-1}}(x^2-1)^n,$$

produces

$$\int_{-1}^{1} [P_n(x)]^2\, dx = \frac{1}{2^n n!} \left[P_n(x) \frac{d^{n-1}}{dx^{n-1}} (x^2-1)^n \Big|_{-1}^{1} - \int_{-1}^{1} P_n'(x) \frac{d^{n-1}}{dx^{n-1}} (x^2-1)^n\, dx \right].$$

The first term on the right vanishes at both limits by virtue of the term $(x^2 - 1)$. Hence after $(n - 1)$ integrations by parts we have

$$\int_{-1}^{1} [P_n(x)]^2\, dx = \frac{(-1)^{n-1}}{2^n n!} \int_{-1}^{1} P_n^{(n-1)}(x) \frac{d}{dx} (x^2 - 1)^n\, dx.$$

One more integration produces

$$\int_{-1}^{1} [P_n(x)]^2\, dx = \frac{(-1)^n}{2^n n!} \int_{-1}^{1} P_n^{(n)}(x)(x^2 - 1)^n\, dx.$$

We now observe that

$$P_n^{(n)}(x) = \frac{(2n)!}{2^n (n!)^2}$$

from Eq. (7.3–14). Using the reduction formula found in most integral tables,

$$\int x^m(ax^n + b)^p\, dx = \frac{1}{m + np + 1}\left[x^{m+1}(ax^n + b)^p + npb \int x^m (ax^n + b)^{p-1}\, dx\right],$$

and integrating n times, we have

$$\int_{-1}^{1} (x^2 - 1)^n\, dx = \frac{(-1)^n 2^{2n+1}(n!)^2}{(2n + 1)!}.$$

Thus, putting everything together (Exercise 5), we have

$$\int_{-1}^{1} [P_n(x)]^2\, dx = \frac{(-1)^n}{2^n n!} \frac{(2n)!}{2^n n!} \frac{(-1)^n 2^{2n+1}(n!)^2}{(2n + 1)!}$$

or

$$\|P_n\|^2 = \frac{2}{2n + 1}, \qquad n = 0, 1, 2, \ldots. \tag{7.3–20}$$

Hence the set

$$\left\{ \frac{P_0(x)}{\sqrt{2}}, \frac{P_1(x)}{\sqrt{2/3}}, \frac{P_2(x)}{\sqrt{2/5}}, \ldots \right\}$$

is an *orthonormal* set on the interval $-1 \leq x \leq 1$ with weight function one. We can also update Eq. (7.3–17) to read

$$A_n = \frac{2n + 1}{2} \int_{-1}^{1} f(x)P_n(x)\, dx, \qquad n = 0, 1, 2, \ldots. \tag{7.3–21}$$

Legendre series have something in common with Fourier series. If a function is defined only on the interval (0, 1), it can be represented by a series of Legendre polynomials of even degree. This is accomplished by making an *even extension* of the function as we did to obtain Fourier cosine series in Section 4.3. We illustrate the procedure for this and for an *odd extension* in the following example.

EXAMPLE 7.3–1 Define

$$f(x) = 2(1 - x), \qquad 0 < x < 1.$$

Obtain the first two terms of the Legendre series representation of this function using (a) polynomials of even degree and (b) polynomials of odd degree.

Solution For (a) we make an even extension and modify Eq. (7.3–21) to

$$A_{2n} = (4n + 1) \int_0^1 f(x)P_{2n}(x)\,dx, \qquad n = 0, 1, 2, \ldots. \qquad \textbf{(7.3–22)}$$

Note that the integrand is an even function; hence we can use symmetry and change the limits as shown. Then

$$A_0 = 2\int_0^1 (1 - x)\,dx = 1,$$

$$A_2 = 5\int_0^1 (1 - x)(3x^2 - 1)\,dx = -\frac{5}{4};$$

hence

$$f(x) \sim P_0(x) - \frac{5}{4}P_2(x) + \cdots.$$

For (b) we make an odd extension and modify Eq. (7.3–21) to

$$A_{2n+1} = (4n + 3) \int_0^1 f(x)P_{2n+1}(x)\,dx, \qquad n = 0, 1, 2, \ldots. \qquad \textbf{(7.3–23)}$$

Observe that the integrand is an even function in this case also. Thus

$$A_1 = 6\int_0^1 (1 - x)x\,dx = 1,$$

$$A_3 = 7\int_0^1 (1 - x)(5x^3 - 3x)\,dx = -\frac{7}{4},$$

so that

$$f(x) \sim P_1(x) - \frac{7}{4}P_3(x) + \cdots.$$

The even and odd extensions are shown in Fig. 7.3–3. Note that $f(0) = 0$ in part (b). (Why?) ■

Another way of looking at the problems in Example 7.3–1 does not require making even and odd extensions of the given function f. Since $f(x)$ is

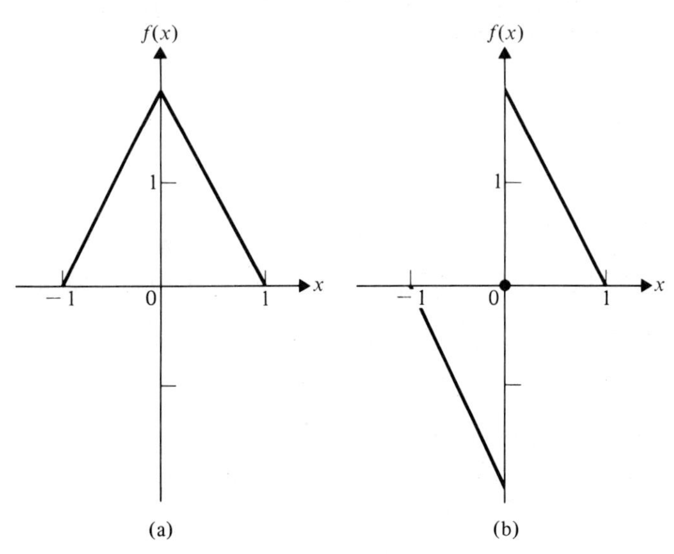

Figure 7.3-3
(a) Even extension. (b) Odd extension.

defined only on (0, 1) and since even and odd Legendre polynomials separately are orthogonal on this interval, we can proceed immediately to Eqs. (7.3-22) and (7.3-23) to obtain the appropriate coefficients.

Legendre Functions of the Second Kind

We close this section with a brief discussion of the Legendre functions of the second kind. Linear combinations of these and the Legendre polynomials comprise the general solution of Legendre's differential equation (7.3-12).

One solution of Legendre's differential equation with $n = 0, 1, 2, \ldots,$

$$(1 - x^2)y'' - 2xy' + n(n + 1)y = 0, \qquad -1 \leq x \leq 1,$$

is $u = P_n(x)$, the Legendre polynomials of degree n. A second linearly independent solution can be obtained by the method known as variation of parameters. It is given by

$$y_n(x) = B_n P_n(x) + A_n P_n(x) \int \frac{dx}{(1 - x^2)[P_n(x)]^2}. \tag{7.3-24}$$

The **Legendre function of the second kind**, $Q_n(x)$, is obtained from Eq. (7.3-24) by putting $A_n = 1$ and $B_n = 0$. Thus

$$Q_0(x) = \int^x \frac{dx}{1 - x^2} = \frac{1}{2} \log\left(\frac{1 + x}{1 - x}\right),$$

$$Q_1(x) = x \int^x \frac{dx}{x^2(1 - x^2)} = \frac{x}{2} \log\left(\frac{1 + x}{1 - x}\right) - 1.$$

Continuing in this way, we have

$$Q_2(x) = \frac{1}{4}(3x^2 - 1)\log\left(\frac{1+x}{1-x}\right) - \frac{3}{2}x,$$

$$Q_3(x) = \frac{x}{4}(5x^2 - 3)\log\left(\frac{1+x}{1-x}\right) - \frac{5}{2}x^2 + \frac{2}{3}.$$

Graphs of these functions are shown in Fig. 7.3–4.

From the definition of the Legendre Q-function,

$$Q_n(x) = P_n(x)\int \frac{dx}{(1 - x^2)[P_n(x)]^2}, \tag{7.3–25}$$

it follows that

$$Q_{2n}(-x) = P_{2n}(-x)\int \frac{-dx}{(1 - x^2)[P_{2n}(-x)]^2}$$

$$= -P_{2n}(x)\int \frac{dx}{(1 - x^2)[P_{2n}(x)]^2} = -Q_{2n}(x)$$

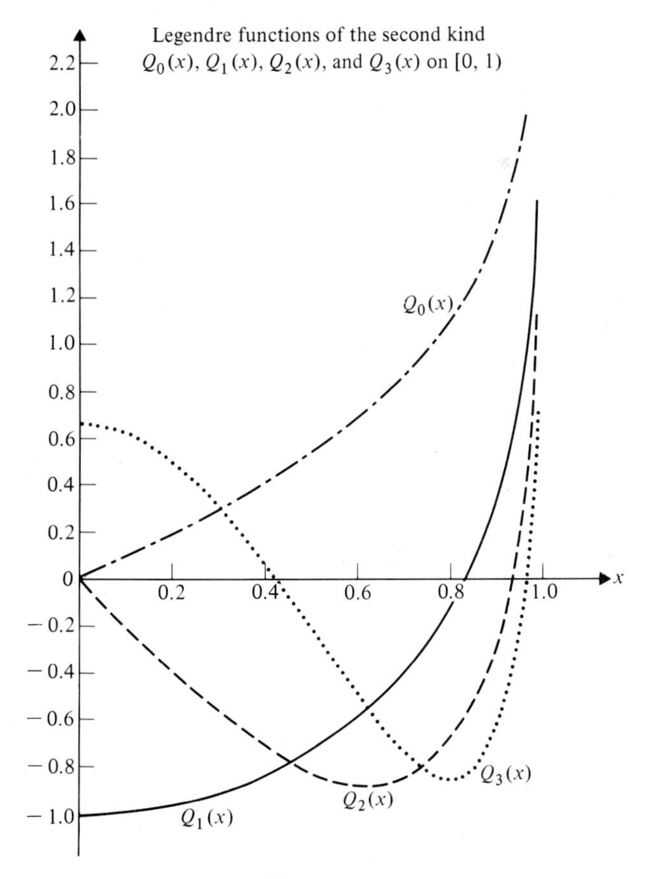

Figure 7.3–4

and also

$$Q_{2n+1}(-x) = P_{2n+1}(-x) \int \frac{-dx}{(1 - x^2)[P_{2n+1}(-x)]^2}$$

$$= P_{2n+1}(x) \int \frac{dx}{(1 - x^2)[P_{2n+1}(x)]^2} = Q_{2n+1}(x).$$

These results may be combined so that we have

$$Q_n(-x) = (-1)^{n+1}Q_n(x).$$

It is for this reason that the graphs in Fig. 7.3–4 are shown only on the interval $0 \leq x < 1$.

The term

$$\log \left(\frac{1 + x}{1 - x}\right)$$

in the Q-functions shows that these functions have a *singularity* at $x = \pm 1$. In spherical coordinates (r, ϕ, θ) this singularity is translated into singularities at $\theta = 0$ and $\theta = \pi$ by virtue of the relation $x = \cos \theta$.

We have obtained the Legendre functions of the second kind (Q-functions) in closed form. These functions can also be expressed as infinite series by expanding $\log (1 + x)/(1 - x)$ into a Maclaurin's series, that is, by using

$$\log \left(\frac{1 + x}{1 - x}\right) = 2\left(x + \frac{x^3}{3} + \frac{x^5}{5} + \frac{x^7}{7} + \cdots\right), \qquad -1 < x < 1.$$

Key Words and Phrases

spherical coordinates
spherical harmonics
Legendre's associated differential equation
associated Legendre functions of the first and second kind
Legendre's differential equation
Legendre polynomials
Legendre series
complete orthonormal set
square of the norm
Rodrigues' formula
Legendre functions of the second kind

7.3 Exercises

• **1.** Show that Eqs. (7.3–3) and (7.3–4) are equivalent forms of Laplace's equation in spherical coordinates.

2. Verify that r^{-n} and $r^{-(n+1)}$ are linearly independent solutions of Eq. (7.3–8). (*Hint*: Compute their Wronskian.)

3. Prove the following properties of the Legendre polynomials (compare Eq. (7.3–13)):

(a) $P_{2n+1}(0) = 0$

(b) $P_n(-1) = (-1)^n$

(c) $xP_n'(x) - P_{n-1}'(x) = nP_n(x), \qquad n = 1, 2, \ldots$

(d) $P_{2n}'(0) = 0$

(e) $P_{2n}(0) = (-1)^n \dfrac{(2n)!}{2^{2n}(n!)^2}$

4. Prove that the Legendre polynomials of odd degree are orthogonal on $0 \le x \le 1$ with weight function 1.

5. Carry out the details needed to arrive at Eq. (7.3–20).

6. Verify Eq. (7.3–20) for $n = 0, 1, 2, 3$, by computing the square of the norm directly from $P_n(x)$.

•• **7.** By direct computation show that $P_0(x)$, $P_1(x)$, and $P_2(x)$ are orthogonal on $-1 \le x \le 1$ with weight function 1.

8. Show that the interval of convergence of $Q_n(x)$ is $-1 < x < 1$.

9. Show that

$$\int_0^1 P_{2n}(x)\,dx = 0, \qquad n = 1, 2, \ldots .$$

(*Hint*: $P_0(x) = 1$.)

10. Show that

$$\int_{-1}^1 P_n'(x)P_m(x)\,dx = 1 - (-1)^{n+m},$$

where $0 \le m \le n$. (*Hint*: Use integration by parts.)

11. Show that

$$\int_{-1}^1 x[P_n(x)]^2\,dx = 0.$$

12. Express each of the following polynomials in terms of Legendre polynomials:

(a) $ax + b$

(b) $ax^2 + bx + c$

(c) $ax^3 + bx^2 + cx + d$

13. Prove that

$$P_n'(1) = \frac{n}{2}(n + 1).$$

14. **(a)** In Example 7.3–1, obtain the coefficients A_0, A_2, A_1, A_3.

(b) Compute A_4 and A_5.

(c) Obtain three coefficients in the Legendre series representation of the function

$$f(x) = \begin{cases} 0, & -1 < x < 0, \\ 2(1 - x), & 0 < x < 1. \end{cases}$$

(d) Evaluate $f(0)$ in the representation of part (c).

15. Obtain the first three nonzero coefficients in the Legendre series representation of each of the following functions.

(a) $f(x) = \begin{cases} 0, & -1 < x < 0, \\ 1, & 0 < x < 1 \end{cases}$

(*Hint*: Use the result in Exercise 3(e).)

(b) $f(x) = \begin{cases} 0, & -1 < x < 0, \\ x, & 0 < x < 1 \end{cases}$

(c) $f(x) = |x|, \quad -1 < x < 1$

16. Verify Eq. (7.3–18). Explain why the last term in the summation is a constant.

17. Obtain the formula

$$\int_x^1 P_n(s)\, ds = \frac{1}{2n+1}[P_{n-1}(x) - P_{n+1}(x)], \qquad n = 1, 2, \ldots.$$

(*Hint*: Use Eq. 7.3–13(f).)

18. If $f(x)$ is a polynomial of degree $m < n$, show that

$$\int_{-1}^1 f(x)\, P_n(x)\, dx = 0.$$

19. An integral representation of $P_n(x)$ called *Laplace's integral* is given by

$$P_n(x) = \frac{1}{\pi}\int_0^\pi [x + (x^2 - 1)^{1/2}\cos\phi]^n\, d\phi.$$

Verify this representation for $n = 0, 1, 2$.

20. (a) Expand the function

$$f(x) = \begin{cases} 0, & -1 < x < 0, \\ 2x + 1, & 0 < x < 1 \end{cases}$$

in a Legendre series, writing only the first four terms.

(b) To what value will the series (not just the four terms) converge when $x = 0$? When $x = -1/2$? When $x = 1/2$?

*(c) Use the four terms obtained in part (a) to compute $f(0)$, $f(-1/2)$, and $f(1/2)$.

21. Expand the function

$$f(x) = \begin{cases} -1, & -1 < x < 0, \\ 1, & 0 < x < 1 \end{cases}$$

in a Legendre series.

22. (a) Use Rodrigues' formula (Eq. 7.3–19) to obtain

$$2^n n! P_{n+1}(x) = (2n + 1)\frac{d^{(n-1)}u^n}{dx^{(n-1)}} + 2n\frac{d^{(n-1)}u^{n-1}}{dx^{(n-1)}},$$

where $u = x^2 - 1$.

(b) In the relation obtained in part (a), make the substitution

$$2^{n-1}(n-1)!P_{n-1}(x) = \frac{d^{(n-1)}u^{n-1}}{dx^{(n-1)}}$$

to obtain

$$P_{n+1}(x) - P_{n-1}(x) = \frac{2n+1}{2^n n!}\frac{d^{(n-1)}u^n}{dx^{(n-1)}}.$$

23. Show that if $f(x)$ is n times differentiable on $-1 < x < 1$, then

$$\int_{-1}^{1} f(x)P_n(x)\,dx = \frac{(-1)^n}{2^n n!}\int_{-1}^{1} (x^2-1)^n f^{(n)}(x)\,dx. \qquad \textbf{(7.3–26)}$$

(*Hint*: Use Rodrigues' formula (Eq. 7.3–19) and integrate by parts n times.)

24. In Example 23, replace $f(x)$ by $P_m(x)$, $m < n$, and obtain the orthogonality relation for the Legendre polynomials.

25. **(a)** Write Legendre's differential equation (7.3–12) in the form of the Sturm–Liouville equation (2.2–6).
(b) Using the notation of Section 2.2, show that $r(b) = 0$ and that the first boundary condition of Eq. (2.2–7) holds.

••• **26.** The associated Legendre functions of the first kind $P_n^m(x)$ are particular solutions of Eq. (7.3–11) defined as follows:

$$P_n^m(x) = (-1)^m(1-x^2)^{m/2}\frac{d^mP_n(x)}{dx^m}.$$

(a) Show that $P_n^0(x) = P_n(x)$, $n = 0, 1, 2, \ldots$
(b) Show that $P_n^m(x) = 0$ if $m > n$
(c) Write out P_1^1, P_2^1, P_3^1, P_2^2, and P_3^2
(d) Verify that P_2^1 and P_3^1 are orthogonal on $(-1, 1)$ using the orthogonality relation

$$\int_{-1}^{1} P_n^m(x)P_k^m(x)\,dx = \begin{cases} 0, & n \neq k \\ \dfrac{2(n+m)!}{(2n+1)(n-m)!}, & n = k. \end{cases}$$

(e) Find

$$\int_{-1}^{1} [P_2^2(x)]^2\,dx.$$

27. Show that solutions to Laplace's equation in spherical coordinates can be written as

$$u(r\,\phi, \theta) = \sum_{m=0}^{\infty}\sum_{n=0}^{\infty}(C_nr^n + D_nr^{-(n+1)})(A_m\cos m\phi$$
$$+ B_m\sin m\phi)(c_{n,m}P_n^m(\cos\theta) + d_{n,m}Q_n^m(\cos\theta)).$$

28. Obtain the recurrence formula

$$P_{n+1}(x) = \left(\frac{2n+1}{n+1}\right)xP_n(x) - \left(\frac{n}{n+1}\right)P_{n-1}(x), \qquad n = 1, 2, \ldots.$$

29. In Example 7.3–1, what is the relation of A_0 to the function being represented by the Legendre series? Generalize.

7.4 APPLICATIONS

We are now ready to solve some boundary-value problems in three-space using cylindrical and spherical coordinates. In the following examples we will use many of the results developed in Sections 7.2 and 7.3.

EXAMPLE 7.4–1 Find the steady-state, bounded temperatures in the interior of a solid sphere of radius b if the temperatures on the surface are given by $f(\cos\theta)$.

Solution Since the surface temperatures are known as a function of θ alone, the temperatures are independent of ϕ. Hence we have the following problem (see Eq. (7.3–3)):

$$\text{P.D.E.:}\qquad \nabla^2 u(r,\theta) = \frac{\partial^2 u}{\partial r^2} + \frac{2}{r}\frac{\partial u}{\partial r} + \frac{1}{r^2}\frac{\partial^2 u}{\partial \theta^2} + \frac{\cot\theta}{r^2}\frac{\partial u}{\partial \theta} = 0,$$
$$0 < r < b, \qquad 0 < \theta < \pi;$$

$$\text{B.C.:}\qquad u(b,\theta) = f(\cos\theta), \qquad 0 < \theta < \pi.$$

The portion of the sphere in the first octant is shown in Fig. 7.4–1. Although the sphere appears to have only one boundary, namely, the surface $r = b$, the P.D.E. shows that the entire vertical axis must be omitted from the interior. (Why?) As a result, additional boundary conditions will come from the fact that the temperatures must be bounded. Using separation of variables and assuming a product solution,

$$u(r,\theta) = R(r)\Theta(\theta),$$

we have

$$\Theta\frac{d^2R}{dr^2} + \frac{2}{r}\Theta\frac{dR}{dr} + \frac{R}{r^2}\frac{d^2\Theta}{d\theta^2} + \frac{R\cot\theta}{r^2}\frac{d\Theta}{d\theta} = 0$$

or

$$\frac{r^2}{R}\frac{d^2R}{dr^2} + \frac{2r}{R}\frac{dR}{dr} = -\frac{1}{\Theta}\frac{d^2\Theta}{d\theta^2} - \frac{\cot\theta}{\Theta}\frac{d\Theta}{d\theta} = \lambda.$$

This results in

$$r^2R'' + 2rR' - \lambda R = 0,$$

which, for $\lambda = n(n+1)$, has solutions

$$R_n(r) = C_n r^n + D_n r^{-(n+1)},$$

as given by Eq. (7.3–10). The second equation can be shown (Exercise 1) to be equivalent to Legendre's differential equation,

$$\frac{1}{\sin\theta}\frac{d}{d\theta}\left(\sin\theta\frac{d\Theta}{d\theta}\right) + n(n+1)\Theta = 0,$$

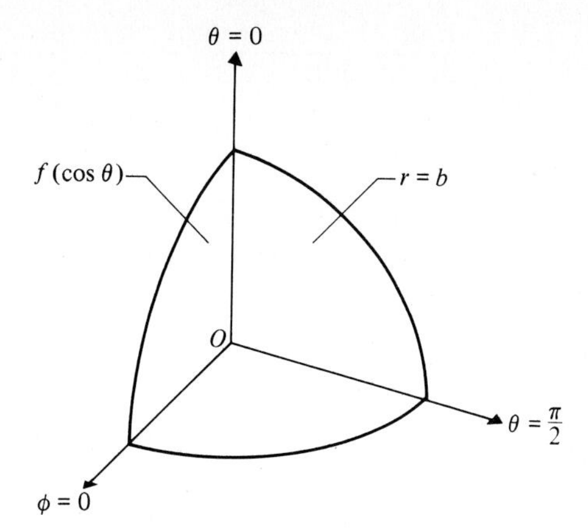

Figure 7.4-1

which has the general solution given by

$$\Theta_n(\theta) = E_n P_n(\cos\theta) + F_n Q_n(\cos\theta).$$

To keep u bounded, we must have* $n = 0, 1, 2, \ldots$ and take $D_n = 0$ and $F_n = 0$ (Exercise 2). Thus $u(r, \theta)$ consists of products,

$$r^n P_n(\cos\theta),$$

and we take a linear combination of these in order to satisfy the remaining nonhomogeneous boundary condition. Then,

$$u(r, \theta) = \sum_{n=0} A_n r^n P_n(\cos\theta),$$

with the A_n to be determined. Using the given surface temperatures,

$$u(b, \theta) = \sum_{n=0} A_n b^n P_n(\cos\theta) = f(\cos\theta),$$

which shows that $f(\cos\theta)$ is to be expressed as a Legendre series. From Eq. (7.3-21),

$$A_n b^n = \frac{2n+1}{2} \int_{-1}^{1} f(x) P_n(x)\, dx, \qquad n = 0, 1, 2, \ldots,$$

so that the solution can be written as

$$u(r, \theta) = \frac{1}{2} \sum_{n=0} (r/b)^n P_n(\cos\theta)(2n+1) \int_{-1}^{1} f(x) P_n(x)\, dx. \quad \blacksquare \tag{7.4-1}$$

*Recall that for $\lambda = n(n+1)$, $n = 0, 1, 2, \ldots$, we were able to solve the equation in θ (see Example 1.5-6) and obtain *bounded* solutions for $0 \le \theta \le \pi$.

EXAMPLE 7.4–2 Find the steady-state, bounded temperatures in the interior of a solid cylinder of radius c and height b, given that the temperature of the curved lateral surface is kept at zero, the base is insulated, and the top is kept at 100°.

Solution We take the axis of the cylinder along the z-axis and use cylindrical coordinates as shown in Fig. 7.4–2. Then we have the following problem (Exercise 3):

$$\text{P.D.E.:} \quad u_{\rho\rho} + \frac{1}{\rho} u_\rho + u_{zz} = 0, \qquad 0 < \rho < c, \qquad 0 < z < b;$$

$$\begin{aligned} \text{B.C.:} \quad & u(c, z) = 0, && 0 < z < b, \\ & u_z(\rho, 0) = 0, && 0 < \rho < c, \\ & u(\rho, b) = 100, && 0 < \rho < c. \end{aligned}$$

If we assume a product $u(\rho, z) = R(\rho)Z(z)$, the partial differential equation becomes

$$ZR'' + \frac{1}{\rho} ZR' + RZ'' = 0$$

or

$$\frac{R''}{R} + \frac{R'}{\rho R} = -\frac{Z''}{Z} = -\lambda^2.$$

We have assumed a *negative* separation constant in order that Z (and u) will *not* be periodic in z. (Why?) The resulting ordinary differential equations with their boundary conditions (only the *homogeneous* conditions may be used here) are (Exercise 4)

$$Z'' - \lambda^2 Z = 0, \qquad Z'(0) = 0,$$

and

$$R'' + \frac{1}{\rho} R' + \lambda^2 R = 0, \qquad R(c) = 0.$$

These differential equations are similar to Eqs. (7.2–2) and (7.2–4), respectively; hence we have the general solutions

$$Z(\lambda z) = A \cosh \lambda z + B \sinh \lambda z$$

and

$$R_0(\lambda\rho) = EJ_0(\lambda\rho) + FY_0(\lambda\rho).$$

To keep the solution bounded, we take $F = 0$. The condition $R(c) = 0$ results in $J_0(\lambda c) = 0$, that is, λc is a zero of the Bessel function $J_0(x)$. Call these positive zeros, $\lambda_j c$, $j = 1, 2, \ldots$, that is, $\lambda_1 c \doteq 2.405$, $\lambda_2 c \doteq 5.520$, etc. The equations in z have solutions (Exercise 5)

$$Z(\lambda_j z) = \cosh(\lambda_j z), \qquad j = 1, 2, \ldots.$$

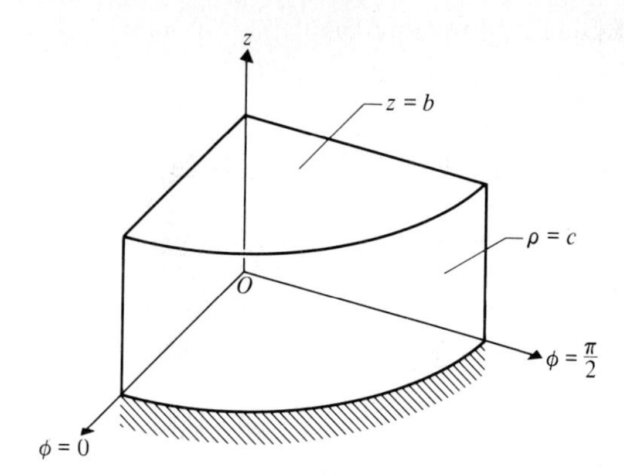

Figure 7.4-2

Thus

$$u(\rho, z) = \sum_{j=1} a_j \cosh(\lambda_j z)\, J_0(\lambda_j \rho),$$

and applying the nonhomogeneous boundary condition, we have

$$u(\rho, b) = \sum_{j=1} a_j \cosh(\lambda_j b)\, J_0(\lambda_j \rho) = 100,$$

which shows that $f(\rho) = 100$ must be expressed as a Fourier–Bessel series on the interval $0 < \rho < c$. We use Eq. (7.2–19) to write

$$A_j = a_j \cosh(\lambda_j b) = \frac{2}{c^2[J_0'(\lambda_j c)]^2} \int_0^c 100 x J_0(\lambda_j x)\, dx$$

$$= \frac{200}{c^2[J_0'(\lambda_j c)]^2} \frac{1}{\lambda_j^2} \int_0^{\lambda_j c} s J_0(s)\, ds$$

on making the substitution $s = \lambda_j x$. Then, using Eq. (7.2–10), we have

$$A_j = a_j \cosh(\lambda_j b) = \frac{200}{\lambda_j c J_1(\lambda_j c)},$$

noting that $J_0'(x) = -J_1(x)$ and simplifying (Exercise 6). Thus

$$u(\rho, z) = \frac{200}{c} \sum_{j=1} \frac{J_0(\lambda_j \rho) \cosh(\lambda_j z)}{\lambda_j \cosh(\lambda_j b) J_1(\lambda_j c)}, \tag{7.4-2}$$

where the $\lambda_j c$ are positive roots of $J_0(\lambda) = 0$. ■

In the next example we consider the two-dimensional wave equation over a circular region.

EXAMPLE 7.4–3 Solve the following boundary-value problem.

$$\text{P.D.E.:}\quad z_{tt} = \frac{a^2}{\rho}(\rho z_\rho)_\rho, \qquad 0 < \rho < c, \qquad t > 0;$$

$$\text{B.C.:}\quad z(c, t) = 0, \qquad t > 0;$$

$$\text{I.C.:}\quad z_t(\rho, 0) = 0, \qquad 0 < \rho < c,$$

$$z(\rho, 0) = f(\rho), \qquad 0 < \rho < c.$$

Solution Here we have a homogeneous membrane (compare Example 6.2–1) of radius c fastened in a frame along its circular edge (see Fig. 7.4–3). The frame is given an initial displacement $f(\rho)$ in the z-direction, and we seek the displacements at an arbitrary point of the membrane at any time t. The fact that the initial displacement is a function of ρ alone indicates that z is independent of ϕ. Using separation of variables, we arrive (Exercise 7) at the following ordinary differential equations and homogeneous conditions:

$$T'' + \lambda^2 a^2 T = 0, \qquad T'(0) = 0,$$

$$\rho^2 R'' + \rho R' + \lambda^2 \rho^2 R = 0, \qquad R(c) = 0.$$

The separation constant was chosen so that T (and z) will be periodic in t consistent with physical fact. The second equation has solutions $J_0(\lambda_j \rho)$ where $J_0(\lambda_j c) = 0$, $j = 1, 2, \ldots$. The first equation has solutions $\cos(\lambda_j at)$, $j = 1, 2, \ldots$; hence we take a linear combination of products of these, that is,

$$z(\rho, t) = \sum_{j=1} A_j J_0(\lambda_j \rho) \cos(\lambda_j at).$$

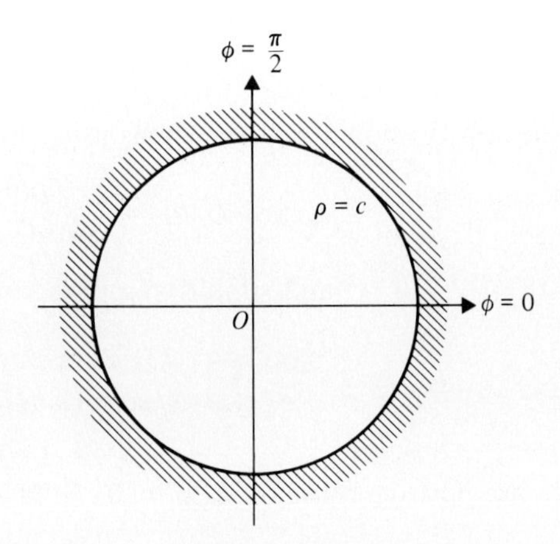

Figure 7.4–3

To satisfy the nonhomogeneous initial condition, we again use Eq. (7.2–19) to determine the A_j. Thus

$$A_j = \frac{2}{c^2[J_1(\lambda_j c)]^2} \int_0^c x f(x) J_0(\lambda_j x)\, dx \qquad j = 1, 2, \ldots,$$

and

$$z(\rho, t) = \frac{2}{c^2} \sum_{j=1} \frac{J_0(\lambda_j \rho) \cos(\lambda_j a t)}{[J_1(\lambda_j c)]^2} \int_0^c x f(x) J_0(\lambda_j x)\, dx, \tag{7.4-3}$$

where $\lambda_j c$ are positive roots of $J_0(x) = 0$. ■

In the next example we investigate the two-dimensional heat equation over a circular region.

EXAMPLE 7.4–4 Solve the following boundary-value problem:

$$\text{P.D.E.:} \quad u_t = \frac{k}{\rho}(\rho u_\rho)_\rho, \qquad 0 < \rho < c, \qquad t > 0;$$

$$\text{B.C.:} \quad u_\rho(c, t) = 0, \qquad t > 0;$$

$$\text{I.C.:} \quad u(\rho, 0) = f(\rho), \qquad 0 < \rho < c.$$

Solution Here we have a homogeneous circular disk of radius c whose outer edge is insulated (see Fig. 7.4–4). We assume that the heat flow is two-dimensional, that is, that the disk is thin and its top and bottom circular faces are also insulated. Moreover, the temperature is independent of ϕ since the initial temperature distribution is a function of ρ alone. We seek the temperatures in the disk at any time t. Using separation of variables, we obtain (Exercise 8) the following ordinary differential equations:

$$T' + k\lambda^2 T = 0$$

$$R'' + \frac{1}{\rho} R' + \lambda^2 R = 0, \qquad R'(c) = 0.$$

Again the separation constant was chosen to be negative, this time in order that $u_\rho(\rho, t)$ have a zero limit as $t \to \infty$ (Exercise 9).

The second equation is Bessel's differential equation of order zero having bounded solution $J_0(\lambda\rho)$. Applying the condition shown, we have

$$J_0'(\lambda c) = -\lambda J_1(\lambda c) = 0,$$

using Exercise 4(d) in Section 7.2, thus showing that λc is a zero of $J_1(x) = 0$. Call these nonnegative zeros $\lambda_j c$, $j = 1, 2, \ldots$, that is $\lambda_1 c = 0$, $\lambda_2 c \doteq 3.832$, $\lambda_3 c \doteq 7.016$, etc. Then we have a solution of the form

$$u(\rho, t) = \sum_{j=1} A_j \exp(-k\lambda_j^2 t) J_0(\lambda_j \rho),$$

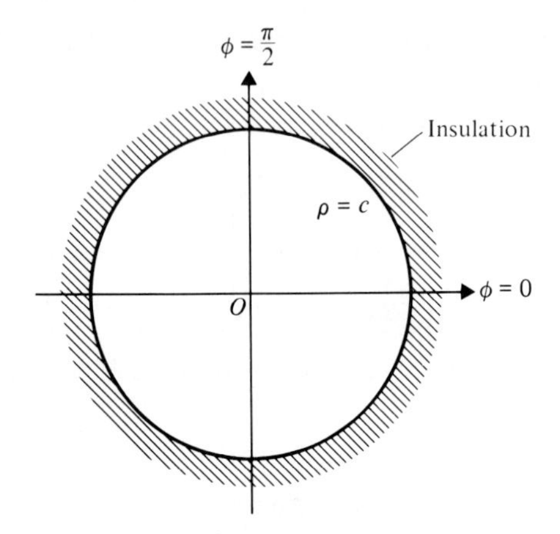

Figure 7.4–4

and to satisfy the nonhomogeneous boundary condition, we compute the A_j using Eq. (7.2–21) with $h = n = 0$ and Eq. (7.2–22). Hence

$$A_1 = \frac{2}{c^2} \int_0^c xf(x)\, dx,$$

$$A_j = \frac{2}{c^2[J_0(\lambda_j c)]^2} \int_0^c xf(x) J_0(\lambda_j x)\, dx, \qquad j = 2, 3, \ldots,$$

so that the final result is

$$u(\rho, t) = A_1 + \sum_{j=2} A_j \exp(-k\lambda_j^2 t) J_0(\lambda_j \rho), \tag{7.4–4}$$

with A_1 and A_j as defined above and with $J_1(\lambda_j c) = 0$, $j = 1, 2, 3, \ldots$. ■

We call attention to a similarity between Eq. (7.4–4) and solutions in Chapter 6 in which Fourier series played a role. In the case of Fourier series we had a constant term $a_0/2$, which represented the **average value*** of the function being represented by the series. Equation (7.4–4) also contains a constant term, namely, A_1. A question arises as to the significance of this term, and we answer this question in the next example.

EXAMPLE 7.4–5 Determine the **steady-state solution** of the problem of Example 7.4–4.

Solution The required solution may be obtained by solving the following problem:

$$\frac{d}{d\rho}\left(\rho \frac{du}{d\rho}\right) = 0, \qquad \frac{du(c)}{d\rho} = 0. \tag{7.4–5}$$

*Also called *mean value*.

We leave it as an exercise to show that the solution of (7.4–5) is a *constant* (Exercise 10).

On the other hand, from Eq. (7.4–4) we have

$$\lim_{t \to \infty} u(\rho, t) = A_1,$$

which is also a *constant*. From physical considerations, however, it is clear that the equilibrium temperature of the region must be a *constant* that, in some sense, is the *average value* of the initial temperature distribution $f(\rho)$.

Recall from calculus that the average value of $g(x, y)$ over a region R in the xy-plane with respect to the area of the region is given by

$$g_{av} = \frac{\iint_R g(x, y)\, dx\, dy}{\iint_R dx\, dy}. \tag{7.4–6}$$

In terms of polar coordinates (ρ, ϕ), Eq. (7.4–6) becomes

$$f_{av} = \frac{\iint_R f(\rho, \phi)\, \rho\, d\rho\, d\phi}{\iint_R \rho\, d\rho\, d\phi}. \tag{7.4–7}$$

In the present example the region R is a circle of radius c, and f is independent of ϕ. Hence the right-hand member of Eq. (7.4–7) is

$$\frac{\int_0^c \int_0^{2\pi} f(\rho)\, \rho\, d\phi\, d\rho}{\pi c^2} = \frac{2}{c^2} \int_0^c \rho f(\rho)\, d\rho,$$

which is exactly the definition of A_1. Thus A_1 gives the average value of $f(\rho)$ over the circular region, and this value is also the steady-state solution of the problem of Example 7.4–4. ■

In connection with the last example, see also Theorem 7.1–1, which gives the mean-value property of a harmonic function.

EXAMPLE 7.4–6 A solid hemisphere of radius b has its plane face perfectly insulated, while the temperature of its curved surface is given by $f(\cos\theta)$. Find the steady-state, bounded temperatures at any point in the interior of the hemisphere.

Solution Figure 7.4–5 shows one-quarter of the hemisphere. Since the temperature of the surface is independent of ϕ, we have the following mathematical formulation of the problem.

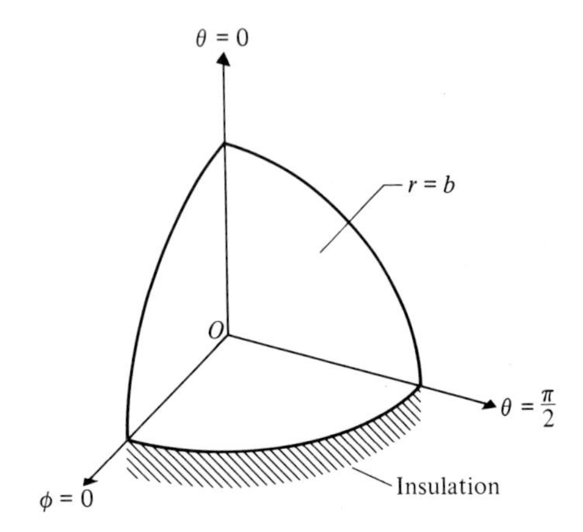

Figure 7.4-5

P.D.E.: $\nabla^2 u = 0$ in spherical coordinates, independent of ϕ with $0 < r < b, \quad 0 < \theta < \pi/2;$

B.C.: $u_z(r, \pi/2) = 0, \quad 0 < r < b,$
$u(b, \theta) = f(\cos\theta), \quad 0 < \theta < \pi/2,$
$|u(r, \theta)| < \infty, \quad 0 < r < b, \quad 0 < \theta < \pi/2.$

The last boundary condition states that u must be bounded. Since the variable z is not one of the coordinates in the spherical coordinate system, we need to alter the homogeneous boundary condition. Referring to Fig. 7.3-1, we have $z = r\cos\theta$, so that

$$\frac{\partial u}{\partial \theta} = \frac{\partial u}{\partial z}\frac{\partial z}{\partial \theta} = -r\sin\theta\,\frac{\partial u}{\partial z}.$$

Accordingly, at $\theta = \pi/2$ we have

$$\frac{\partial u}{\partial z} = -\frac{1}{r}\frac{\partial u}{\partial \theta},$$

and the condition $u_z = 0$ implies that $u_\theta = 0$ at $\theta = \pi/2$.

Referring to Example 7.4-1, we have

$$u(r, \theta) = \sum_{n=0} A_n r^n P_n(\cos\theta),$$

which is a bounded solution of Laplace's equation as stated in this example. Now

$$u_\theta(r, \theta) = \sum_{n=0} A_n r^n(-\sin\theta)P_n'(\cos\theta)$$

so that

$$u_\theta(r, \pi/2) = -\sum_{n=0} A_n r^n P_n'(0) = 0,$$

from which it follows that n is *even* (see Exercise 3(d) in Section 7.3); call it $2m$. **Updating** the solution gives us

$$u(r, \theta) = \sum_{m=0} A_{2m} r^{2m} P_{2m}(\cos\theta).$$

Finally, we apply the nonhomogeneous boundary condition to obtain

$$u(b, \theta) = \sum_{m=0} A_{2m} b^{2m} P_{2m}(\cos\theta) = f(\cos\theta),$$

that is, $f(\cos\theta)$ must be represented on the interval $0 < \theta < \pi/2$ by a series of Legendre polynomials of even degree. This is possible for appropriate functions f because the polynomials are orthogonal on the given interval, and we can use Eq. (7.3–22) to compute the coefficients. Hence

$$A_{2m} b^{2m} = (4m + 1) \int_0^{\pi/2} f(\cos\theta)\, P_{2m}(\cos\theta) \sin\theta\, d\theta,$$

and the final solution becomes

$$u(r, \theta) = \sum_{m=0} (4m + 1) \left(\frac{r}{b}\right)^{2m} P_{2m}(\cos\theta) \int_0^1 f(x) P_{2m}(x)\, dx. \quad \blacksquare \tag{7.4–8}$$

We recommend the procedure used in the last example of updating the solution every time new information is obtained about it. Generally, it is better to apply the *homogeneous* conditions when the separate *ordinary* differential equations are solved, that is, before the product solutions are formed. In Example 7.4–6, however, we could safely depart from this procedure. (Why?)

EXAMPLE 7.4–7 Solve the following problem:

P.D.E.: (a) $u_{\rho\rho} + \frac{1}{\rho} u_\rho + u_{zz} = 0, \quad 0 < \rho < c, \quad z > 0;$

B.C.: (b) $u_\rho(c, z) + hu(c, z) = 0, \quad z > 0, \quad h > 0,$

(c) $u(\rho, 0) = f(\rho), \quad 0 < \rho < c,$

(d) $|u(\rho, z)| < \infty, \quad 0 < \rho < c, \quad z > 0.$

Solution As shown in Fig. 7.4–6, the mathematical statements may be interpreted in the following way.

(a) Find the steady-state (independent of t) temperatures at any point within a solid, semi-infinite right circular cylinder of radius c.

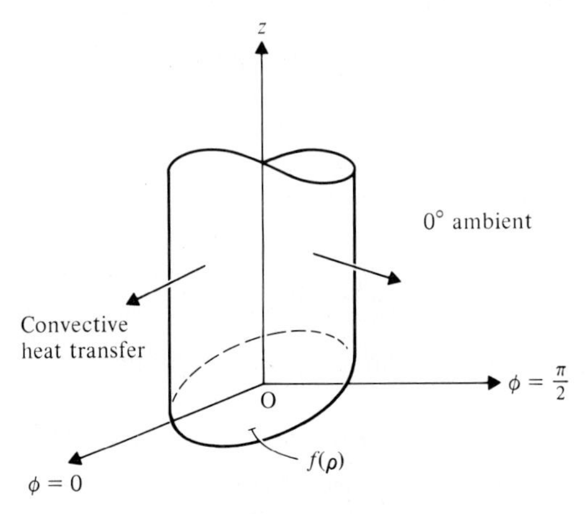

Figure 7.4-6

(b) There is **convective heat transfer** from the lateral surface into a surrounding medium at temperature zero. Another way of saying this is that the lateral surface is cooling in accordance with **Newton's law of cooling.**

(c) The temperature of the base is a given function of ρ, that is, $f(\rho)$.

(d) We seek a **bounded solution**, which means that

$$\lim_{z \to \infty} u(\rho, z) = 0$$

and $u(\rho, z)$ is bounded in the neighborhood of the z-axis, that is, as $\rho \to 0$.

Other interpretations are equally valid, since we are solving the **potential equation** (Laplace's equation).

We use separation of variables, assuming that the desired solution can be written as a product, that is,

$$u(\rho, z) = R(\rho)Z(z).$$

Then we have

$$\frac{R''}{R} + \frac{R'}{\rho R} = -\frac{Z''}{Z} = -\lambda^2,$$

where we have chosen the separation constant as $-\lambda^2$ so that Z (and u) will not be periodic in z. Thus we are led to the following two ordinary differential equations:

$$\rho R'' + R' + \lambda^2 \rho R = 0, \qquad R'(c) + hR(c) = 0,$$
$$Z'' - \lambda^2 Z = 0$$

The first equation is Bessel's differential equation of order zero; its general solution is

$$R(\lambda\rho) = AJ_0(\lambda\rho) + BY_0(\lambda\rho).$$

We take $B = 0$ because $Y_0(\lambda\rho)$ is undefined for $\rho = 0$. Applying condition (b) yields

$$\lambda J_0'(\lambda c) + hJ_0(\lambda c) = 0, \qquad \textbf{(7.4-9)}$$

which defines the positive λ's, call them λ_j, $j = 1, 2, \ldots$. The second equation has the general solutions

$$Z(\lambda_j z) = C_j \exp(\lambda_j z) + D_j \exp(-\lambda_j z),$$

in which we take $C_j = 0$ to keep the solution bounded for $z > 0$.

Hence the solution may be written as

$$u(\rho, z) = \sum_{j=1} A_j \exp(-\lambda_j z) J_0(\lambda_j \rho).$$

We apply condition (c) to obtain

$$u(\rho, 0) = \sum_{j=1} A_j\, J_0(\lambda_j \rho) = f(\rho),$$

which shows that $f(\rho)$ must be expressed as a Fourier–Bessel series. In view of Eq. (7.4–9) the coefficients A_j are obtained from Eq. (7.2–21), that is,

$$A_j = \frac{2\lambda_j^2}{(\lambda_j^2 c^2 + h^2 c^2) J_0^2(\lambda_j c)} \int_0^c x\, f(x)\, J_0(\lambda_j x)\, dx, \qquad j = 1, 2, \ldots.$$

Thus the updated solution can be written as

$$u(\rho, z) = \frac{2}{c^2} \sum_{j=1} \frac{\lambda_j^2 \exp(-\lambda_j z) J_0(\lambda_j \rho)}{(\lambda_j^2 + h^2) J_0^2(\lambda_j c)} \int_0^c x\, f(x)\, J_0(\lambda_j x)\, dx, \qquad \textbf{(7.4–10)}$$

where the λ_j are positive roots of

$$\lambda_j J_0'(\lambda_j c) + hJ_0(\lambda_j c) = 0, \qquad h > 0. \quad \blacksquare$$

We have presented a few examples to show how circular symmetry in a problem can lead to Bessel functions and spherical symmetry to Legendre polynomials. We have used separation of variables to solve the boundary-value problems, since this technique allows us to transfer *homogeneous* conditions to the separate *ordinary* differential equations. We have also used our knowledge of the physical situation wherever possible to assign particular values to the separation constants.

In conclusion we point out that the method of separation of variables is not limited to problems phrased in rectangular, polar, cylindrical, and spherical coordinates. Other coordinate systems in which the method can be used include the following:* elliptic cylindrical coordinates, conical coordinates, parabolic coordinates, elliptic coordinates (both prolate and oblate), ellipsoidal coordinates, and paraboloidal coordinates.

Key Words and Phrases

average value
steady-state solution
updating
convective heat transfer
Newton's law of cooling
bounded solution
potential equation

7.4 Exercises

• **1.** In Example 7.4–1, show that

$$\frac{1}{\Theta}\frac{d^2\Theta}{d\theta^2} + \frac{\cot\theta}{\Theta}\frac{d\Theta}{d\theta} = -n(n+1)$$

is equivalent to

$$\frac{1}{\sin\theta}\frac{d}{d\theta}\left(\sin\theta\,\frac{d\Theta}{d\theta}\right) + n(n+1)\,\Theta = 0.$$

2. Show that we must take $D_n = F_n = 0$ in Example 7.4–1 even though the region in which we seek a solution does *not* include the points $r = 0$, $(b, 0)$ and (b, π). Why must these points be excluded? (*Hint*: Look at the partial differential equation being solved.)

3. Translate each of the given boundary conditions of Example 7.4–2 into the mathematical statements shown.

4. Carry out the details in separating the variables in Example 7.4–2.

5. Solve

$$Z'' - \lambda_j^2 Z = 0, \qquad Z'(0) = 0.$$

(Compare Example 7.4–2.)

6. Fill in the necessary details to obtain the A_j in Example 7.4–2.

7. Use separation of variables to obtain the ordinary differential equations in Example 7.4–3.

8. Use separation of variables to obtain the two ordinary differential equations in Example 7.4–4.

---

*See Section 5.4 of the author's *Advanced Engineering Mathematics* (Reading, Mass.: Addison-Wesley, 1982).

9. In Example 7.4–4, explain why we must have

$$\lim_{t\to\infty} u_\rho(\rho, t) = 0.$$

10. Obtain a solution to the following problem:

$$\frac{d}{d\rho}\left(\rho \frac{du}{d\rho}\right) = 0, \qquad \frac{du(c)}{d\rho} = 0.$$

•• **11.** Obtain the solution to the problem in Example 7.4–1, given that the surface temperature is a constant 100°. Does your result agree with physical fact and with Theorem 6.1–1?

12. Obtain the solution to the problem in Example 7.4–1 if the surface temperature is given by $f(\cos\theta) = \cos\theta$. (*Hint*: Recall from Section 7.3 that $P_1(\cos\theta) = \cos\theta$.)

***13.** **(a)** In Eq. (7.4–2), put $b = c = 1$, and write out the first three terms of the sum.
(b) Use the result in part (a) to compute $u(0, 0)$.
(c) Is the result in part (b) what you would expect? Explain.

14. Solve the problem of Example 7.4–2, given that the base and curved lateral surface are both kept at zero and the top is kept at 100°.

15. What would the result be if the separation constant in Example 7.4–3 were of the opposite sign? Do these results agree with physical fact? Explain.

16. In Example 7.4–3, change $f(\rho)$ to 1, and obtain the solution corresponding to Eq. (7.4–3). Is such an initial displacement physically possible? Explain.

17. Solve the problem of Example 7.4–4, given that the outer edge of the disk is kept at temperature zero instead of being insulated and all other conditions remain the same.

18. Find the bounded, steady-state temperatures inside a solid hemisphere of radius b if its plane surface is kept at temperature zero and the remaining surface has a temperature distribution given by $f(\cos\theta)$.

19. Modify Exercise 18 so that $f(\cos\theta) = 100$, and obtain the solution.

20. A solid hemisphere of radius b has its plane surface kept at a constant 100°, and its curved surface is insulated. Find the steady-state temperature at an arbitrary point in the interior.

21. Use the condition $f(\cos\theta) = 100$ in Example 7.4–6, and obtain the solution.

22. Modify condition (c) in Example 7.4–7 to $u(\rho, 0) = 100$, and obtain the solution.

23. **(a)** Change the homogeneous boundary condition in Example 7.4–7 to $u(c, z) = 0$, and obtain the solution.
(b) Give a physical interpretation of the problem in part (a).

24. Concentric spheres of radius a and b ($a < b$) are held at constant potentials V_1 and V_2, respectively. Determine the potential at any point between the spheres.

25. A dielectric sphere* of radius b is placed in a uniform electric field of intensity E in the z-direction. Determine the potential inside the sphere and outside the sphere. (*Hint*: Both the potential inside (v) and the potential outside (V) must satisfy the

*This is *not* a solid sphere.

potential equation. Continuity conditions are given by

$$v(b, \theta) = V(b, \theta), \qquad 0 < \theta < \pi,$$

and

$$Kv_r(b, \theta) = V_r(b, \theta), \qquad 0 < \theta < \pi, \qquad K > 0.$$

Moreover,

$$\lim_{r \to \infty} V(r, \theta) = -Ez = -Er \cos \theta.$$

26. Find the potential of a grounded, conducting sphere of radius b placed in a uniform electric field of intensity E in the z-direction. (*Hint*: Again $\nabla^2 v = 0$ with $v(b, \theta) = 0$ and

$$\lim_{r \to \infty} v(r, \theta) = -Ez = -Er \cos \theta.)$$

27. Show that the positive roots of

$$\lambda J_0'(\lambda c) + hJ_0(\lambda c) = 0, \qquad h > 0,$$

are the same as those of

$$KJ_0(\lambda c) + \lambda c J_0'(\lambda c) = 0, \qquad K > 0.$$

28. Using the fact that $r = (x^2 + y^2 + z^2)^{1/2}$, show that

$$\frac{\partial}{\partial z}\left(\frac{1}{r}\right) = -\frac{1}{r^2} P_1 (\cos \theta).$$

29. Show that the functions

$$u(\rho, \phi, z) = \exp(-\lambda z)\, J_n(\lambda \rho) \cos n\phi$$

are potential functions for $n = 1, 2, \ldots$. (*Hint*: Use the recurrence relations in Exercises 4(e) and 26 of Section 7.2.)

30. A homogeneous membrane of radius b is fastened along its circumference. It is given an initial displacement $C(b^2 - \rho^2)$, causing it to vibrate from rest. Show that the displacements $z(\rho, t)$ are given by

$$z(\rho, t) = 8Cb^2 \sum_{j=1}^{\infty} \frac{J_0(\rho\lambda_j/b) \cos(a\lambda_j t/b)}{\lambda_j^3 J_1(\lambda_j)},$$

where the λ_j are roots of $J_0(\lambda) = 0$.

••• **31.** **(a)** Solve the following problem.

$$\begin{aligned}
&\text{P.D.E.:} && v_t = k\left(v_{\rho\rho} + \frac{1}{\rho} v_\rho\right), && 0 < \rho < a, \quad t > 0;\\
&\text{B.C.:} && v(a, t) = 0, \quad t > 0, &&\\
& && \lim_{t \to \infty} v(\rho, t) = 0, && 0 < \rho < a,\\
& && |v(\rho, t)| < \infty, && 0 < \rho < a, \quad t > 0;\\
&\text{I.C.:} && v(\rho, 0) = u_1 - u_0, && 0 < \rho < a.
\end{aligned}$$

(b) Explain how the problem of part (a) arises from the following one. A long circular cylinder of radius a is initially heated to a uniform temperature u_1. Its surface is then maintained at a constant temperature u_0. Show that the temperature $u(\rho, t)$ is given by

$$u(\rho, t) = u_0 + 2(u_1 - u_0) \sum_{j=1} \frac{\exp(-\lambda_j^2 kt/a^2)\, J_0(\lambda_j \rho/a)}{\lambda_j J_1(\lambda_j)},$$

where the λ_j are positive zeros of $J_0(\lambda) = 0$.

32. An infinitely long solid cylinder of radius b is initially at temperature $f(\rho)$.For $t > 0$ the boundary surface $\rho = b$ dissipates heat by convection into a medium at temperature zero (see Example 7.4–7).

(a) State the problem in mathematical terms.

(b) Obtain the solution

$$u(\rho, t) = \frac{2}{b^2} \sum_{j=1} \frac{\lambda_j^2 \exp(-k\lambda_j^2 t)\, J_0(\lambda_j \rho)}{(h^2 + \lambda_j^2)\, J_0^2(\lambda_j b)} \int_0^b x\, f(x)\, J_0(\lambda_j x)\, dx,$$

where the λ_j are the positive roots of

$$\lambda_j J_0'(\lambda_j b) + hJ_0(\lambda_j b) = 0, \qquad h > 0.$$

(c) For the special case $f(\rho) = 100$, show that the solution becomes

$$u(\rho, t) = \frac{200h}{b^2} \sum_{j=1} \frac{\exp(-k\lambda_j^2 t)\, J_0(\lambda_j \rho)}{(\lambda_j^2 + h^2)\, J_0(\lambda_j b)},$$

with λ_j defined as in part (b).

33. **(a)** If the temperatures in a solid sphere are independent of ϕ and θ, show that the heat equation can be written as

$$u_t = k\left(u_{rr} + \frac{2}{r} u_r\right).$$

(b) If in the result of part (a) a new variable is defined as

$$U(r, t) = ru(r, t),$$

show that the heat equation becomes the familiar one-dimensional equation discussed in Section 6.3.

34. In Example 7.4–1 we solved the interior Dirichlet problem for a sphere of radius b. Solve the exterior Dirichlet problem, that is, assume that $r > b$, all other conditions remaining the same.

35. Solve the following problem.

P.D.E.: $u_{\rho\rho} + \frac{1}{\rho} u_\rho + u_{zz} = 0, \quad 0 < \rho < a, \quad 0 < z < L;$

B.C.: $u(a, z) = 0, \quad 0 < z < L,$
$u(\rho, L) = 0, \quad 0 < \rho < a,$
$u(\rho, 0) = 100, \quad 0 < \rho < a,$
$|u(\rho, z)| < \infty, \quad 0 < \rho < a, \quad 0 < z < L.$

36. In analyzing antenna radiation patterns for a system with a circular aperture the equation

$$f(z) = \int_0^1 g(\rho)J_0(\rho z)\rho \, d\rho$$

arises. Solve this equation for the particular case when $g(\rho) = 1 - \rho^2$.

REFERENCES

Bowman, F., *Introduction to Bessel Functions*. London: Longmans, Green, 1938.

Gray, A., and G. B. Mathews, *A Treatise on Bessel Functions and Their Applications to Physics*, 2nd ed. New York: Dover, 1966.

McLachlan, N. W., *Bessel Functions for Engineers*. London: Oxford University Press, 1934.

Sneddon, I. N., *Special Functions of Mathematical Physics and Chemistry*, 3rd ed. London: Longman, 1980.

Wyld, H. W., *Mathematical Methods for Physics*. Reading, Mass.: Benjamin, 1976.

8 Numerical Methods

8.1 INTRODUCTION

Up to this point we have presented various analytical methods for solving boundary-value problems. In all cases, however, the equations were *linear*, and the boundaries were those of *regular regions*, such as rectangles, circles, half-planes, semi-infinite strips, cylinders, spheres, and the like.

In practice it is a rarity when a problem falls neatly into a class that can be solved by the methods we have discussed. Classical approaches may fail for one or more of the following reasons.

1. The partial differential equation is nonlinear and cannot be linearized without seriously affecting the result.
2. The boundary is irregular.
3. Boundary conditions are of mixed types.
4. Boundary values are time-dependent.
5. Materials must be considered that are not homogeneous and isotropic.

Some of the above can cause complexities that make any method except a numerical one completely impractical. Of course, numerical methods also have a number of shortcomings, as we will see later.

As an example of the difficulties that may arise in boundary-value problems, consider the following one involving a **free-boundary**.

$$\begin{aligned}
&\text{P.D.E.:} && ku_{xx} - u_t = 0, \quad a < x < s(t), \quad t > 0;\\
&\text{B.C.:} && u(a, t) = T, \quad t > 0,\\
& && u[s(t), t] = 0, \quad t > 0,\\
& && \frac{ds(t)}{dt} = -\alpha u_x[s(t), t], \quad t > 0, \quad \alpha > 0.
\end{aligned}$$

Here the right-hand boundary is time-dependent, the left-hand boundary is constant, and the third condition expresses the law of conservation of energy.

The above problem is called a one-phase **Stefan problem.*** Moving interfaces occur in problems involving water and melting ice or in the crystallization of liquids. In such problems the interface between solid and liquid phases moves as latent heat is absorbed or released there. For substances like water the solidification takes place at a fixed temperature so that the liquid and solid phases are separated by a sharp interface. For alloys, mixtures, and impure materials the solidification or crystallization takes place over a range of temperatures. Thus instead of a sharp demarcation between the phases, there is a moving region consisting of two phases.

Further complications can occur due to the fact that the diffusivity coefficient k may vary with temperature (among other things) and the quantity α in the energy balance equation may be a function of density and the latent heat of melting (or solidification). Thus Stefan problems are nonlinear, and techniques such as superposition cannot be used. Under certain simplified conditions, analytic solutions can be obtained,† but, in general, numerical methods are required.

8.2 A MONTE CARLO METHOD

We present first a numerical method that is unusual in that it uses a principle from probability theory. Suppose that we need to solve a **Dirichlet problem** ($\nabla^2 u = 0$ with the values of u known on the boundary) in the plane over a region R that has an irregular boundary C. In particular, we seek the value of u at some given point P. This is accomplished in the following way.

We subdivide the region R by means of a **grid** (Fig. 8.2–1a). Beginning at P, we move to one of the four neighboring points (Fig. 8.2–1b) with equal probability. At the second stage we move again to one of the new four neighboring points with equal probability. Such a chain of moves is called a **random walk.**

Starting at P, a random walk will eventually take us to a boundary point; call it C_1. An example of such a walk is shown in Fig. 8.2–2. Since the value of u is known at C_1, we record it; call it $u(C_1)$. We begin at P again and start a second random walk to obtain $u(C_2)$, and so on.

The theory of probability tells us that the probability of reaching a part of the boundary that is *near* to P is greater than the probability of reaching a part of the boundary that is *far* from P. But this is just another way of saying that the boundary values near to P have a greater influence in determining $u(P)$

*After Josef Stefan (1835–1893), an Austrian physicist.

†See M. N. Ozisik's, *Heat Conduction* (New York: John Wiley, 1980), Chapter 10.

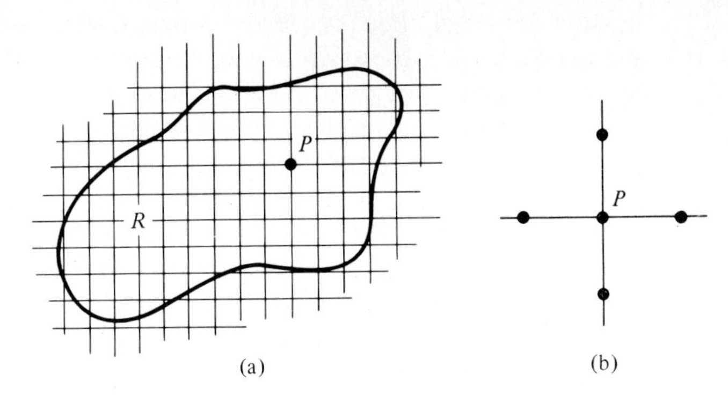

Figure 8.2-1
(a) *R* subdivided by a grid. (b) The neighboring points of *P*.

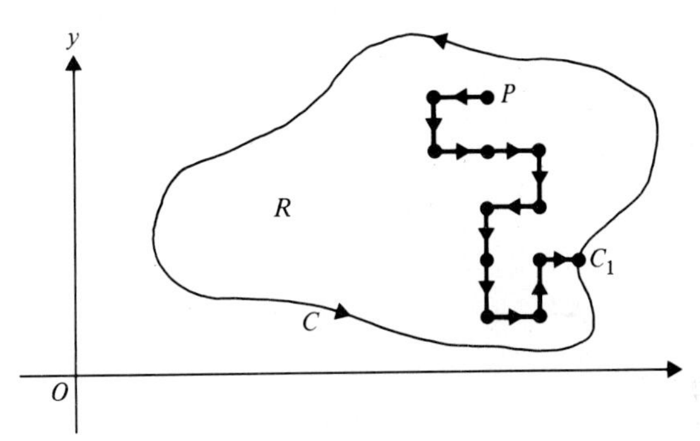

Figure 8.2-2
A random walk.

than those farther away. This statement can be readily verified by experiment in the case where R is a metal plate, u is temperature, and we seek the steady-state temperature at the point P.

Hence after a large number of walks (say, 1,000) we would expect that the *average* value,

$$\frac{1}{1000}\sum_{i=1}^{1000} u(C_i),$$

would be close to the desired value $u(P)$. In the language of statistics this average value is an **unbiased estimate** of the temperature at the point P. Thus we could conjecture that

$$u(P) = \lim_{n\to\infty} \frac{1}{n}\sum_{i=1}^{n} u(C_i). \tag{8.2-1}$$

The method described above is called a **Monte Carlo method** for solving a Dirichlet problem. It is especially useful if solutions at a few isolated points are desired. Usually, however, solutions at many points are required, and of special importance are curves that connect points having the same value of u. In a steady-state temperature problem these curves are called **isothermals**.

Probabilistic methods such as the Monte Carlo method have enjoyed varying degrees of popularity. One of the earliest problems involving probability to generate some interest was *Buffon's needle problem** in which the value of π was determined by repeatedly dropping a needle on a board ruled with parallel lines.

Today, probabilistic methods with their questionable accuracy and their requirement of large storage capacity are not popular with workers utilizing high-speed computers. This seems somewhat contradictory in view of the fact that such computers are ideally suited to these methods.

Key Words and Phrases

free boundary
Stefan problem
grid
random walk
unbiased estimate
Monte Carlo method
isothermals

8.2 Exercises

• 1. What are the units of α in the one-phase Stefan problem of the text?

2. Comment on ways in which the Monte Carlo method of the text can be made more accurate.

3. If the boundary C in the Monte Carlo method for solving a Dirichlet problem is an irregular one, it can be expected that the probability of a random walk terminating on C is small. Suggest how this difficulty may be overcome.

•• 4. Generate a set of 100 pseudorandom numbers in the following way. Using a telephone directory, start at some page and record the last two digits of each telephone number at the top and bottom of each column. Divide each number in this set repeatedly† (if necessary) by four and record the *remainder*. This will produce a set like {2, 1, 0, 3, 1, 2, 2, 0, 3, . . .}.

5. Consider the Dirichlet problem over a rectangular plate measuring 10 cm by 20 cm with one of the shorter sides held at temperature 100° and the other three sides held at zero. Choose coordinate axes as shown in Fig. 8.2–3. Find the temperature

*After Comte Georges Louis Leclerc de Buffon (1707–1788), a French naturalist.

†The number 25 divided by 4 is 6, which can again be divided by 4, so the remainder is 2.

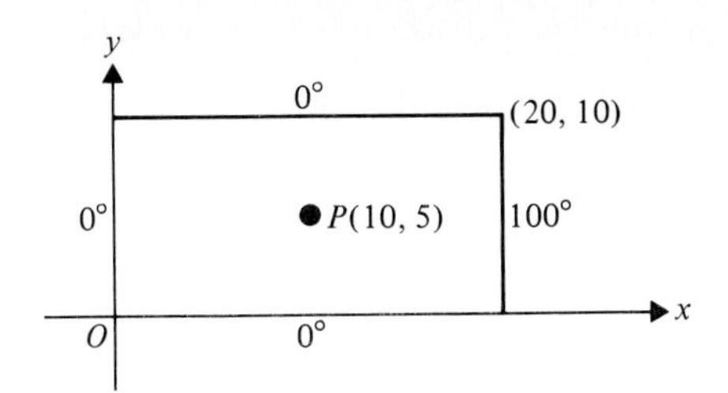

Figure 8.2-3

at the center of the plate by making ten random walks using the numbers generated in Exercise 4. The numbers can be interpreted as follows:

0	move 2 cm to the right,
1	move 2 cm up,
2	move 2 cm to the left,
3	move 2 cm down.

When one random walk is finished, continue in the set of numbers for the second random walk, and so on. If more numbers are needed, enlarge your set.

6. Solve the problem in Exercise 5 by the method of separation of variables. Then compare the analytical result with that found in Exercise 5.

7. Given the following Dirichlet problem:

$$\begin{aligned} \text{P.D.E.:} \quad & v_{xx} + v_{yy} = 0, \quad 0 < x < 1, \quad 0 < y < 1; \\ \text{B.C.:} \quad & v(0, y) = v(1, y) = 0, \quad 0 < y < 1, \\ & v(x, 1) = 0, \quad v(x, 0) = 1, \quad 0 < x < 1. \end{aligned}$$

Find $v(1/4, 3/4)$ by the Monte Carlo method.

8.3 FINITE-DIFFERENCE APPROXIMATIONS

A widely used method for solving a Dirichlet problem begins with a transformation of the partial differential equation to a number of **difference equations**. The latter, being algebraic equations, are solved by one of the many methods available for solving a system of linear equations. One of these methods, called a **relaxation method**,* is described next.

We begin by noting that the derivative

$$\frac{du}{dx} = \lim_{h \to 0} \frac{u(x + h) - u(x)}{h} = \lim_{h \to 0} \frac{u(x) - u(x - h)}{h}$$

*The method is also known as *Liebmann's method* after Karl O. H. Liebmann (1874–1939), a German mathematician.

can be approximated at a given point x by the difference quotient,

$$\frac{u(x + h) - u(x)}{h} \quad \text{or} \quad \frac{u(x) - u(x - h)}{h},$$

provided that h is sufficiently small. Similarly,

$$\begin{aligned}\frac{d^2u}{dx^2} &= \lim_{h \to 0} \frac{1}{h}\left[\frac{du(x + h)}{dx} - \frac{du(x)}{dx}\right] \\ &= \lim_{h \to 0} \frac{1}{h}\left[\frac{u(x + h) - u(x)}{h} - \frac{u(x) - u(x - h)}{h}\right].\end{aligned}$$

Thus

$$\frac{d^2u}{dx^2} \doteq \frac{1}{h^2}[u(x + h) - 2u(x) + u(x - h)].$$

If u is a function of two variables, x and y, then

$$\frac{\partial^2 u(x, y)}{\partial x^2} \doteq \frac{1}{h^2}[u(x + h, y) - 2u(x, y) + u(x - h, y)]$$

and

$$\frac{\partial^2 u(x, y)}{\partial y^2} \doteq \frac{1}{h^2}[u(x, y + h) - 2u(x, y) + u(x, y - h)];$$

hence the potential equation can be approximated as

$$\begin{aligned}u_{xx} + u_{yy} \doteq \frac{1}{h^2}&[u(x + h, y) + u(x, y + h) \\ &+ u(x - h, y) + u(x, y - h) - 4u(x, y)] = 0,\end{aligned}$$

so that

$$u(x, y) = \frac{1}{4}[u(x + h, y) + u(x, y + h) + u(x - h, y) + u(x, y - h)] \tag{8.3-1}$$

is a finite-difference approximation to *Laplace's equation* at the point (x, y). In terms of a grid over the region R (Fig. 8.3-1) Eq. (8.3-1) shows that the value of u at the point P is approximately the *average* of the values at the four neighboring points numbered 1, 2, 3, and 4. In other words,

$$u(P) \doteq \frac{1}{4}(u_1 + u_2 + u_3 + u_4). \tag{8.3-2}$$

This relation must hold at *every* point of the grid so that the method produces a system of linear algebraic equations that can be solved with the aid of a computer. Note that some of the grid points fall on the boundary and hence are known. If the boundary lies between two grid points, then linear interpolation can be used. In the relaxation method, initial values at interior points are *assumed*, then corrected as the computations progress. It should be pointed

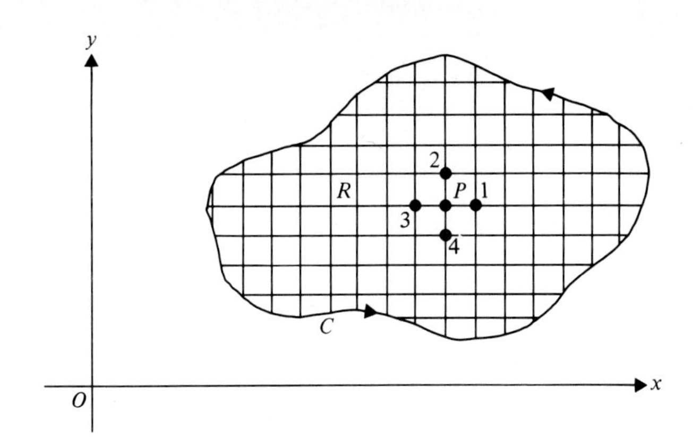

Figure 8.3-1
Finite-difference approximation of Laplace's equation.

out that the better the initial approximations, the more rapid the convergence. We will apply the foregoing method to some problems in magnetic induction heating in Section 8.4.

When using a finite-difference method to solve a problem involving a *parabolic equation*, additional complexities arise. In the one-dimensional diffusion equation, for example, we not only have to approximate u_{xx} but we also have to deal with u_t. Although we are working in xt-space here, we do not necessarily want to subdivide these two variables in the same way as we did in the xy-space of the Dirichlet problem.

If the diffusion problem is given as

$$u_t = ku_{xx}, \qquad 0 < x < L, \qquad t > 0, \tag{8.3-3}$$

then we can take space and time coordinates subdivided as follows:

$$\begin{aligned} x &= i\Delta x, & i &= 0, 1, 2, \ldots, N, \\ t &= j\Delta t, & j &= 0, 1, 2, \ldots. \end{aligned}$$

This leads to

$$u(x, t) = u(i\Delta x, j\Delta t) \equiv u_i^j,$$

that is, subscripts will denote position and superscripts will denote time. With this notation the finite-difference approximation of u_{xx} at the (i, j)th node written in terms of a **central-difference formula** is (see Fig. 8.3-2)

$$u_{xx}\Big|_{i,j} = \frac{u_{i-1}^j - 2u_i^j + u_{i+1}^j}{(\Delta x)^2} + O(\Delta x)^2, \tag{8.3-4}$$

where $O(\Delta x)^2$ represents terms* that are no greater than $(\Delta x)^2$.

*Omitting these terms leads to a **truncation error**.

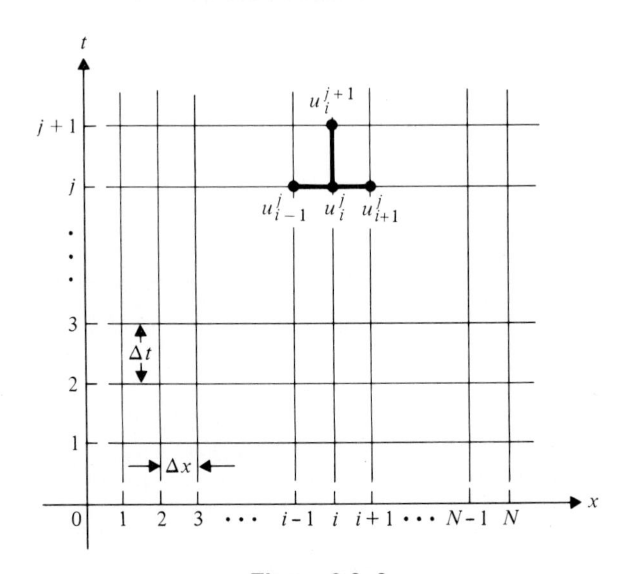

Figure 8.3-2
Finite-difference approximation of diffusion equation.

For the finite-difference representation of u_t we use a **forward-difference formula** to obtain

$$u_t\Big|_{i,j} = \frac{u_i^{j+1} - u_i^j}{\Delta t} + O(\Delta t) \tag{8.3-5}$$

at the (i, j)th node. By using Eqs. (8.3–4) and (8.3–5) the diffusion equation in (8.3–3) becomes

$$\frac{u_i^{j+1} - u_i^j}{\Delta t} = k\frac{u_{i-1}^j - 2u_i^j + u_{i+1}^j}{(\Delta x)^2},$$

with an error of $O(\Delta x)^2 + O(\Delta t)$. Solving this last equation for u_i^{j+1} gives us

$$u_i^{j+1} = ru_{i-1}^j + (1 - 2r)u_i^j + ru_{i+1}^j, \tag{8.3-6}$$

where

$$r = \frac{k\Delta t}{(\Delta x)^2}. \tag{8.3-7}$$

Equation (8.3–6) is called the **explicit form** of the finite-difference representation of the diffusion equation because the temperature can be computed at the $(j + 1)$st time step from a knowledge of the temperatures at the previous time step. For $j = 0$ we use the initial condition.

If in Eq. (8.3–7) Δt and Δx are chosen so that $r = 1/2$, then Eq. (8.3–6) can be further simplified. It can be shown* that the value of r *must* not exceed

*See C. F. Gerald's *Applied Numerical Analysis,* 2nd ed. (Reading, Mass.: Addison-Wesley, 1978), Chapter 8.

1/2 in order that the method remain **stable**, that is, errors do not become larger. If we arrange things so that $r = 1/2$, then Eq. (8.3–6) takes on the particularly simple form given by

$$u_i^{j+1} = \frac{1}{2}(u_{i-1}^j + u_{i+1}^j). \tag{8.3-8}$$

We illustrate this with an example.

EXAMPLE 8.3–1 A large flat steel plate is 2 cm thick. The plate is initially at 100°C. Its faces are then held at 0°C. Find the internal temperatures as a function of position and time.

Solution Since the plate is large, we can assume heat flow in one dimension. From tables we find that $k = 0.1515$ cm²/sec. In order to use Eq. (8.3–8) we will take $\Delta x = 0.2$ cm and $\Delta t = 0.132$ sec. We observe that there is symmetry about the line $x = 1$, allowing us to restrict our computations to the interval $0 \leq x \leq 1$. In mathematical terms the problem can be stated as follows:

$$\begin{aligned} \text{P.D.E.:} \quad & u_t = 0.1515 u_{xx}, \quad 0 < x < 2, \quad t > 0; \\ \text{B.C.:} \quad & u(0, t) = u(2, t) = 0, \quad t > 0; \\ \text{I.C.:} \quad & u(x, 0) = 100, \quad 0 < x < 2. \end{aligned}$$

The computations can be arranged in a table from which we can read discrete values of $u(x, t)$. ■

$t =$	$x = 0$	$x = 0.2$	$x = 0.4$	$x = 0.6$	$x = 0.8$	$x = 1.0$	$x = 1.2$
0	0	20	40	60	80	100	80
0.132	0	20	40	60	80	80	80
0.264	0	20	40	60	70	80	70
0.396	0	20	40	55	70	70	70
0.528	0	20	37.5	55	62.5	70	62.5
0.660	0	18.75	37.5	50	62.5	62.5	62.5
0.792	0	18.75	34.38	50	56.25	62.5	56.25
0.924	0	17.19	34.38	45.32	56.25	56.25	56.25
1.056	0	17.19	31.26	45.32	50.79	56.25	50.79
1.188	0	15.63	31.26	41.03	50.79	50.79	50.79

In contrast to the last finite-difference method we present next an **implicit method**. If we write the partial differential equation $u_t = ku_{xx}$ in the form

$$\frac{u_i^{j+1} - u_i^j}{\Delta t} = k\,\frac{u_{i-1}^{j+1} - 2u_i^{j+1} + u_{i+1}^{j+1}}{(\Delta x)^2}, \tag{8.3-9}$$

we can no longer solve for one of the temperatures in terms of *known* quantities. Equation (8.3–9) is obtained by representing u_{xx} in terms of finite differences at the ith node for the $(j + 1)$st time step by using a central-difference formula, whereas u_t is represented at the $(j + 1)$st time step by using a **backward-difference formula**. In order to use Eq. (8.3–9) it is necessary to solve a set of simultaneous algebraic equations for the u_i^{j+1}. The advantage of the method, however, lies in the fact that it is stable for all values of Δt and thus removes one of the restrictions in the explicit method.

A modification of the above implicit method uses the *average value* of the finite-difference expressions for u_{xx} in Eqs. (8.3–4) and (8.3–9). This modification is known as the **Crank–Nicolson method**. It has the advantages of being stable for all values of r in Eq. (8.3–7) and having a truncation error (or discretization error) of the order of $(\Delta t)^2 + (\Delta x)^2$.

One serious disadvantage of a finite-difference method is that interpolation of some kind must be used to obtain solutions at points that are not grid points. It would appear that this difficulty could be overcome by using a finer grid, but there is a limit to how fine a grid can be. At some stage in the computation the truncation and round-off errors can deteriorate the accuracy of the calculations. Moreover, the finer the grid, the greater the number of computations that have to be performed and the larger the amount of storage required. Limitations on the fineness of the grid also cause serious difficulties in the vicinity of a boundary that is irregular.

In an attempt to correct these shortcomings of finite-difference methods a *variational form* of the method, called the **finite-element method**, has been developed. The finite-element method utilizes the relationship between Euler's equation in the calculus of variations and a partial differential equation. For example, Euler's equation for the functional*

$$F[u] = \iint_R (u_x^2 + u_y^2)\, dx\, dy \qquad \textbf{(8.3–10)}$$

is $u_{xx} + u_{yy} = 0$, the potential equation. In other words, the function that minimizes the functional in Eq. (8.3–10) is harmonic. The significance of this is that the integrand in Eq. (8.3–10) can be written as

$$|\text{grad } u|^2$$

and interpreted as kinetic energy. Thus it seems natural that the finite-element method has its origin in the field of solid mechanics and structural analysis.

The Dirichlet problem over a region R having a boundary C can be solved by finding a function u that minimizes the integral in Eq. (8.3–10) and assumes the given values on C. This latter problem, while more complicated than a direct approach, has a number of aspects that lead to a more accurate solution. Instead of using square or rectangular grids, it is possible to use other configurations—for example, triangles of various sizes as shown in Fig. 8.3–3. In this way the node lines may be a better approximation to the boundary. Furthermore, the sides of the triangles may be polynomials (called **spline func-**

*Given a set $\mathcal{F}$ of functions defined on an interval $[a, b]$, we define a **functional** $F[u]$ to be a mapping from $\mathcal{F}$ into the set of real or complex numbers that assigns a unique number to each function u of $\mathcal{F}$. (See Exercise 16.)

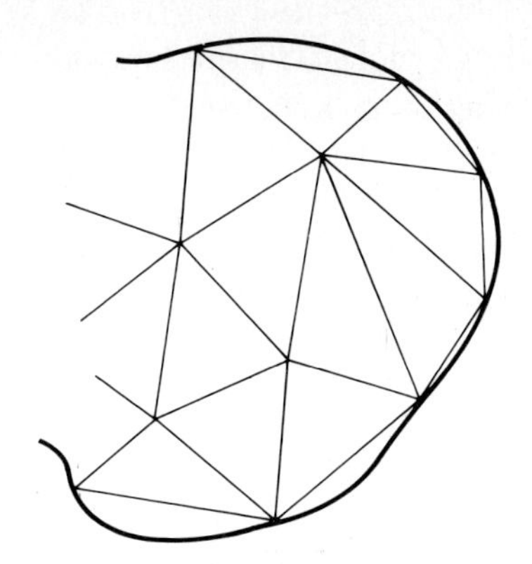

Figure 8.3-3
A triangular grid.

tions) so that even closer fits to an irregular boundary may be achieved and differentiability at the nodes may be preserved as well.

Details of the finite-element method and the variational principle on which it is based can be found in books listed at the end of this chapter.

Key Words and Phrases

difference equation
relaxation method
central-difference formula
truncation error
forward-difference formula
explicit and implicit difference equations
stable method
backward-difference formula
Crank–Nicolson method
finite-element method
functional
spline function

8.3 Exercises

• **1.** The formula

$$\frac{d^2u}{dx^2} = \frac{1}{h^2}[u(x + 2h) - 2u(x + h) + u(x)]$$

is known as a *forward-difference formula*, whereas the formula

$$\frac{d^2u}{dx^2} = \frac{1}{h^2}[u(x + h) - 2u(x) + u(x - h)]$$

is known as a *central-difference formula*. Explain the meaning of these terms.

2. Verify Eq. (8.3–6).

3. Explain the units of k in Example 8.3–1.

4. Explain why there is symmetry about the line $x = 1$ in Example 8.3–1.

•• 5. Consider the Dirichlet problem over a rectangular plate measuring 10 cm by 20 cm with one of the shorter sides held at temperature 100° and the other three sides held at zero. Choose coordinate axes as shown in Fig. 8.2–3 in the exercises of Section 8.2. Divide the region into eight subregions by lines 5 cm apart parallel to the axes so that *three* interior points will be produced.
   (a) Use Eq. (8.3–2) to write a system of three algebraic equations.
   (b) Solve the system of equations in part (a).
   (c) Compare the results in part (a) with those of Exercises 5 and 6 in Section 8.2.

6. In Exercise 5, change the grid spacing from 5 cm to 2.5 cm. How many interior points (and equations) are there now?

7. Change the grid spacing of Exercise 5 to $3\frac{1}{3}$ cm, and solve the resulting system of equations. (*Hint*: Use symmetry.)

8. Show that the one-dimensional wave equation and initial conditions $u_{tt} = u_{xx}$, $u(x, 0) = f(x)$, $u_t(x, 0) = g(x)$ can be approximated by the finite-difference equations

$$U(x, t + k) = 2U(x, t) - U(x, t - k) + \lambda^2[U(x + h, t) - 2U(x, t) + U(x - h, t)],$$
$$U(x, 0) = f(x), \qquad U(x, k) = kg(x) + f(x),$$

where $\lambda = k/h$. (*Note*: It can be shown that the convergence of U to u requires that $\lambda < 1$.)

9. Use the relaxation method to find the potential at the interior nodal points, given that the potential on the boundary is as shown in Fig. 8.3–4.

10. Solve $\nabla^2 u = xy$ on the square region bounded by $x = 0$, $x = 1$, $y = 0$, and $y = 1$. Use a spacing of 1/3, and take $u = 0$ on the boundary.

11. **(a)** A guitar string is 80 cm long. At a point 20 cm from one end it is raised 0.6 cm from its equilibrium position, then released from rest. Find the displacements of points along the string as a function of time. Let $\Delta x = 10$ cm, $\Delta t = 179\ \mu$s, and $a^2 = 3.136 \times 10^9$.
    (b) By determining the time required for one complete cycle of motion from part (a), compute the frequency with which the string vibrates.

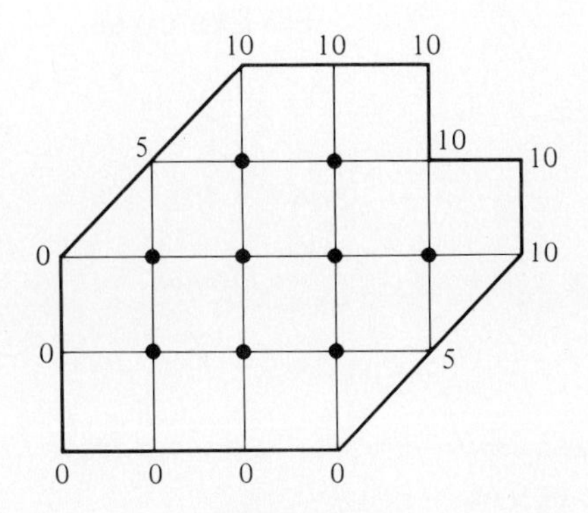

Figure 8.3–4

12. Consider the following problem:

$$\begin{aligned}
&\text{P.D.E.:} && U_{xx} = U_{tt}, \quad 0 < x < 1, \quad t > 0;\\
&\text{B.C.:} && \left.\begin{aligned} U(0, t) &= 0,\\ U(1, t) &= 0, \end{aligned}\right\} \quad t > 0;\\
&\text{I.C.:} && \left.\begin{aligned} U_t(x, 0) &= 0,\\ U(x, 0) &= \sin \pi x, \end{aligned}\right\} \quad 0 < x < 1.
\end{aligned}$$

(a) Solve the problem by the finite-difference method.
(b) Compare the solution in part (a) with the analytical solution

$$u(x, t) = \sin \pi x \cos \pi t.$$

13. Solve the following problem by the relaxation method:

$$\begin{aligned}
&\text{P.D.E.:} && u_{xx} + u_{yy} = 0, \quad 0 < x < 1, \quad 0 < y < 1;\\
&\text{B.C.:} && u(0, y) = u(1, y) = 0, \quad 0 < y < 1,\\
& && u(x, 1) = 0, \quad u(x, 0) = 1, \quad 0 < x < 1.
\end{aligned}$$

Divide the region so that there are nine interior points assumed to be zero initially, and carry the results to three decimals. (*Hint*: Use symmetry where possible.)

14. A better approximation for the starting values at the interior points of Exercise 13 can be obtained by using the following:

$$u(0.5, 0.25) = 0.75, \qquad u(0.5, 0.5) = 0.5, \qquad u(0.5, 0.75) = 0.25.$$

Use these values and use linear interpolation and symmetry to obtain the remaining six starting values. Then solve the problem again, and contrast the number of iterations required with that in Exercise 13.

15. Show that the finite-difference method can be extended to Poisson's equation

$$u_{xx} + u_{yy} = f(x, y)$$

by changing Eq. (8.3–2) to

$$u(P) = \frac{1}{4}(u_1 + u_2 + u_3 + u_4) - \frac{1}{4}h^2 f(P).$$

16. Show that each of the following are functionals.

(a) $\mathscr{I}[f] = \int_a^b f(x)\, dx$

(b) $\mathscr{D}[f] = f'(x_0), \quad a < x_0 < b$

(c) $\mathscr{S}[f] = f(x_0), \quad a \le x_0 \le b$

17. In Example 8.3–1, change the material from steel to copper ($k = 1.14$ cm²/sec), and explain how the solution changes.

18. Repeat Exercise 17 using firebrick ($k = 0.516$ m²/sec).

19. When using a finite-difference representation of Laplace's equation in plane *polar coordinates* (Eq. 7.1–2), we can represent the coordinates (ρ, ϕ) of a point P at a node by

$$\rho = i\Delta\rho \qquad \text{and} \qquad \phi = j\Delta\phi.$$

Then

$$u(\rho, \phi)\big|_P = u(i\Delta\rho, j\Delta\phi) \equiv u_i^j,$$

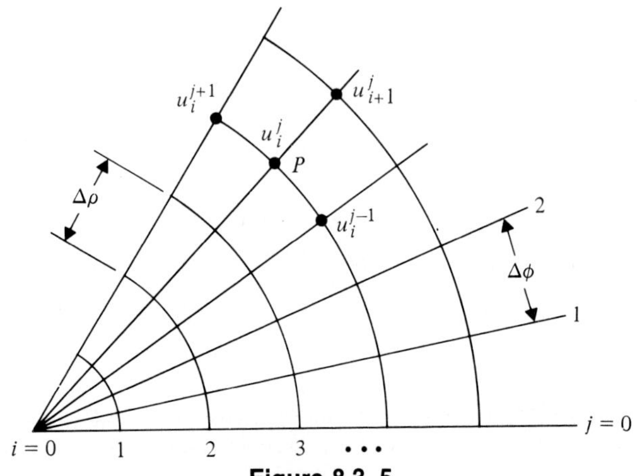

Figure 8.3-5
A grid in polar coordinates.

so that the subscripts denote *radial* position and superscripts denote *angular* position. Using central-difference formulas and referring to Fig. 8.3-5, obtain the following approximations.

(a) $u_{\rho\rho}\Big|_{i,j} = \dfrac{u_{i-1}^j - 2u_i^j + u_{i+1}^j}{(\Delta\rho)^2}$

(b) $u_{\phi\phi}\Big|_{i,j} = \dfrac{u_i^{j-1} - 2u_i^j + u_i^{j+1}}{(\Delta\phi)^2}$

(c) $u_{\rho}\Big|_{i,j} = \dfrac{u_{i+1}^j - u_{i-1}^j}{2\Delta\rho}$

(d) $$\nabla^2 u\Big|_{i,j} = \frac{1}{(\Delta\rho)^2}\left[\left(1 - \frac{1}{2i}\right)u_{i-1}^j - 2u_i^j + \left(1 + \frac{1}{2i}\right)u_{i+1}^j\right] + \frac{1}{i^2(\Delta\rho\Delta\phi)^2}\left[u_i^{j-1} - 2u_i^j + u_i^{j+1}\right]$$

8.4 APPLICATION TO MAGNETIC INDUCTION

There are many industrial applications of magnetic induction heating. Some of these are the following: surface hardening of iron or steel parts subject to excessive wear; bonding of one metal on another; heating for soldering and brazing; through-heating for forging and upsetting. The sources of power for induction heating are high-frequency* *ac*-generators. The high-frequency *emf*

*These vary from 3,000 to 450,000 Hz.

is applied to an induction coil, which may be a multiturn helically wound coil or a copper forging, the equivalent of a single turn.

In heating a cylindrical steel bar by induction the bar is placed within the induction coil and acts as a one-turn secondary of a high-frequency transformer. When the high-frequency sinusoidal voltage is impressed across the primary, a changing magnetic field is produced that cuts the bar. This field induces an *emf* that causes a flow of current through the resistance of the bar and thus produces heat. This heating is called **eddy-current heating**.

The changing magnetic field also causes heating due to a hysteresis effect. The theory here is that the molecules of the bar, under the influence of the strong magnetic field, behave like tiny magnets and tend to change their positions so that their axes are parallel to the field in much the same way that a compass needle lines itself up with the direction of the earth's magnetic field. Since the field is built up and collapsed thousands of times a second, a considerable amount of molecular motion takes place in the bar, and this motion also produces heat. Heating produced in this manner is called **hysteresis heating**. Above the **Curie temperature*** of 746°C, steel is nonmagnetic, so this heating effect vanishes at this temperature.

In air the magnetic potential Ω satisfies Laplace's equation, and the solution of this equation in the region within the coil results in equipotential surfaces. Of more interest, however, is the determination of the stream or flow surfaces, since these represent **magnetic lines**. If a longitudinal section is taken through the axis of the coil, then the problem resolves into the solution of a *plane-harmonic* equation subject to certain boundary conditions.

Accordingly, if Ω is considered as a function of a complex variable $x + iy$, we can define a **conjugate harmonic function**† ψ such that

$$-H_x = \frac{\partial \Omega}{\partial x} = \frac{\partial \psi}{\partial y} \quad \text{and} \quad -H_y = \frac{\partial \Omega}{\partial y} = -\frac{\partial \psi}{\partial x}. \qquad \textbf{(8.4-1)}$$

Thus ψ is harmonic (Exercise 1), and lines of constant ψ are magnetic lines.

We transform the partial differential equation

$$\frac{\partial^2 \psi}{\partial x^2} + \frac{\partial^2 \psi}{\partial y^2} = 0$$

into a finite-difference equation by means of Eq. (8.3-2). In Fig. 8.4-1 we show the interior portion of a typical induction coil. Only one fourth of the region is shown, since the remainder can be obtained from the symmetry of the figure. The magnetic lines in gausses are shown in air, the **permeability** μ being unity.

As Fig. 8.4-1 indicates, the region of interest was covered by a square grid, boundary values were calculated, and values at grid points were successively adjusted until Eq. (8.3-2) was satisfied at each point.

*After Pierre Curie (1859–1906), a French physicist whose early research was in the area of magnetism and piezoelectricity.

†See Section 10.3 of the author's *Advanced Engineering Mathematics* (Reading, Mass.: Addison-Wesley, 1982).

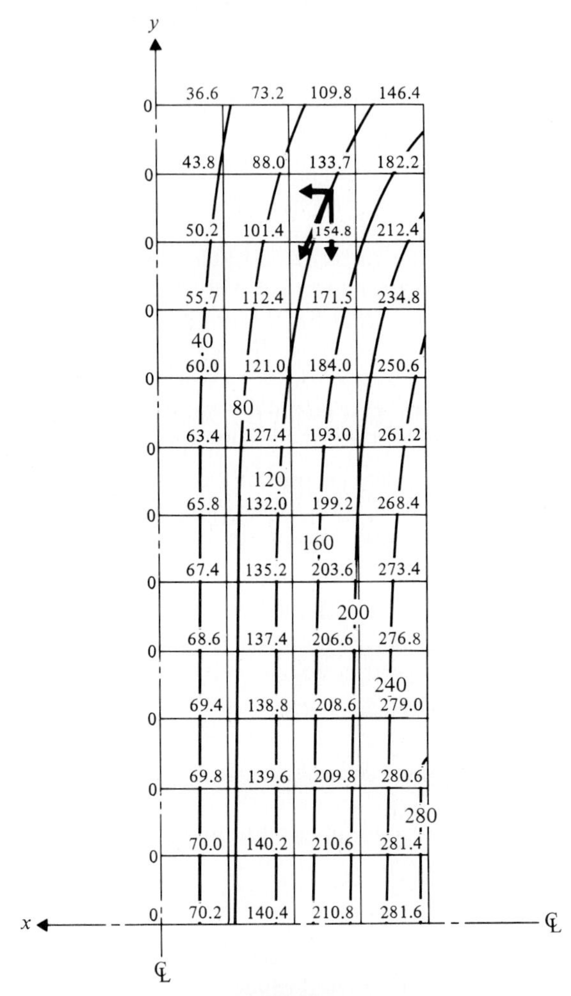

Figure 8.4-1
Magnetic induction lines (μ = 1).

Now, if a cylindrical steel bar is placed inside the induction coil, the value of μ will change at the interface between steel and air. At the boundary between the two media, Gauss' theorem* holds. This theorem states that if any closed surface is taken in the magnetic field, and if $\mu \partial\Omega/\partial n$ denotes the component of magnetic induction at any point of this surface in the direction of the outward normal, then

$$\iint_S \mu \frac{\partial \Omega}{\partial n} dS = -4\pi \sum m, \qquad \textbf{(8.4-2)}$$

*After Karl Friedrich Gauss (1777–1855), a German mathematician and physicist.

where $\Sigma\, m$ is the sum of the strengths of all the poles inside this surface. Since the surface does not cut through any magnetized matter, $\Sigma\, m$ is the aggregate strength of the poles of complete magnetic particles and is therefore equal to zero. At the boundary at which the value of μ changes abruptly we may take a closed surface formed of two areas fitting closely about an element dS of the boundary with the two areas being on opposite sides of the boundary. Then, if μ_1 is the permeability of air, μ_2 the permeability of steel, and n_1 and n_2 are the respective normals to the surfaces drawn into the two media, respectively, we have, from Eq. (8.4–2),

$$\mu_1 \frac{\partial \Omega}{\partial n_1} + \mu_2 \frac{\partial \Omega}{\partial n_2} = 0. \tag{8.4–3}$$

Corresponding to Eqs. (8.4–1), we have

$$-B_x = \mu \frac{\partial \Omega}{\partial x} = \frac{\partial \psi}{\partial y} \quad \text{and} \quad -B_y = \mu \frac{\partial \Omega}{\partial y} = -\frac{\partial \psi}{\partial x}.$$

Referring to Fig. 8.4–2, we have for medium 1 (Exercise 2)

$$\begin{aligned} \mu_1 \frac{\partial \Omega}{\partial n_1} &= \mu_1 \frac{\partial \Omega}{\partial x} \cos(x, n_1) + \mu_1 \frac{\partial \Omega}{\partial y} \sin(x, n_1) \\ &= \frac{\partial \psi}{\partial y} \cos(x, n_1) - \frac{\partial \psi}{\partial x} \sin(x, n_1) \\ &= \frac{\partial \psi}{\partial s}. \end{aligned}$$

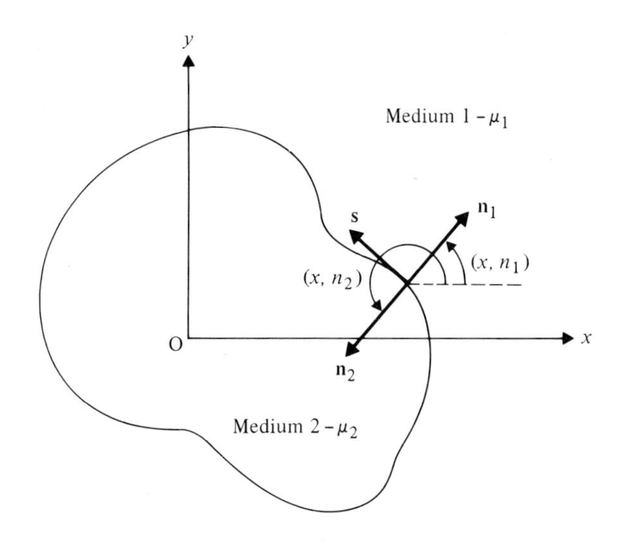

Figure 8.4–2
Refraction at an interface.

Similarly, for medium 2,

$$\mu_2 \frac{\partial \Omega}{\partial n_2} = \frac{\partial \psi}{\partial s}.$$

We can write similar equations for the function ψ, that is,

$$\begin{aligned}\frac{1}{\mu_1}\frac{\partial \psi}{\partial n_1} &= \frac{1}{\mu_1}\left[\frac{\partial \psi}{\partial x}\cos(x, n_1) + \frac{\partial \psi}{\partial y}\sin(x, n_1)\right] \\ &= -\frac{\partial \Omega}{\partial y}\cos(x, n_1) + \frac{\partial \Omega}{\partial x}\sin(x, n_1) \\ &= -\frac{\partial \Omega}{\partial s}\end{aligned}$$

and also

$$\frac{1}{\mu_2}\frac{\partial \psi}{\partial n_2} = -\frac{\partial \Omega}{\partial s}.$$

Thus the boundary condition expressed by Eq. (8.4–3) implies that

$$\frac{1}{\mu_1}\left(\frac{\partial \psi}{\partial n}\right)_1 = \frac{1}{\mu_2}\left(\frac{\partial \psi}{\partial n}\right)_2. \qquad \textbf{(8.4–4)}$$

It can be shown* that Eq. (8.4–4) can be implemented as follows: for nodal points at the boundary between air and steel, values of ψ are multiplied by 1 in air, $1/\mu_2$ in steel, and $(\mu_2 + 1)/2\mu_2$ along the interface. At other points the computations are carried out as in the previous example.

In order to determine the effect of permeability on magnetic induction, computations were carried out for permeabilities of 2, 100, 300, and 1,000. The results are shown in Figs. 8.4–3, 8.4–4, 8.4–5, and 8.4–6, respectively. For $\mu = 300$, values of ψ were also calculated at intermediate points of the net, a process called "advance to a finer net." This technique is useful for obtaining a closer view in regions of particular interest.

Up to this point the permeability of steel has been considered constant. It is known, however, that permeability varies with temperature, a fact that further complicates the computation. We next indicate how variable permeability can be taken into account. We begin with

$$\frac{\partial}{\partial x}\left(\mu \frac{\partial \Omega}{\partial x}\right) + \frac{\partial}{\partial y}\left(\mu \frac{\partial \Omega}{\partial y}\right) = 0$$

rather than with Laplace's equation. Introducing a conjugate function ψ defined by

$$\mu \frac{\partial \Omega}{\partial x} = \frac{\partial \psi}{\partial y} \quad \text{and} \quad \mu \frac{\partial \Omega}{\partial y} = -\frac{\partial \psi}{\partial x}$$

*See R. V. Southwell, *Relaxation Methods in Theoretical Physics,* Vol. 1 (Oxford: Clarendon Press, 1946).

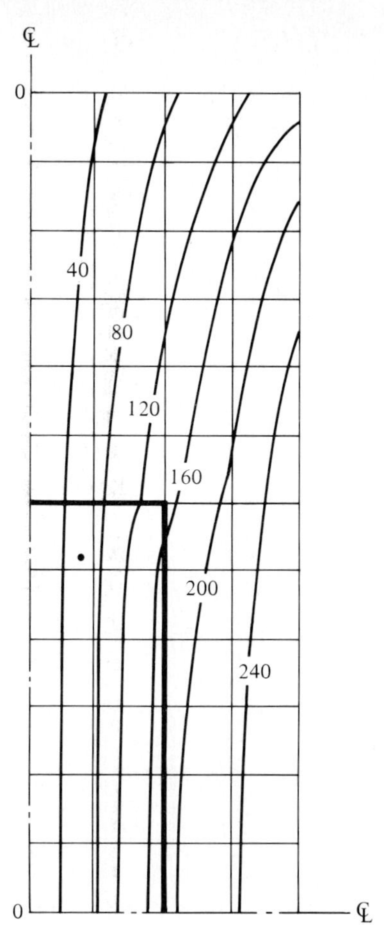

Figure 8.4-3
Magnetic induction lines (μ = 2).

and eliminating Ω, we obtain (Exercise 3)

$$\frac{\partial}{\partial x}\left(\frac{1}{\mu}\frac{\partial \psi}{\partial x}\right) + \frac{\partial}{\partial y}\left(\frac{1}{\mu}\frac{\partial \psi}{\partial y}\right) = 0. \tag{8.4-5}$$

Referring to Fig. 8.4-7, we can write the following finite difference approximations:

$$\left(\frac{\partial \psi}{\partial x}\right)_{1'} = \psi_1 - \psi_0 \quad \text{and} \quad \left(\frac{\partial \psi}{\partial x}\right)_{3'} = \psi_0 - \psi_3.$$

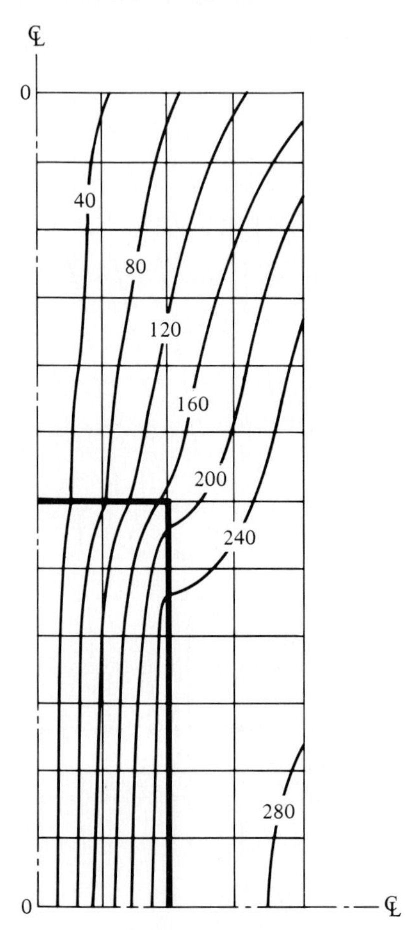

Figure 8.4-4
Magnetic induction lines (μ = 100).

Hence

$$\left[\frac{\partial}{\partial x}\left(\frac{1}{\mu}\frac{\partial \psi}{\partial x}\right)\right]_0 = \left(\frac{1}{\mu}\frac{\partial \psi}{\partial x}\right)_{1'} - \left(\frac{1}{\mu}\frac{\partial \psi}{\partial x}\right)_{3'}$$

$$= \frac{1}{\mu_{1'}}(\psi_1 - \psi_0) - \frac{1}{\mu_{3'}}(\psi_0 - \psi_3).$$

Similarly,

$$\left[\frac{\partial}{\partial y}\left(\frac{1}{\mu}\frac{\partial \psi}{\partial y}\right)\right]_0 = \frac{1}{\mu_{2'}}(\psi_2 - \psi_0) - \frac{1}{\mu_{4'}}(\psi_0 - \psi_4).$$

Thus the approximation to Eq. (8.4–5) is (Exercise 4)

$$\frac{1}{\mu_{1'}}\psi_1 + \frac{1}{\mu_{2'}}\psi_2 + \frac{1}{\mu_{3'}}\psi_3 + \frac{1}{\mu_{4'}}\psi_4 - \psi_0\left(\frac{1}{\mu_{1'}} + \frac{1}{\mu_{2'}} + \frac{1}{\mu_{3'}} + \frac{1}{\mu_{4'}}\right) = 0. \tag{8.4–6}$$

In Eq. (8.4–6), values of the permeability must be known at points 1′, 2′, 3′, and 4′, which are points midway between the nodal points of the net.

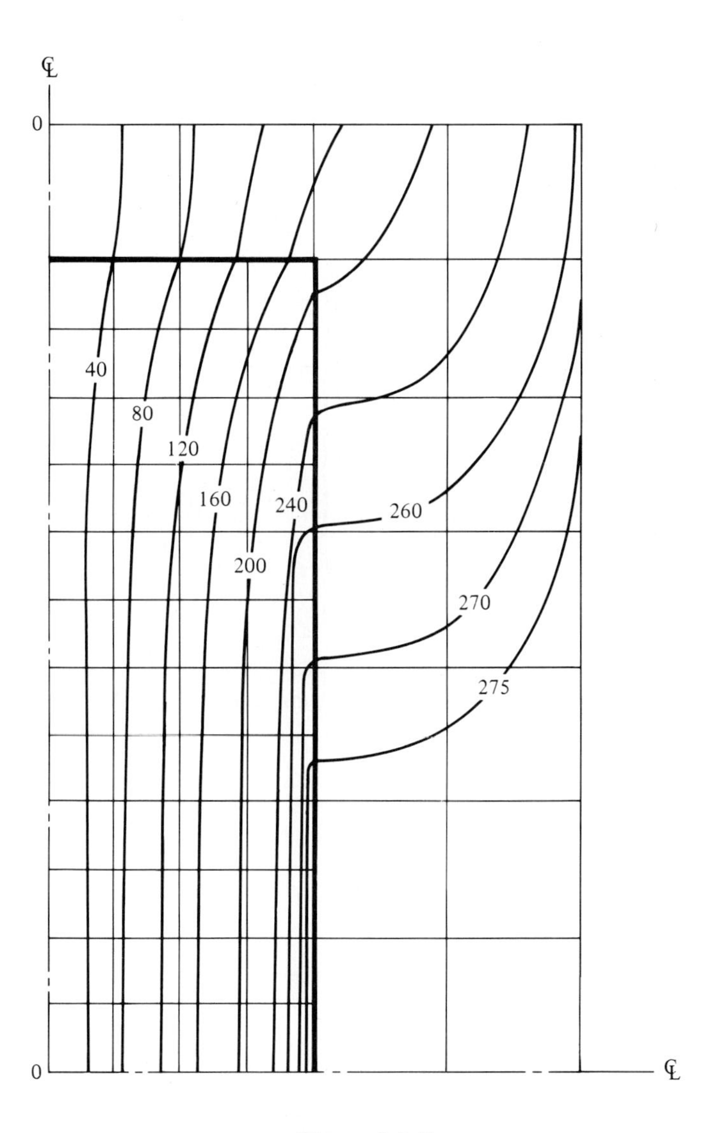

Figure 8.4–5
Magnetic induction lines (μ = 300).

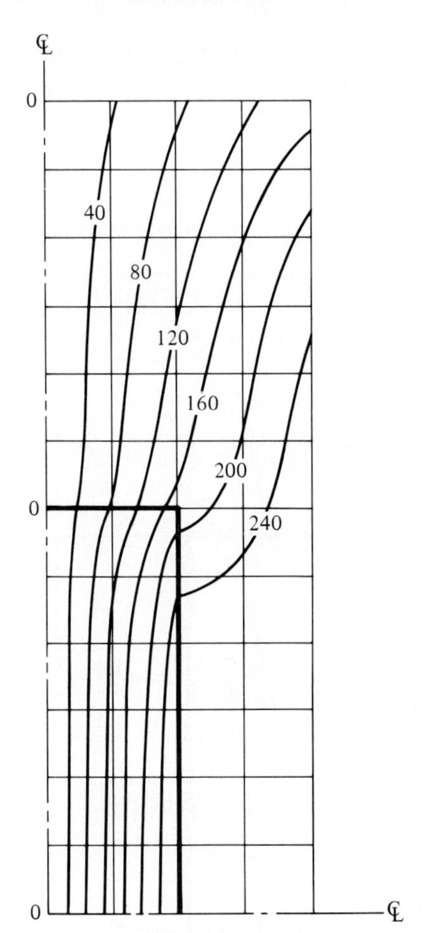

Figure 8.4-6
Magnetic induction lines (μ = 1000).

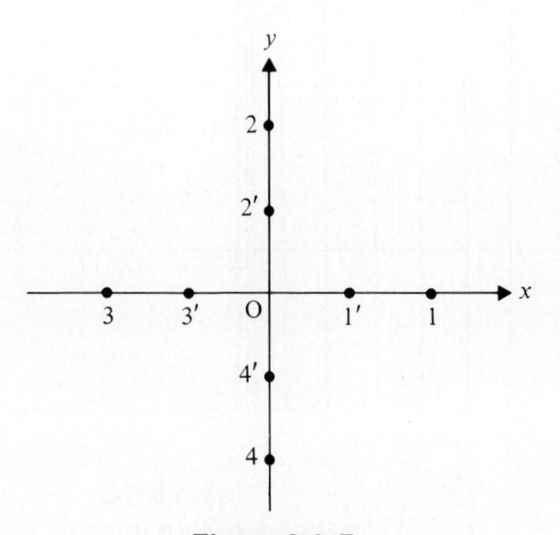

Figure 8.4-7

Our final example deals with the determination of the induced magnetic field in a cylindrical steel bar that is being heated by induction and whose surface temperature is approximately 740°C. With the permeability varying as shown, the results are given in Fig. 8.4–8.

We have presented these examples of magnetic induction heating in order to illustrate how departures from idealized conditions can be treated numerically. There are, of course, other difficulties that we have ignored. For example, at high frequencies the induced eddy-currents are confined to a thin layer on the surface of the material, and this results in rapid surface heating accompanied by a corresponding increase in resistivity and a decrease in permeability.

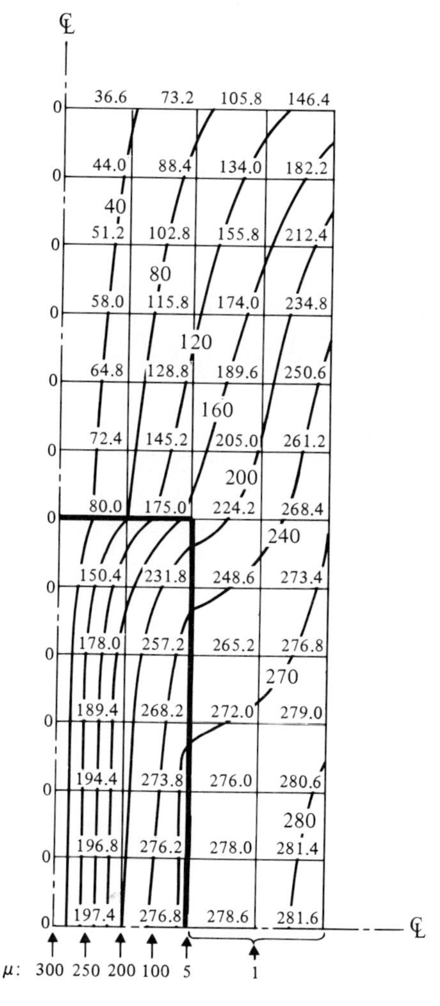

Figure 8.4–8
Magnetic induction lines (variable μ).

Key Words and Phrases

eddy-current heating
hysteresis heating
Curie temperature
magnetic lines
conjugate harmonic function

8.4 Exercises

• **1.** Use Eq. (8.4–1) to show that ψ is harmonic as a consequence of Ω being harmonic.

2. Referring to Fig. 8.4–2, verify that

$$\mu_1 \frac{\partial \Omega}{\partial n_1} = \frac{\partial \psi}{\partial s}.$$

3. Obtain Eq. (8.4–5).

4. Show that Eq. (8.4–6) follows from the previous steps.

REFERENCES

Monte Carlo

Buslenko, N. P., D. I. Golenko, Y. A. Shreider, I. M. Sobol', and V. G. Sragovich, *Monte Carlo Method.* New York: Pergamon, 1966.

Hammersley, J. M., and D. C. Handscomb, *Monte Carlo Methods.* London: Methuen, 1967.

Finite-Difference Methods

Ames, W. F., *Numerical Methods for Partial Differential Equations,* 2nd ed. New York: Academic Press, 1977.

Ferziger, J. H., *Numerical Methods for Engineering Application.* New York: Wiley, 1981.

Forsythe, G. E., and W. R. Wasow, *Finite-Difference Methods for Partial Differential Equations.* New York: Wiley, 1960.

Gerald, C. F., *Applied Numerical Analysis,* 2nd ed. Reading, Mass.: Addison-Wesley, 1978.

Gustafson, K. E., *Introduction to Partial Differential Equations and Hilbert Space Methods.* New York: Wiley, 1980.

Ozisik, M. N., *Heat Conduction.* New York: Wiley, 1980.

Smith, G. D., *Numerical Solution of Partial Differential Equations: Finite Difference Methods,* 2nd ed. Oxford: Clarendon Press, 1978.

Zachmanoglou, E. C., and D. W. Thoe, *Introduction to Partial Differential Equations with Applications.* Baltimore: Williams and Wilkins, 1976.

Finite-Element Method

Gladwell, I., and R. Wait (eds.), *A Survey of Numerical Methods for Partial Differential Equations.* Oxford: Clarendon Press, 1979.

Huebner, K. H., *Finite Element Method for Engineers.* New York: Wiley, 1975.

Johnson, L. W., and R. D. Riess, *Numerical Analysis,* 2nd ed. Reading, Mass.: Addison-Wesley, 1982.

Lapidus, L., and G. F. Pinder, *Numerical Solution of Partial Differential Equations in Science and Engineering.* New York: Wiley, 1982.

Martin, H. C., and G. F. Carey, *Introduction to Finite-Element Analysis.* New York: McGraw-Hill, 1973.

Ural, O., *Finite Element Method: Basic Concepts and Applications.* Scranton, Pa.: International Textbook, 1973.

Zienkiewicz, O. C., and I. K. Cheung, *The Finite Element Method in Engineering Science.* New York: McGraw-Hill, 1971.

Variational Principles

Chester, C. R., *Techniques in Partial Differential Equations.* New York: McGraw-Hill, 1971.

Friedman, A., *Variational Principles and Free-Boundary Problems.* New York: Wiley, 1982.

Hildebrand, F. B., *Advanced Calculus for Applications,* 2nd ed. Englewood Cliffs, N.J.: Prentice-Hall, 1976.

Lebedev, N. N., I. P. Skalskaya, and Y. S. Uflyand, *Worked Problems in Applied Mathematics.* New York: Dover, 1979.

Sagan, H., *Boundary and Eigenvalue Problems in Mathematical Physics.* New York: Wiley, 1961.

Sneddon, I. N., *Elements of Partial Differential Equations.* New York: McGraw-Hill, 1957.

General References

Brigham, E. O., *The Fast Fourier Transform.* Englewood Cliffs, N.J.: Prentice-Hall, 1974.

Broman, A., *Introduction to Partial Differential Equations from Fourier Series to Boundary-Value Problems.* Reading, Mass.: Addison-Wesley, 1970.

Dettman, J. W., *Mathematical Methods in Physics and Engineering,* 2nd ed. New York: McGraw-Hill, 1969.

Duff, G. F. D., and D. Naylor, *Differential Equations of Applied Mathematics.* New York: John Wiley, 1966.

Gladwell, I., and R. Wait (eds.), *A Survey of Numerical Methods for Partial Differential Equations.* London: Clarendon Press, 1979.

Hanna, J. R., *Fourier Series and Integrals of Boundary Value Problems.* New York: John Wiley, 1982.

Henrici, P., *Discrete Variable Methods in Ordinary Differential Equations.* New York: John Wiley, 1962.

Hochstadt, H., *The Functions of Mathematical Physics.* New York: John Wiley, 1971.

Jackson, D., *Fourier Series and Orthogonal Polynomials* (Carus Mathematical Monograph No. 6). New York: Mathematical Association of America, 1941.

John, F., *Partial Differential Equations,* 4th ed. New York: Springer Verlag, 1981.

Kovach, L. D., *Advanced Engineering Mathematics.* Reading, Mass.: Addison-Wesley, 1982.

Li, Wen-Hsiung, *Engineering Analysis.* Englewood Cliffs, N.J.: Prentice-Hall, 1960.

Maron, M. J., *Numerical Analysis: A Practical Approach.* New York: Macmillan, 1982.

Mikhlin, S. G. (ed.), *Linear Equations of Mathematical Physics.* New York: Holt, Rinehart and Winston, 1967.

Miles, J. W., *Integral Transforms in Applied Mathematics.* Cambridge: Cambridge University Press, 1971.

Schechter, M., *Modern Methods in Partial Differential Equations: An Introduction.* New York: McGraw-Hill, 1977.

Stromberg, K. R., *An Introduction to Classical Real Analysis.* Belmont, Calif.: Wadsworth, 1981.

Wilf, H. S., *Mathematics for the Physical Sciences.* New York: John Wiley, 1962.

Appendix

Table I Exponential and hyperbolic functions

x	e^x	e^{-x}	$\sinh x$	$\cosh x$
0.00	1.0000	1.00000	0.0000	1.0000
0.05	1.0513	0.95123	0.0500	1.0013
0.10	1.1052	0.90484	0.1002	1.0050
0.15	1.1618	0.86071	0.1506	1.0113
0.20	1.2214	0.81873	0.2013	1.0201
0.25	1.2840	0.77880	0.2526	1.0314
0.30	1.3499	0.74082	0.3045	1.0453
0.35	1.4191	0.70469	0.3572	1.0619
0.40	1.4918	0.67032	0.4108	1.0811
0.45	1.5683	0.63763	0.4653	1.1030
0.50	1.6487	0.60653	0.5211	1.1276
0.55	1.7333	0.57695	0.5782	1.1551
0.60	1.8221	0.54881	0.6367	1.1855
0.65	1.9155	0.52205	0.6967	1.2188
0.70	2.0138	0.49659	0.7586	1.2552
0.75	2.1170	0.47237	0.8223	1.2947
0.80	2.2255	0.44933	0.8881	1.3374
0.85	2.3396	0.42741	0.9561	1.3835
0.90	2.4596	0.40657	1.0265	1.4331
0.95	2.5857	0.38674	1.0995	1.4862
1.00	2.7183	0.36788	1.1752	1.5431
1.05	2.8577	0.34994	1.2539	1.6038
1.10	3.0042	0.33287	1.3356	1.6685
1.15	3.1582	0.31664	1.4208	1.7374
1.20	3.3201	0.30119	1.5095	1.8107
1.25	3.4903	0.28650	1.6019	1.8884
1.30	3.6693	0.27253	1.6984	1.9709
1.35	3.8574	0.25924	1.7991	2.0583

Table I (*continued*)

x	e^x	e^{-x}	*sinh x*	*cosh x*
1.40	4.0552	0.24660	1.9043	2.1509
1.45	4.2631	0.23457	2.0143	2.2488
1.50	4.4817	0.22313	2.1293	2.3524
1.55	4.7115	0.21225	2.2496	2.4619
1.60	4.9530	0.20190	2.3756	2.5775
1.65	5.2070	0.19205	2.5075	2.6995
1.70	5.4739	0.18268	2.6456	2.8283
1.75	5.7546	0.17377	2.7904	2.9642
1.80	6.0496	0.16530	2.9422	3.1075
1.85	6.3598	0.15724	3.1013	3.2585
1.90	6.6859	0.14957	3.2682	3.4177
1.95	7.0287	0.14227	3.4432	3.5855
2.00	7.3891	0.13534	3.6269	3.7622
2.05	7.7679	0.12873	3.8196	3.9483
2.10	8.1662	0.12246	4.0219	4.1443
2.15	8.5849	0.11648	4.2342	4.3507
2.20	9.0250	0.11080	4.4571	4.5679
2.25	9.4877	0.10540	4.6912	4.7966
2.30	9.9742	0.10026	4.9370	5.0372
2.35	10.4856	0.09537	5.1951	5.2905
2.40	11.0232	0.09072	5.4662	5.5569
2.45	11.5883	0.08629	5.7510	5.8373
2.50	12.1825	0.08208	6.0502	6.1323
2.55	12.8071	0.07808	6.3645	6.4426
2.60	13.4637	0.07427	6.6947	6.7690
2.65	14.1540	0.07065	7.0417	7.1123
2.70	14.8797	0.06721	7.4063	7.4735
2.75	15.6426	0.06393	7.7894	7.8533
2.80	16.4446	0.06081	8.1919	8.2527
2.85	17.2878	0.05784	8.6150	8.6728
2.90	18.1741	0.05502	9.0596	9.1146
2.95	19.1060	0.05234	9.5268	9.5791
3.00	20.0855	0.04979	10.018	10.068
3.05	21.1153	0.04736	10.534	10.581
3.10	22.1980	0.04505	11.076	11.122
3.15	23.3361	0.04285	11.647	11.690
3.20	24.5325	0.04076	12.246	12.287
3.25	25.7903	0.03877	12.876	12.915
3.30	27.1126	0.03688	13.538	13.575
3.35	28.5027	0.03508	14.234	14.269
3.40	29.9641	0.03337	14.965	14.999
3.45	31.5004	0.03175	15.734	15.766

Table I (*continued*)

x	e^x	e^{-x}	*sinh x*	*cosh x*
3.50	33.1155	0.03020	16.543	16.573
3.55	34.8133	0.02872	17.392	17.421
3.60	36.5982	0.02732	18.286	18.313
3.65	38.4747	0.02599	19.224	19.250
3.70	40.4473	0.02472	20.211	20.236
3.75	42.5211	0.02352	21.249	21.272
3.80	44.7012	0.02237	22.339	22.362
3.85	46.9931	0.02128	23.486	23.507
3.90	49.4024	0.02024	24.691	24.711
3.95	51.9354	0.01925	25.958	25.977
4.00	54.5982	0.01832	27.290	27.308
4.10	60.3403	0.01657	30.162	30.178
4.20	66.6863	0.01500	33.336	33.351
4.30	73.6998	0.01357	36.843	36.857
4.40	81.4509	0.01227	40.719	40.732
4.50	90.0171	0.01111	45.003	45.014
4.60	99.4843	0.01005	49.737	49.747
4.70	109.9472	0.00910	54.969	54.978
4.80	121.5104	0.00823	60.751	60.759
4.90	134.2898	0.00745	67.141	67.149
5.00	148.4132	0.00674	74.203	74.210

Table II Laplace transforms

$$F(s) = \int_0^\infty f(t) \exp(-st)\, dt$$

$F(s)$	$f(t)$
1. $\frac{a}{s}$	a
2. $\frac{1}{s^2}$	t
3. $\frac{2}{s^3}$	t^2
4. $\frac{n!}{s^{n+1}}$	t^n, n a positive integer
5. $\frac{\Gamma(\alpha + 1)}{s^{\alpha+1}}$	t^α, α a real number, $\alpha > -1$
6. $\frac{a}{s^2 + a^2}$	$\sin at$
7. $\frac{s}{s^2 + a^2}$	$\cos at$

Table II (*continued*)

$F(s)$	$f(t)$
8. $\dfrac{1}{s-a}$, $s > a$	$\exp(at)$
9. $\dfrac{1}{(s-a)^2}$	$t \exp(at)$
10. $\dfrac{2}{(s-a)^3}$	$t^2 \exp(at)$
11. $\dfrac{n!}{(s-a)^{n+1}}$	$t^n \exp(at)$, n a positive integer
12. $\dfrac{a}{s^2-a^2}$	$\sinh at$
13. $\dfrac{s}{s^2-a^2}$	$\cosh at$
14. $\dfrac{s^2+2a^2}{s(s^2+4a^2)}$	$\cos^2 at$
15. $\dfrac{s^2-2a^2}{s(s^2-4a^2)}$	$\cosh^2 at$
16. $\dfrac{2a^2}{s(s^2+4a^2)}$	$\sin^2 at$
17. $\dfrac{2a^2}{s(s^2-4a^2)}$	$\sinh^2 at$
18. $\dfrac{2a^2 s}{s^4+4a^4}$	$\sin at \sinh at$
19. $\dfrac{a(s^2-2a^2)}{s^4+4a^4}$	$\cos at \sinh at$
20. $\dfrac{a(s^2+2a^2)}{s^4+4a^4}$	$\sin at \cosh at$
21. $\dfrac{s^3}{s^4+4a^4}$	$\cos at \cosh at$
22. $\dfrac{2as}{(s^2+a^2)^2}$	$t \sin at$
23. $\dfrac{s^2-a^2}{(s^2+a^2)^2}$	$t \cos at$
24. $\dfrac{2as}{(s^2-a^2)^2}$	$t \sinh at$
25. $\dfrac{s^2+a^2}{(s^2-a^2)^2}$	$t \cosh at$
26. $\dfrac{a}{(s-b)^2+a^2}$	$\exp(bt) \sin at$
27. $\dfrac{s-b}{(s-b)^2+a^2}$	$\exp(bt) \cos at$
28. $\dfrac{1}{(s-a)(s-b)}$	$\dfrac{\exp(at) - \exp(bt)}{a-b}$

Table II (*continued*)

$F(s)$	$f(t)$
29. $\dfrac{s}{(s-a)(s-b)}$	$\dfrac{a\exp(at)-b\exp(bt)}{a-b}$
30. $\dfrac{1}{s(s-a)(s-b)}$	$\dfrac{1}{ab}+\dfrac{b\exp(at)-a\exp(bt)}{ab(a-b)}$
31. $\dfrac{1}{(s-a)(s-b)(s-c)}$	$\dfrac{(c-b)e^{at}+(a-c)e^{bt}+(b-a)e^{ct}}{(a-b)(a-c)(c-b)}$
32. $\dfrac{s}{(s-a)(s-b)(s-c)}$	$\dfrac{a(b-c)e^{at}+b(c-a)e^{bt}+c(a-b)e^{ct}}{(a-b)(b-c)(a-c)}$
33. $\dfrac{1}{s(s^2+a^2)}$	$\dfrac{1}{a^2}(1-\cos at)$
34. $\dfrac{1}{s(s^2-a^2)}$	$\dfrac{1}{a^2}(\cosh at-1)$
35. 1	$\delta(t)$
36. $\exp(-as)$	$\delta(t-a)$
37. $\dfrac{1}{s}\exp(-as)$	$u_a(t)=\begin{cases}0, & \text{for } t<a,\\ 1, & \text{for } t>a,\end{cases}\quad a\geq 0$
38. $(s^2+\lambda^2)^{-1/2}$	$J_0(\lambda t)$
39. $\dfrac{1}{s}\log s$	$-\log t - C$
40. $\log\dfrac{s-a}{s}$	$\dfrac{1}{t}(1-\exp(at))$
41. $\log\dfrac{s-a}{s-b}$	$\dfrac{1}{t}(\exp(bt)-\exp(at))$
42. $\log\dfrac{s+a}{s-a}$	$\dfrac{2}{t}\sinh at$
43. $\arctan\left(\dfrac{a}{s}\right)$	$\dfrac{1}{t}\sin at$
44. $\dfrac{\sqrt{\pi}}{2}\exp\left(\dfrac{s^2}{4}\right)\operatorname{erfc}\left(\dfrac{s}{2}\right)$	$\exp(-t^2)$
45. $\dfrac{1}{b}F\left(\dfrac{s}{b}\right),\quad b>0$	$f(bt)$
46. $F(s-b)$	$\exp(bt)f(t)$
47. $s\sqrt{k}(s+k)^{-1/2}$	$\operatorname{erf}\sqrt{kt},\quad k>0$
48. $\dfrac{1}{s}[1-\operatorname{erf}(s/2k)]\exp(s^2/4k^2)$	$\operatorname{erf}(kt)\quad k>0$
49. $\dfrac{1}{s}\exp(-k\sqrt{s})$	$\operatorname{erfc}(k/2\sqrt{t}),\quad k>0$
50. $2a(\sqrt{s+2a}+\sqrt{s})^{-2}$	$\dfrac{1}{t}I_1(at)\exp(-at)$
51. $\lambda^{-\nu}(\sqrt{\lambda^2+s^2}-s)^{\nu}(\lambda^2+s^2)^{-1/2}$	$J_\nu(\lambda t),\quad \nu>-1$

Table III Fourier transforms

$$\bar{f}(\alpha) = \int_{-\infty}^{\infty} f(x) \exp(i\alpha x)\, dx$$

$\bar{f}(\alpha)$	$f(x)$
1. $\dfrac{2 \sin \alpha c}{\alpha}$	$\begin{cases} 1, & \lvert x\rvert < c, \\ 0, & \lvert x\rvert > c, \\ 1/2, & \lvert x\rvert = c \end{cases}$
2. $\pi a \exp(-a\lvert\alpha\rvert)$	$\dfrac{a^2}{a^2 + x^2}$
3. $\dfrac{2ih}{\alpha}(1 - \cos \alpha c)$	$\begin{cases} -h, & -c < x < 0, \\ h, & 0 < x < c, \\ 0, & \lvert x\rvert \geq c \end{cases}$
4. $2a\sqrt{\pi} \exp(-a^2\alpha^2)$	$\exp(-x^2/4a^2)$
5. $[2 \cos(a\pi/2)]/(1 - \alpha^2)$	$\begin{cases} \cos x, & \lvert x\rvert \leq \pi/2, \\ 0, & \lvert x\rvert > \pi/2 \end{cases}$
6. $\dfrac{2i \sin n\pi\alpha}{\alpha^2 - 1}$	$\begin{cases} \sin x, & \lvert x\rvert \leq n\pi \\ 0, & \lvert x\rvert > n\pi \end{cases}$
7. $-i\alpha\bar{f}(\alpha)$	$f'(x)$
8. $-\alpha^2\bar{f}(\alpha)$	$f''(x)$
9. $\dfrac{1}{a}\bar{f}(\alpha/a)$	$f(ax)$
10. $\bar{f}(a\alpha)$	$\dfrac{1}{a} f(x/a)$
11. $\bar{f}(\alpha) \exp(i\alpha b)$	$f(x - b)$
12. $(\cos \alpha b)\bar{f}(\alpha)$	$\frac{1}{2}[f(x + b) + f(x - b)]$
13. $2\pi f(-\alpha)$	$\bar{f}(x)$
14. $\dfrac{2}{1 + \alpha^2}$	$\exp(-\lvert x\rvert)$
15. $\dfrac{1}{\sqrt{1 + \alpha^2}}$	$\dfrac{1}{\pi} K_0(\lvert x\rvert)$

Answers and Hints to Selected Exercises

CHAPTER 1

Section 1.1, p. 5

12. **(a)** $y = c_1 \exp t + c_2 \exp(2t)$ **(c)** $y = (c_1 \cos 4t + c_2 \sin 4t) \exp(3t)$

14. **(b)** $y = [c_1 \cos(\sqrt{7}t/2) + c_2 \sin(\sqrt{7}t/2)] \exp(-t/2)$
(d) $x = -(3/2) \cos 2t + \sin 2t$

15. **(a)** $y = 2 \exp(-x) - 2 \exp(-2x)$ **(c)** $y = 3(1 - 4x) \exp(2x)$

16. **(a)** $y = c_1 \cosh 2t + c_2 \sinh 2t$ **(c)** $y = c_1 + c_2 \exp(-5r)$

17. **(b)** $u = 5 \exp(-2t) - 4 \exp(-3t)$
(d) $y = (1/2)(2 + 9t) \exp(-5t/2)$

18. The Wronskian is $\beta \exp(\alpha x)$, which is different from zero for all x. Note that β cannot be zero.

25. Remove common factors: 3 from row one, 2 from row two, 4 from row three; then -3 from column three. Then interchange rows two and three; multiply row one by -1, and add the result to row two; multiply row one by 1, and add the result to row three. The final result is $(3)(2)(4)(-3)(1)(5)(2) = 720$.

Section 1.2, p. 10

6. **(a)** $(A \cos x + B \sin x) \exp x + (C \cos 2x + D \sin 2x) \exp(2x)$
(b) $Ax^3 + Bx^2 + Cx + D) \exp x$ **(c)** $A + B \cos x + C \sin x$

7. **(a)** $y = c_1 \exp(-x) + c_2 \exp(3x) - (1/2) \exp x + \exp(2x)$
(c) $y = c_1 \exp x + c_2 \exp(-4x) + (3x/5) \exp x$
(e) $y = [c_1 + c_2x + x^2 + (1/6)x^3] \exp(-2x)$
(g) $y = (c_1 + c_2x) \exp(-2x) + x^2 - 4x + (7/2)$

9. $y_p(x) = (-3x/4) \cos 2x$

11. **(a)** $y = c_1 \cos x + [c_2 + (x/2)] \sin x - \sin 2x$
(c) $y = (c_1 + c_2 x) \exp(-x) + (1/50)(4 \sin 2x - 3 \cos 2x + 25)$

12. **(a)** $y = (3 \cos 2x + 4 \sin 2x) \exp(-x) + 2 \cos x + \sin x$
(c) $y = -3 \exp x + x^2 + 4x + 5$

13. **(d)** $u'(x) = \cos x - \sec x$; $v'(x) = \sin x$
(e) $u(x) = \sin x - \log |\sec x + \tan x|$;

$v(x) = -\cos x$ (*Note*: Constants of integration are not required, since we are seeking a *particular* solution.)

(f) $y_p(x) = -\cos x \log |\sec x + \tan x|$

14. **(b)** $y = [c_1 + c_2 x - \log(1 - x)] \exp x, x < 1$
(d) $y = [c_1 + \log(\cos x)] \cos x + (c_2 + x) \sin x, |x| < \pi/2$
(f) $y = c_1 \cos 2x + c_2 \sin 2x + (1/4) \sin 2x \log |\csc 2x - \cot 2x|, |x| < \pi/2$

15. $y = (2 + 3x - \log |1 - x|) \exp x$

17. $y = c_1 \exp(-x) + c_2 \exp x - (1/2)(x \sin x + \cos x)$

Section 1.3, p. 18

2. $y = (c_1/x) + (c_2/x) \log x + x^3/16$

4. $y = (3/x)(x^3 + 1)$

6. $y = t^{-2}[c_1 \cos(\log t) + c_2 \sin(\log t)]$

8. $y = c_1 x^{-3} + c_2 x^3 + (x/4) - (x^2/5)$

10. $y = (x/2)[3 \cos(\log x) + 5 \sin(\log x)] + (5/2) - 4x$

11. **(a)** $y = 1 + 2 \log x$

12. **(a)** Let $y_2(x) = xu(x)$ to obtain $xu'' + u' = 0$; then put $u = v$ to obtain $vdx + xdv = 0$; $y_2(x) = x \log x$.
(c) $y_2(x) = x^{-2}$

13. xy' becomes $cx\, dy/d(cx) = xy'$

15. **(a)** $y = x^{1/2}(x - 1)$

16. $y = c_1 x^{1-a} + c_2, a \neq 1$; $y = c_1 \log x + c_2, a = 1$

Section 1.4, p. 24

1. **(c)** $\{1, 2/3, 13/15, 76/105, \ldots\}$
(d) $\{1, 4/3, 13/9, 40/27, \ldots\}$

2. This is a geometric series with $a = 1$, $r = 1/3$, and sum $3/2$.

3. For $x = 0$ we have $-(1 + 1/2 + 1/3 + 1/4 + \cdots)$, a divergent harmonic series. For $x = 2$ we have $1 - 1/2 + 1/3 - 1/4 + - \cdots$, which is a convergent alternating series.

7. Three terms are sufficient for 4D accuracy.

8. **(a)** $(2^n - 1)/2^n$

10. **(a)** This is a geometric series with first term 1/10 and common ratio 1/10.
(b) 1/9; note that the series can also be written as 0.111 . . . .

11. **(a)** 1 **(b)** ∞ **(c)** 1 **(d)** e **(e)** 3 **(f)** ∞

12. **(a)** $-2 < x < 2$ **(c)** $0 \le x < 2$ **(e)** $-\infty < x < \infty$

14. **(d)** Use the integral test.

15. The series diverges if $p = 1$ and converges if $p > 1$.

16. Evaluate $\int_1^\infty x^{-p}\,dx$.

17. **(a)** $S_n = 1/2[1/n - 1/(n + 2)]$ **(b)** 3/4

18. The sum of the series is $(1/p)(1 + 1/2 + 1/3 + \cdots + 1/p)$.

Section 1.5, p. 39

1. **(b)** $y_2(x) = a_1 \sum_{n=0} \frac{2\cdot 5\cdot 8 \cdots (3n - 1)}{(3n + 1)!} x^{3n+1}$

(*Note*: Here and in part (a), use 1 for the numerator when $n = 0$.)

6. **(a)** $x = 0$ is a regular singular point.
(d) $x = 1$ is a regular singular point, $x = 0$ is an irregular singular point.

7. **(a)** $y = a_0 \cos x + a_1 \sin x$ **(c)** $y = c \exp x - x^2 - 2x - 2$
(e) $y = a_0(1 + x \arctan x) + a_1 x$

8. **(a)** $y = a_0 J_0(x) = a_0\left(1 - \frac{x^2}{2^2} + \frac{x^4}{2^2\cdot 4^2} - \frac{x^6}{2^2\cdot 4^2\cdot 6^2} + - \cdots\right)$
(c) $y = a_0 x + a_1 x^{-2}$

10. **(a)** $y = a_0[1 + \frac{1}{2}(x - 1)^2 + \frac{1}{6}(x - 1)^3 + \frac{1}{6}(x - 1)^4 + \cdots]$
$+ a_1[(x - 1) + \frac{1}{2}(x - 1)^2 + \frac{1}{2}(x - 1)^3 + \frac{1}{4}(x - 1)^4 + \cdots]$

12. $y = a_0\left(x + x^2 + \frac{x^3}{3} + \frac{x^4}{18} + \cdots\right)$

15. $y = 1 + x + x^2 - \frac{x^3}{6} - \frac{x^4}{12} - \frac{x^5}{20} + \frac{x^6}{180} + \cdots$

17. $-\infty < x < \infty$ and $0 < x < \infty$

19. $y = a_0 x\left(1 + \frac{x^2}{10} + \frac{x^4}{360} + \frac{x^6}{28{,}080} + \cdots\right)$
$+ a_1 x^{1/2}\left(1 + \frac{x^2}{6} + \frac{x^4}{168} + \frac{x^6}{11{,}088} + \cdots\right)$

20. $y = a_0 \sum_{n=0} \frac{(-1)^n(n + 1)}{(n + 3)!} x^{n+2} + a_1 \sum_{n=0} \frac{(-1)^n(n - 2)}{n!} x^{n-1}$

Section 1.6, p. 46

1. (a) $N = 3$

2. The integral converges to zero for $x = 0$ and to one for $0 < x \leq 1$.

4. The integral defining $G(a)$ can be shown to be uniformly convergent when $a \geq 0$. From this it can be proved that $G(a)$ is continuous when $a \geq 0$.

8. $N = a/\epsilon$ for uniform convergence.

9. (b) $$\lim_{n\to\infty} \cos n^2x = \begin{cases} \text{does not exist if } x \neq 0 \\ 1 \text{ if } x = 0 \end{cases}$$

10. (b) For example, the series diverges for $x = 0$.

CHAPTER 2

Section 2.1, p. 53

4. (a) Has only the trivial solution.

(b) The eigenvalues are all positive real numbers; eigenfunctions are

$$y_\lambda(x) = (\sinh \lambda x)/\sinh \lambda a.$$

(c) $$\lambda_n = \frac{(2n - 1)\pi}{2a}, \ n = 1, 2, \ldots$$

$$y_n(x) = \sin \frac{(2n - 1)\pi x}{2a}$$

(d) λ^2 is a positive real number,

$$\lambda \neq (2n - 1)\frac{\pi}{2}, \ n = 1, 2, \ldots$$

$$y_\lambda(x) = \cos \lambda x + \tan \lambda a \sin \lambda x$$

5. (b) $y_0 = x/a$ (d) $y_0 = 1$

7. $y_0(t) = t$

8. If $\lambda_n = n\pi$, $n = 1, 2, \ldots$, then $y_n = \sin n\pi x$; if $\lambda_n = \dfrac{(2n - 1)\pi}{2}$, $n = 1, 2, \ldots$, then $y_n = \cos\left(\dfrac{2n - 1}{2}\right)\pi x$. These results may be combined into

$$\lambda_n = \frac{n\pi}{2}, \ y_n = \sin \frac{n\pi}{2}(x + 1), \qquad n = 1, 2, \ldots.$$

9. $\lambda_n = n^2$, $y_n = \sin nx \exp(-x)$, $n = 1, 2, \ldots$

11. $y_0(x) = x - 1$

14. (a) $y = 2 \cos x + (1/2)(2 - \pi) \sin x + x$

(c) $y = 0$

Section 2.2, p. 61

6. $\lambda_n = (2n - 1)^2/4$, $n = 1, 2, \ldots$; $y_n(x) = \cos[(2n - 1)x/2]$

7. $\lambda_n = n^2$, $n = 0, 1, 2, \ldots$; $y_n(x) = \cos nx$; also $\lambda_n = (2n + 1)^2/4$, $n = 0, 1, 2, \ldots$; $y_n(x) = \sin[(2n + 1)x/2]$. These results can be combined to give $y_n(x) = \cos[(n/2)(x + \pi)]$.

9. $\lambda_n = n^2 - 1$, $n = 1, 2, \ldots$; $y_n(x) = \sin nx$

10. $\lambda_n = -n^2\pi^2$, $n = 1, 2, \ldots$; $y_n(x) = \sin n\pi x \exp(-x)$

11. $\lambda_n = -n^2$, $n = 1, 2, \ldots$; $y_n(x) = (n \cos nx + \sin nx)\exp(-x)$; also $\lambda = 1$; $y_1(x) = 1$

13. $\lambda = -1$; $y(x) = x \exp(-2x)$

14. no solution

16. **(b)** $\exp(-x)$

17. **(b)** $\exp(-x^2)$

18. **(b)** $(1 - x^2)^{-1/2}$

Section 2.3, p. 66

4. $\{\sqrt{2/\pi} \sin nx, n = 1, 2, \ldots\}$

6. $$\int_0^1 (\sin\sqrt{\lambda_n}x - \sqrt{\lambda_n}\cos\sqrt{\lambda_n}x)(\sin\sqrt{\lambda_m}x - \sqrt{\lambda_m}\cos\sqrt{\lambda_m}x)\,dx = 0$$

if $m \neq n$ and λ's are solutions of $\sqrt{\lambda} = \tan\sqrt{\lambda}$.

8. $\{\sqrt{2/\pi} \cos[(2n - 1)x/2], n = 1, 2, \ldots\}$

9. **(a)** $\{\sqrt{2/\pi} \sin[(2n - 1)x/2], n = 1, 2, \ldots\}$
(c) $\{\sqrt{c}/c, \sqrt{2/c}\cos(n\pi x/c), n = 1, 2, \ldots\}$

10. **(b)** The appropriate factor $\mu(x)$ is $\exp(2x)$.
(c) The weight function $w(x)$ is $\exp(2x)$.

Section 2.4, p. 71

1. $\mathbf{e}_k = 1\mathbf{e}_k$

3. **(a)** not defined **(c)** -1 **(e)** -1 **(g)** $1/2$

7. **(a)** $1, -1, 1, -1$ **(c)** π, 0, not defined, not defined

8. **(a)** $1/2$ **(c)** 0

9. Each successive pair of functions is "more nearly orthogonal" on the interval (0, 1) with weight function $w(x) = 1$. The pair in part (c) *is* orthogonal on the interval.

11. $\{\sqrt{2}/2, \sqrt{3/2}x, \sqrt{5/8}(1 - 3x^2)\} \doteq \{0.707, 1.225x, 3.162(1 - 3x^2)\}$

13. $\{\sqrt{3}/2, \sqrt{15}x/2, \sqrt{21/32}(5x^2 - 1)\} \doteq \{0.866, 1.936x, 0.810(5x^2 - 1)\}$

14. $c_1 = 4/3,\ c_2 = 0,\ c_3 = -1/3$

16. **(a)** $$f(x) = \begin{cases} 1 \text{ if } x \text{ is a rational number} \\ 0 \text{ if } x \text{ is an irrational number} \end{cases}$$

(c) $f(x) = 1/(x - a)$

18. **(b)** a_1 and a_2 can be any nonzero constants; $a_4 = 0$; $a_3 = -3a_5$, with a_5 nonzero but otherwise arbitrary.

Section 2.5, p. 78

1. First, the given quantity is a minimum when its square is a minimum. Second, the only quantity in the expansion of Eq. (2.5–5) over which we have any control is the one containing c_j, and this quantity is nonnegative. Hence the minimum occurs when this nonnegative quantity assumes its minimum value, namely, zero.
2. The inequality (2.5–8) holds for all N; hence the sequence of partial sums consisting of *positive terms* is bounded above by some positive number, namely, the maximum value of the square of the norm of f on (a, b). Consequently, the given series converges.

5. $$\int_0^{2\pi} \cos x \sin jx\, dx = 0 \text{ for } j = 1, 2, 3, \ldots$$

Section 2.6, p. 84

7. **(a)** $\begin{bmatrix} 0 & 1 \\ -\lambda^2 & 0 \end{bmatrix}$ **(c)** $\begin{bmatrix} 0 & 1 \\ \lambda & -1 \end{bmatrix}$

8. **(c)** $\mathbf{v}' = -A^T\mathbf{v}$, where

$$-A^T = \begin{bmatrix} 0 & -\lambda \\ -1 & 1 \end{bmatrix}$$

9. **(c)** $v'' - v' - \lambda v = 0$

14. $$\frac{d}{dx}\left[\frac{dy}{dx}\exp(-x^2)\right] + \frac{y \exp(-x^2)}{x} = 0$$

15. $$y_1Ly_2 - y_2My_1 = a_2y_2''y_1 + a_1y_2'y_1 + a_0y_2y_1 - y_2(a_2y_1)'' + y_2(a_1y_1)' - a_0y_2y_1$$
$$= \frac{d}{dx}\left[a_2y_2'y_1 - y_2(a_2y_1)' + y_2a_1y_1\right]$$

Section 2.7, p. 89

5. $\cos(n\pi x/c),\ n = 0, 1, 2, \ldots$, and $\sin(n\pi x/c),\ n = 1, 2, \ldots$
7. $\cos(2n - 1)x$ and $\sin(2n - 1)x$, $n = 1, 2, \ldots$, are linearly independent solutions

8. **(a)** $\sin \sqrt{\lambda}x$, $\lambda > 0$

10. The suggested substitution produces

$$\int_{-1}^{1} \frac{T_m(x)T_n(x)}{\sqrt{1-x^2}}\,dx = \int_0^{\pi} T_m(\cos\alpha)T_n(\cos\alpha)\,d\alpha.$$

Now, DeMoivre's formula $(\cos\alpha + i\sin\alpha)^n = \cos n\alpha + i\sin n\alpha$ shows that $\cos n\alpha$ is a polynomial of degree n in $\cos\alpha$. Finally, we know that

$$\int_0^{\pi} \cos m\alpha \cos n\alpha\,d\alpha = 0 \qquad \text{if } m \neq n.$$

Section 2.8, p. 98

3. **(a)** $y = 3\sinh x - 2x$

7. $y_3 = 0.910$, $y_4 = 0.840$, $y_5 = 0.750$, $y_6 = 0.641$, $y_7 = 0.514$, $y_8 = 0.368$, $y_9 = 0.207$, $y_{10} = 0.032$

8. $y_1 = -0.005$, $y_2 = -0.020$, $y_3 = -0.045$, $y_4 = -0.079$, $y_5 = -0.122$, $y_6 = -0.174$, $y_7 = -0.234$, $y_8 = -0.300$, $y_9 = -0.371$, $y_{10} = -0.445$

13. **(a)** The general solution is $y = c_1\cosh x + c_2\sinh x$. Boundary conditions transform this into the system

$$\begin{cases} 1.5431c_1 + 1.1752c_2 = 1.5431, \\ 3.7622c_1 + 3.6269c_2 = 3.7622, \end{cases}$$

which is easily solved by Cramer's rule to obtain $c_1 = 1$, $c_2 = 0$. Hence the solution is $y = \cosh x$.

(c) In finite-difference form the differential equation becomes

$$\frac{y_{i+1} - 2y_i + y_{i-1}}{h^2} = y_i$$

or $y_{i+1} + y_{i-1} = (h^2 + 2)y_i$. Using $h = 0.25$, $y_1 = 1.5431$, and $y_5 = 3.7622$ produces the system

$$\begin{cases} -2.0625y_2 + y_3 = -1.5431 \\ y_2 - 2.0625y_3 + y_4 = 0 \\ y_3 - 2.0625y_4 = -3.7622, \end{cases}$$

which has solutions $y_2 = 1.8894$, $y_3 = 2.3538$, $y_4 = 2.9653$.

16. Let $\bar{y}$ be the approximate value of y at the midpoint $x = 1/2$. Then

$$\frac{4 - 2\bar{y} + 1}{0.25} = \frac{3}{2}(\bar{y})^2,$$

which has approximate solutions 1.855 and -7.188. We choose the first of these, and with $h = 0.2$ we try $y_1 = 3$ and use

$$y_{i+1} = 0.06y_i^2 + 2y_i - y_{i-1}, \qquad i = 1, 2, 3, 4.$$

This yields the values of $y_2 = 2.540$, $y_3 = 2.467$, $y_4 = 2.759$, and $y_5 = 3.509$. Next we try $y_1 = 2.8$ to obtain the values $y_2 = 2.070$, $y_3 = 1.60$, $y_4 = 1.279$, and

$y_5 = 1.058$. Using *linear* extrapolation, we find a better approximation to y_1, namely, $y_1 = 2.7953$. With this value we obtain $y_2 = 2.0594$, $y_3 = 1.5780$, $y_4 = 1.2460$, and $y_5 = 1.0071$. The above values may be refined further. The other solution can be found by starting with $y_1 = -2.5$ and improving this value.

17. **(b)** $\ddot{y} = 4y$, $y(0) = 1.175$, $y(1) = 10.018$
(d) $y = \sinh x$

18. **(e)**
$$\begin{bmatrix} 2+h^2 & -1 & & & \\ -1 & 2+h^2 & -1 & & \\ \cdot & & \cdot & & \\ & \cdot & & \cdot & \\ & & \cdot & & \cdot \\ & & & -1 & 2+h^2 \end{bmatrix} \begin{bmatrix} y_1 \\ y_2 \\ \cdot \\ \cdot \\ \cdot \\ y_n \end{bmatrix} = \begin{bmatrix} -h^2x_1 + y_0 \\ -h^2x_2 \\ \cdot \\ \cdot \\ \cdot \\ -h^2x_n + y_{n+1} \end{bmatrix}$$

CHAPTER 3

Section 3.1, p. 109

3. The given expression satisfies the partial differential equation identically, *and* it contains two arbitrary functions; hence it qualifies as the set of all solutions.

5. **(b)** elliptic when $x > 0$; hyperbolic when $x < 0$; parabolic when $x = 0$
(d) hyperbolic outside the unit circle with center at the origin; elliptic inside this unit circle; parabolic on the unit circle

11. **(b)**, **(d)**, and **(e)** are not linear

15. $f(y - 3x)$ and $g(y + 2x)$ are solutions, where f and g are twice-differentiable functions of x and y.

17. **(b)** hyperbolic **(c)** parabolic if $x = 0$ or $y = 0$, otherwise hyperbolic

19. If the roots of Eq. (3.1–3) are $\alpha \pm \beta i$, then

$$u(x, y) = f[y + (\alpha + \beta i)x] + g[y + (\alpha - \beta i)x].$$

22. If we consider first the term $x\psi_1$, differentiate, and substitute into the biharmonic equation, we have after rearranging terms

$$4\frac{\partial}{\partial x}\left[\frac{\partial^2\psi_1}{\partial x^2} + \frac{\partial^2\psi_1}{\partial y^2}\right] + x\frac{\partial^2}{\partial x^2}\left[\frac{\partial^2\psi_1}{\partial x^2} + \frac{\partial^2\psi_1}{\partial y^2}\right] + x\frac{\partial^2}{\partial y^2}\left[\frac{\partial^2\psi_1}{\partial x^2} + \frac{\partial^2\psi_1}{\partial y^2}\right].$$

Since ψ_1 is harmonic, each bracketed term is zero. The term ψ_2 is simpler.

Section 3.2, p. 119

4. **(a)** Starting with Eq. (3.2–11) and applying the nonhomogeneous boundary condition produces $u_n(x, 0) = b_n \sin nx = 2 \sin 3x$. Hence $n = 3$, $b_3 = 2$, and $u(x, y) = (2/\sinh 3b) \sin 3x \sinh 3(b - y)$.
(c) Using the hint, $\sin 2x \cos x = (\sin 3x + \sin x)/2$. Hence we need two

terms, one with $n = 1$ and $b_1 = 1/2$, the other with $n = 3$ and $b_3 = 1/2$. Thus

$$u(x, y) = \frac{\sin x \sinh (b - y)}{2 \sinh b} + \frac{\sin 3x \sinh 3(b - y)}{2 \sinh 3b}.$$

8. $u(x, y) = \dfrac{3 \sin x \sinh (b - y)}{\sinh b}$

9. **(b)** 0 **(d)** 0.9721

12. $T' - \lambda T = 0,\ kX'' + (a - \lambda)X = 0$

13. $T' - \lambda T = 0,\ kX'' + bX' - \lambda X = 0$

14. $T' - \lambda T = 0,\ kX'' + bX' - \lambda X = -a$

15. $Y'' + \lambda^2 Y = 0$, $Y(0) = Y(b) = 0$, has solutions $Y_n(y) = \sin (n\pi y/b)$. $X'' - (n^2\pi^2/b^2)X = 0$, $X(\pi) = 0$, has solutions

$$X_n(x) = \left[\sinh \frac{n\pi}{b}(x - \pi)\right] / \cosh \frac{n\pi^2}{b}.$$

Hence the solution can be taken as a linear combination of functions of the form

$$u_n(x, y) = \left[\sin \frac{n\pi}{b} y \sinh \frac{n\pi}{b}(x - \pi)\right] / \cosh \frac{n\pi^2}{b}.$$

17. $X'' + \lambda^2 X = 0$, $X(0) = X(\pi) = 0$, has solutions $X_n(x) = \sin nx$. $Y'' - n^2 Y = 0$, $Y(0) = 0$, has solutions $Y_n(y) = \sinh ny$. Hence solutions are linear combinations of $u_n(x, y) = \sin nx \sinh ny$, that is,

$$u(x, y) = \sum_{n=1} b_n \sin nx \sinh ny.$$

The nonhomogeneous condition produces

$$\phi(x) = u(x, b) = \sum_{n=1} b_n \sin nx \sinh nb,$$

showing that

$$b_n \sinh nb = \frac{2}{\pi} \int_0^\pi \phi(x) \sin nx\, dx.$$

21. $X'' - \lambda X = 0$, $Y'' + \lambda y Y = 0$, if $y \neq 0$

Section 3.3, p. 131

5. $\phi(x) = -\psi(2L - x)$, but $\psi(x)$ has already been defined in the interval $-L \le x \le L$, so that the above equation now extends $\phi(x)$ to the interval $-L \le 2L - x \le L$, that is, $-3L \le -x \le -L$ or $L \le x \le 3L$.

7. $u_{tt} = c^2(u_{xx} + u_{yy} + u_{zz})$

8. **(a)** y_{tt} has dimensions of acceleration; $y_{xx} = d(dy/dx)/dx$ has dimensions of cm^{-1}.

9. $y = \sin x \cos at$

10. $y = (\cos x \sin at)/a$

15. $y = \dfrac{2L}{3\pi a} \cos \dfrac{3\pi x}{L} \sin \dfrac{3\pi at}{L}$

17. $-\int_{\alpha}^{x-at} = \int_{x-at}^{\alpha}$

18. Assuming the function of t to be of the form $\exp(i\omega t)$ is equivalent to choosing the sign of the separation constant so that $T(t)$ will be expressed in terms of sines and cosines. The separation constant is ω^2 here.

22. The boundary-value problem becomes (see Exercise 20) $a^2y''(x) - g = 0$, $y(0) = y(L) = 0$, whose solution is $4g(x - L/2)^2 = 8a^2y + gL^2$, which is a parabola. Differentiating, we find $dy/dx = 0$ when $x = L/2$ and the maximum displacement is $-gL^2/8a^2$.

Section 3.4, p. 138

3. k has units of cm^2/sec; K has units of calorie/°C cm.

6. The bar is insulated except for the ends, which are kept at zero for all time; there are no heat sources; and heat flows from a region of higher temperature $[f(x)]$ to one of lower temperature (0°).

9. **(a)** $3 \sin(2\pi x/L) \exp(-4k\pi^2 t/L^2)$

11. **(a)** The problem $u''(x) = 0$, $u(0) = 0$, $u(L) = 100$ has solution $u = 100x/L$.
(c) $u(x) = 100$
(d) $u(x) = 50(1 + x/L)$

13. The substitution $u(x, t) = f(ax + bt)$ leads to the differential equation $a^2kf'' - bf' = 0$ with solutions $f(ax + bt) = c_1 + c_2 \exp[b(ax + bt)/ka^2]$. Substituting this result into $u_t = ku_{xx}$ shows that $b = a^2k$.

15. $c = 1/k$

16. The diffusion of respiratory gases is rendered more rapid by the large surface area of the lungs.

Section 3.5, p. 149

14. **(b)** hyperbolic everywhere except on the x-axis, where it is parabolic;
(d) parabolic on the unit circle, elliptic in its interior, and hyperbolic elsewhere;
(f) parabolic everywhere

15. **(b)** $-y/x^2$ **(d)** $-2y/x$

18. **(a)** $\cot 2\theta = -3/4$, $\theta \doteq -0.4636$; rotation through an angle θ results in $u_{\xi\xi} + 6u_{\eta\eta} = 0$.

CHAPTER 4

Section 4.2, p. 161

4. $g'(1)$ is undefined

6. **(a)** 4π **(c)** $2/3$ **(e)** 4

7. Use mathematical induction.

9. **(a)** $$-\frac{\pi}{4} + \frac{2}{\pi}\left(\frac{\cos x}{1^2} + \frac{\cos 3x}{3^2} + \cdots\right) + \frac{\sin x}{1} - \frac{\sin 2x}{2} + - \cdots$$

(c) $$1 - \frac{4}{\pi}\left(\frac{\sin \pi x}{1} + \frac{\sin 3\pi x}{3} + \cdots\right)$$

(e) $$\frac{1}{24} + \frac{1}{2\pi^2}\sum_{n=1}\frac{(-1)^n \cos 2n\pi x}{n^2} + \frac{1}{4\pi^3}\sum_{n=1}\frac{(2n-1)^2\pi^2 - 4}{(2n-1)^3}\sin 2(2n-1)\pi x - \frac{1}{8\pi}\sum_{n=1}\frac{\sin 4n\pi x}{n}$$

(g) $$\frac{\pi}{4} - \frac{2}{\pi}\sum_{n=1}\frac{\cos(2n-1)x}{(2n-1)^2} + \sum_{n=1}\frac{(-1)^{n+1}\sin nx}{n}$$

(i) $$\frac{8}{\pi^2}\sum_{n=1}\frac{\cos(2n-1)x}{2n-1}$$

10. **(a)** 1 **(c)** 3/2 **(e)** 1 **(g)** 1

13. **(a)** $\pi/2$ **(c)** $\pi/2$ **(e)** $\pi/2$

14. **(b)** 1 **(d)** 1 **(f)** 1

17. Twenty terms give a result that is approximately 1.2 percent low.

26. **(a)** $$\frac{\exp(2\pi) - 1}{\pi}\left(\frac{1}{2} + \sum_{n=1}\frac{\cos nx - n\sin nx}{n^2 + 1}\right)$$

Section 4.3, p. 170

8. $$c_n = \frac{1}{2}(a_n - ib_n) = \frac{1}{2\pi}\int_{-\pi}^{\pi} f(s)\,[\cos ns - i \sin ns]\,ds$$
$$= \frac{1}{2\pi}\int_{-\pi}^{\pi} f(s)\left[\frac{\exp(ins) + \exp(-ins)}{2} - i\,\frac{\exp(ins) - \exp(-ins)}{2i}\right]ds,$$

from which the desired result follows.

9. $$c_n = \frac{1}{2\pi}\int_{-L}^{L} f(s)\exp(-in\pi s/L)\,ds,\ n = 0, \pm 1, \pm 2, \ldots$$

10. **(d)** $F(-x) = (-x)\exp[-(-x)^2] = -x\exp(-x^2) = -F(x)$

11. (e) $F(-x) = -x \tan(-x) = (-x)(-\tan x) = x \tan x = F(x)$

12. (d) $F(-x) = (-x)^2/1 - x = x^2/1 - x$, which is neither $-F(x)$ nor $F(x)$

13. If f is odd and g is even, then $f(-x) = -f(x)$ and $g(-x) = g(x)$; hence $f(-x)g(-x) = -f(x)g(x)$, showing that the product is odd.

14. (a) $\displaystyle \frac{4}{\pi}\sum_{n=1} \frac{\sin(n\pi x/2)}{n}$

15. (b) a

16. (a) $\displaystyle \frac{2}{\pi}\sum_{n=1} \frac{1}{n}\left[(-1)^{n+1} - 2\frac{1+(-1)^{n+1}}{n^2\pi^2}\right]\sin(n\pi x)$

(b) $\displaystyle \frac{1}{3} + \frac{4}{\pi^2}\sum_{n=1} \frac{(-1)^n \cos(n\pi x)}{n^2}$

17. (b) $\displaystyle \frac{1}{2}(e^2 - 1) + 4\sum_{n=1} \frac{(-1)^n e^2 - 1}{4 + n^2\pi^2}\cos(n\pi x/2)$

18. (b) $\displaystyle \frac{2}{\pi} - \frac{4}{\pi}\sum_{n=1} \frac{\cos(2n\pi x)}{4n^2 - 1}$

19. (a) $\displaystyle \frac{8}{\pi}\sum_{n=1} \frac{n \sin(2nx)}{4n^2 - 1}$

23. (a) $\displaystyle \frac{1}{2} + \frac{4}{\pi^2}\sum \frac{\cos(2n-1)\pi x}{(2n-1)^2}$

(b) let $x = 1$

24. $\displaystyle \frac{4}{\pi}\sum_{n=1} \frac{\sin(2n-1)x}{2n-1}$

26. $\displaystyle \frac{1}{2} - \frac{4}{\pi^2}\sum_{n=1} \frac{\cos(2n-1)\pi x}{(2n-1)^2}$

27. $c_n = 0$ if n is even (including $n = 0$) and $c_n = 2/(in\pi)$ if n is odd; thus the series can be written as

$$\frac{2}{i\pi}\sum_{n=-\infty}^{\infty} \frac{\exp[i(2n-1)\pi x/2]}{2n-1}.$$

Note that this can be written in terms of sines and cosines, in which case the cosine terms vanish and the sine terms double so that the final result is the same as when we use Eq. (4.3–6).

28. Make an even periodic extension of the function $g(x) = 1, -1 \le x \le 1$. Then $c_0 = 1$ and all other $c_n = 0$.

31. $f(-x) = -x - x^2 = -x(1 + x) = f(x + 1) = f(x)$

37. If $f(x)$ is even, then $f(-x) = f(x)$. Differentiating this produces

$$f'(-x) = \frac{df(-x)}{d(-x)} = \frac{df(-x)}{dx}\frac{dx}{d(-x)} = -\frac{df(-x)}{dx} = \frac{df(x)}{dx}$$

since $f'(x)$ is even. Hence $df(x) = -df(-x)$, that is, $f(x) = -f(-x) + C$. But $f(-x) = f(x)$ so that $2f(x) = C$, proving the assertion.

39. $\frac{2}{\pi}\left(\sin x - \sin 2x + \frac{1}{3}\sin 3x + \frac{1}{5}\sin 5x - \frac{1}{3}\sin 6x\right)$

Section 4.4, p. 181

7. $V(x, y) = \operatorname{csch}\pi \sin \pi x \sinh \pi(1 - y)$

9. Let $x = 1/2$, and compute $V(x, y)$ for various values of y. Approximate results are shown in the following table:

y	0.1	0.2	0.3	0.4	0.5	0.6	0.7	0.8	0.9
$V(1/2, y)$	0.72	0.52	0.38	0.28	0.20	0.14	0.09	0.06	0.03

Repeat for $x = 0.1, 0.2, 0.3$, and 0.4. Note that by symmetry the results are the same for $x = 0.9, 0.8, 0.7$, and 0.6, respectively. Locate the values of $V(x, y)$ on a suitable square, and connect points of equal V by a smooth curve. It is also helpful to compute $V(x, 0)$ for the above values of x. See figure below.

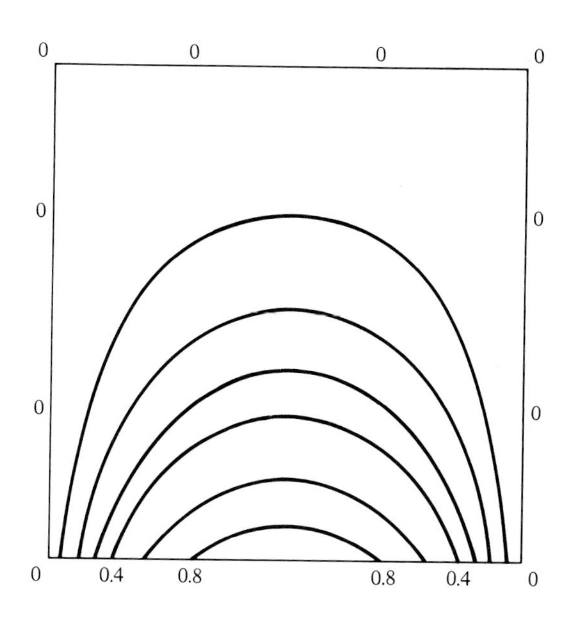

10. (a) P.D.E.: $u_{tt} = a^2 u_{xx}$, $0 < x < 1$, $t > 0$;
B.C.: $u(0, t) = u(1, t) = 0$, $t > 0$;
I.C.: $u(x, 0) = 0.01x$, $u_t(x, 0) = 0$, $0 < x < 1$

(b) $$u(x, t) = \frac{0.02}{\pi} \sum_{n=1} \frac{(-1)^{n+1} \sin n\pi x \cos n\pi a t}{n}$$

12. (a) $y_x(\pi, t) = 0$ implies that there is no vertical force on the string at $x = \pi$; that is, the end is free (see Section 3.3).

(c) $|g(t)|$ must be "small" in comparison to the length of the string.

Section 4.5, p. 190

4. Write $\sin(N + \frac{1}{2})t$ as $\sin Nt \cos \frac{1}{2}t + \cos Nt \sin \frac{1}{2}t$, and then use L'Hôpital's rule.

8. The series converges uniformly by the Weierstrass M-test.

10. $f''(x)$ does not exist because the infinite series is divergent.

13. (a) $$\frac{4}{\pi} \sum_{n=1} \frac{\sin(2n-1)x}{2n-1}$$

(c) $$\frac{\pi}{2} - \frac{4}{\pi} \sum_{n=1} \frac{\cos(2n-1)x}{(2n-1)^2}$$

14. (a) The representation of $F(x) = x$, $0 < x < 2\pi$, $F(x + 2\pi) = F(x)$ is

$$\pi - 2 \sum_{n=1} \frac{\sin nx}{n},$$

hence

$$f(x) = \frac{1}{2}(\pi - x) \sim \sum_{n=1} \frac{\sin nx}{n}.$$

(c) Since $f(x)$ satisfies the hypotheses of Theorem 4.5–4, we can integrate from zero to a to obtain

$$\frac{\pi a}{2} - \frac{a^2}{4} = \int_0^a f(x)\,dx = \sum_{n=1} \frac{1 - \cos na}{n^2}, \qquad 0 \leq a \leq 2\pi.$$

Then, replacing a by x, we have

$$\frac{x(2\pi - x)}{4} = \sum_{n=1} \frac{1}{n^2} - \sum_{n=1} \frac{\cos nx}{n^2}, \qquad 0 \leq x \leq 2\pi.$$

(d) Outside the interval $[0, 2\pi]$, the even periodic extension of the function on the left is given by the series on the right.

15. The series is the $a_0/2$ term in the representation of $g(x)$.

16. Since (see Exercise 3)

$$\frac{\sin (2n + 1)t}{2 \sin t} = \frac{1}{2} + \cos 2t + \cos 4t + \cdots + \cos 2nt$$

and $\int_0^{\pi/2} \cos 2kt\, dt = 0$, the result follows.

CHAPTER 5

Section 5.1, p. 201

2. The integral is finite.

11. Only the function in part (a) is absolutely integrable, the integral being 2.

19. **(a)** $\frac{2}{\pi} \int_0^\infty \frac{\alpha \sin \alpha x}{1 + \alpha^2}\, d\alpha$

(c) $\frac{2}{\pi} \int_0^\infty \frac{\sin \alpha - \alpha}{\alpha^2} \sin \alpha x\, d\alpha$

(e) $\frac{2}{\pi} \int_0^\infty \frac{2\alpha \sin \alpha + 2 \cos \alpha - \alpha^2 \cos \alpha - 2}{\alpha^3} \sin \alpha x\, d\alpha$

20. **(b)** $\frac{2h}{\pi} \int_0^\infty \frac{\sin \alpha L}{\alpha} \cos \alpha x\, d\alpha$

(d) $\frac{2}{\pi} \int_0^\infty \frac{\sin \alpha}{\alpha} (1 + 2 \cos \alpha) \cos \alpha x\, d\alpha$

21. **(a)** $\frac{i}{2\pi} \int_{-\infty}^\infty \frac{\alpha\, [\exp (i\alpha\pi) + 1]}{\alpha^2 - 1} \exp (-i\alpha x)\, d\alpha$

(c) $\frac{1}{2\pi} \int_{-\infty}^\infty \frac{1 + i\alpha}{1 + \alpha^2} \exp (-i\alpha x)\, d\alpha$

23. **(b)** When $x = L$, we have

$$\frac{h}{\pi} \int_0^\infty \frac{\sin 2\alpha L}{\alpha}\, d\alpha = \frac{h}{\pi} \left(\frac{\pi}{2}\right) = \frac{h}{2},$$

using Exercise 14.

(d) When $x = 0$, we have

$$\frac{1}{2\pi} \int_{-\infty}^\infty \frac{d\alpha}{1 + \alpha^2} + \frac{i}{2\pi} \int_{-\infty}^\infty \frac{\alpha d\alpha}{1 + \alpha^2} = \frac{1}{2} + 0 = \frac{1}{2},$$

noting that the first integrand is an even function, whereas the second is odd.

Section 5.2, p. 215

4. $\bar{f}(\alpha) = 2 \int_0^1 (1 - x) \cos \alpha x\, dx = \left(\sin^2 \frac{\alpha}{2}\right) / (\alpha/2)^2;$

$$u(x, t) = \frac{4}{\pi} \int_0^\infty \frac{\sin^2 (\alpha/2) \exp (-\alpha^2 t) \cos \alpha x}{\alpha^2}\, d\alpha$$

5. $\bar{f}(\alpha) = b \int_0^c \exp(i\alpha x)\, dx = \dfrac{bi}{\alpha}[1 - \exp(i\alpha c)];$

$$u(x, t) = \frac{2bi}{\pi} \int_0^\infty \frac{[1 - \exp(i\alpha c)] \exp(-\alpha^2 kt) \cos \alpha x \, d\alpha}{\alpha}$$

7. $u(x, t) = \dfrac{2}{\pi} \displaystyle\int_0^\infty \frac{\exp(-\alpha^2 t) \cos \alpha x}{1 + \alpha^2}\, d\alpha$

9. **(a)** $u(x, t) = \dfrac{2}{\pi} \displaystyle\int_0^\infty \frac{\exp(-\alpha^2 kt) \cos \alpha x}{1 + \alpha^2}\, d\alpha$

10. **(a)** $\bar{f}(\alpha) = [2 \cos(\alpha\pi/2)]/(1 - \alpha^2)$

11. **(b)** $\dfrac{2[\pi - \alpha \sin(\alpha/2)] \exp(i\alpha/2)}{\pi^2 - \alpha^2}$

(d) $\dfrac{4\alpha \cos \alpha + 2(\alpha^2 - 2) \sin \alpha}{\alpha^3}$

(e) $\dfrac{2ih}{\alpha}(1 - \cos \alpha c)$

12. **(a)** $\dfrac{h}{\alpha}[1 - \cos(\alpha/h)]$

(c) $\dfrac{1}{\alpha^2}(\alpha - \sin \alpha)$

13. **(a)** $\dfrac{1 + \sin \alpha\pi}{1 - \alpha^2}$

(c) $\dfrac{2\pi \cos(\alpha/2)[1 - \sin(\alpha/2)]}{\pi^2 - \alpha^2}$

14. **(c)** Using the inverse Fourier transform,

$$\frac{1}{2\pi} \int_{-\infty}^{\infty} \bar{f}(x) \exp(i\alpha x)\, dx = f(-\alpha)$$

15. **(a)** $\displaystyle\int_{-\infty}^{\infty} f(x - b) \exp(i\alpha x)\, dx$

$$= \int_{-\infty}^{\infty} f(x - b) \exp[i\alpha(x - b)] \exp(i\alpha b)\, d(x - b) = \bar{f}(\alpha) \exp(i\alpha b)$$

24. $\dfrac{2i \sin 6\pi\alpha}{\alpha^2 - 1}$

Section 5.3, p. 221

6. Consult the table of Fourier transforms in the Appendix.

7. $u(x, t) = \dfrac{2u_0}{\pi} \displaystyle\int_0^\infty \frac{\sin \alpha x}{\alpha} [1 - \exp(-k\alpha^2 t)]\, d\alpha$

8. **(b)** $u(x, t) = \frac{2a}{\pi} \int_0^\infty \frac{\cos \alpha x \exp(-k\alpha^2 t)}{a^2 + \alpha^2} \, d\alpha$

9. $u(x, t) = \frac{2u_0}{\pi} \int_0^\infty \frac{\sin \alpha x \, (1 - \cos \alpha L) \exp(-k\alpha^2 t)}{\alpha} \, d\alpha$

10. $u(x, t) = \frac{2\beta}{\pi} \int_0^\infty \frac{\cos \alpha x}{\alpha^2} [1 - \exp(-k\alpha^2 t)] \, d\alpha$

12. **(a)** $y(x) = cY_0(x)$

14. **(b)** $u(0, 1) = u(c, 1) = \frac{1}{\pi} \arctan c$; $u(c/2, 1) = \frac{2}{\pi} \arctan \frac{c}{2}$

16. **(a)** $y(x) = -\frac{Ci}{\pi} \int_0^\infty \frac{1}{\alpha} \sin\left(\frac{1}{\alpha} + \alpha x\right) d\alpha$

(c) $y(x) = \frac{C}{\pi} \int_0^\infty \exp(-\alpha^2/2) \cos\left(\frac{\alpha^3}{3} - \alpha x\right) d\alpha$

21. **(a)** $y(x) = \frac{C}{2\pi} \int_{-\infty}^\infty \exp(i\alpha^3 - \alpha^2) \exp(-i\alpha x) \, d\alpha$

CHAPTER 6

Section 6.1, p. 236

2. Use the identity $\sinh(A - B) = \sinh A \cosh B - \cosh A \sinh B$.

4. $B_{mn} = \frac{4ab(-1)^{m+n}}{\pi^2 \, mn \sinh(\pi\omega_{mn})}$

6. $V(x, y) = \frac{2}{\pi} \int_0^\infty \frac{\alpha}{\alpha^2 + 1} \frac{\sinh \alpha(b - y)}{\sinh(\alpha b)} \sin(\alpha x) \, d\alpha$

7. $V(x, y) = \frac{2}{\pi} \int_0^\infty \frac{\sin(\alpha\pi)}{1 - \alpha^2} \frac{\sinh \alpha(b - y)}{\sinh(\alpha b)} \sin(\alpha x) \, d\alpha$

15. $u(x, y) = \frac{2}{\pi^2} (\pi - y) + \frac{4}{\pi} \sum_{n=1} \frac{\cos(2nx)}{1 - 4n^2} \frac{\sinh 2n(\pi - y)}{\sinh(2n\pi)}$

17. $\sum_{n=1} \frac{1}{1 - 4n^2} = \frac{1}{2} \sum_{n=1} \left(\frac{1}{1 + 2n} + \frac{1}{1 - 2n}\right)$

18. $u(x, y) = \frac{2a}{\pi} \int_0^\infty \frac{\sinh[\alpha(1 - x)] \cos(\alpha y)}{(a^2 + \alpha^2) \sinh \alpha} d\alpha$

20. **(c)** $u(x, y) = \frac{2}{\pi} \sum_{n=1} \frac{(-1)^{n+1}(\sinh n\pi y \sin n\pi x + \sinh n\pi x \sin n\pi y)}{n \sinh n\pi}$

24. If $\phi_x(x, y) = -\phi_y(x, y)$

28. **(b)** $u(a, 0) = 0$; $u(0, 0) = 10$

32. **(b)** $f(x, y) = xy = u(x, y)$ as given in the solution to Exercise 20(c).

Section 6.2, p. 249

3. $u(x, y, t) = k \sin(\pi y/b) \sin(\pi x/a) \cos(c\pi\sqrt{a^2 + b^2}t/ab)$

$f = c\sqrt{a^2 + b^2}/2ab \text{ sec}^{-1}$

4. $B_{mn} = \dfrac{64a^2b^2}{\pi^6(2n - 1)^3(2m - 1)^3}$, $n = 1, 2, \ldots, m = 1, 2, \ldots$

7. E has dimensions g cm/cm^2 sec^2; ρ has dimensions g/cm^3

19. T_0 has dimensions g cm/sec^2 cm; w has dimensions g/cm^2; g has dimensions cm/sec^2; G has dimensions g cm/sec^2 cm^2

20. $b_{2n-1} = 4kL/c\pi^2(2n - 1)^2$, $n = 1, 2, \ldots$

21. $$u(x, t) = \frac{8h}{\pi^2} \sum_{n=1} \frac{(-1)^{n+1} \sin[(2n - 1)\pi x/L] \cos[(2n - 1)\pi ct/L]}{(2n - 1)^2}$$

23. $u(1/4, 2) = 3/16$, $u(1/2, 3/2) = -1/2$

24. Compute $U(0, 0)$.

27. **(b)** The difference $ds - dx = \sqrt{1 + u_x^2}\, dx - dx \doteq u_x^2\, dx/2$. Hence the resulting increase in potential energy is $T_0 u_x^2\, dx/2$, where $T_0 = \rho c^2$.

28. $u(x, t) = \sum_{n=1} b_n \sin nx \cos\sqrt{n^2 - 1}\, t$, where

$$b_n = \frac{2}{\pi} \int_0^\pi f(s) \sin ns\, ds, \quad n = 1, 2, \ldots$$

29. **(b)** $u(x, t) = (1/bc)(\cosh bx \sinh bct) + 2xt^2 + (t^3/6)$

Section 6.3, p. 259

4. Because the steady-state solution is zero.

12. $u(x, t) = \sum_{n=1} b_{2n-1} \exp[-\pi^2(2n - 1)^2 t/4L^2] \sin[(2n - 1)\pi x/2L]$, where

$$b_{2n-1} = \frac{2}{L} \int_0^L f(s) \sin[(2n - 1)\pi s/2L]\, ds, \quad n = 1, 2, \ldots$$

15. $u(x, t) = \dfrac{Cx(L - x)}{2} + \sum_{n=1} b_n \sin(n\pi x/L) \exp(-n^2\pi^2 t/L^2)$, where

$$b_n = \frac{2}{L}\int_0^L \left[\frac{Cs(s - L)}{2} + f(s)\right] \sin(n\pi s/L)\, ds$$

17. $u(x, t) = \sum_{n=1} a_n \cos[(2n - 1)\pi x/4] \exp[-(2n - 1)^2\pi^2 t/4]$, where

$$a_n = \frac{8a}{\pi^2}\left[\frac{\pi(2n - 1)(-1)^{n+1} - 2}{(2n - 1)^2}\right], \; n = 1, 2, \ldots$$

20. $u(x, t) = \sum_{n=1} b_n \sin[(2n - 1)\pi x/2] \exp[-(2n - 1)^2\pi^2 kt/4]$, where

$$a_n = \frac{16a}{\pi^3(2n - 1)^3}[(-1)^{n+1}\pi(2n - 1) - 2]$$

21. **(a)** $u(x, t) = \frac{1}{2}[1 - \cos 2x \exp(-4kt)]$ Note that $\sin^2 x = \frac{1}{2}(1 - \cos 2x)$.

22. $u(x, t) = \dfrac{\pi}{2} \sin x \exp(-kt) - \dfrac{16}{\pi} \sum_{n=2} \dfrac{\sin 2nx \exp(-4n^2kt)}{n(4n^2 - 1)^2}$

Section 6.4, p. 272

10. $u(x, t) = 1 - \dfrac{4}{\pi} \sum_{n=1} \dfrac{\sin(2n - 1)\pi x \exp[-(2n - 1)^2\pi^2 t]}{2n - 1}$

14. **(a)** $u(x, t) = \dfrac{2u_0}{\pi} \displaystyle\int_0^\infty \frac{[\exp(-\alpha^2 t) - 1] \cos \alpha x \, d\alpha}{\alpha^2}$

$$= u_0 x + \frac{2u_0}{\pi}\int_0^\infty \frac{\exp(-\alpha^2 t) - 1}{\alpha^2} d\alpha$$

$$- \frac{2u_0}{\pi}\int_0^\infty \int_0^x \frac{\exp(-\alpha^2 t)}{\alpha} \sin \alpha s \, ds \, d\alpha$$

16. **(a)** $v(x, y) = \dfrac{2}{\pi}\displaystyle\int_0^\infty \frac{\sinh(\alpha x) \cos(\alpha y)}{\alpha \cosh(\alpha c)} d\alpha \int_0^\infty f(s) \cos(\alpha s)\, ds$

17. **(a)** $v(x, y) = \dfrac{2}{\pi}\displaystyle\int_0^\infty \frac{\cos(\alpha x) \cosh(\alpha y)}{\cosh(\alpha b)} d\alpha \int_0^\infty f(s) \cos(\alpha s)\, ds$

22. $u(1, 1/4) \doteq 0.61u_0$; $u(1, 4) \doteq 0.89u_0$

30. Use the relation

$$\frac{\cos \alpha x}{\alpha} = \frac{1}{\alpha} - \int_0^x \sin \alpha s \, ds.$$

Section 6.5, p. 283

9. $$\frac{\cosh \omega_n(b - y)}{\sinh \omega_n b} = \frac{\exp[\omega_n(b - y)] + \exp[-\omega_n(b - y)]}{\exp(\omega_n b) - \exp(-\omega_n b)}$$

$$= \frac{\exp(-\omega_n y)[1 + \exp[-2\omega_n(b - y)]]}{1 - \exp(-2\omega_n b)}$$

$$\leq \frac{2\exp(-\omega_n y)}{1 - \exp(-2b)}$$

16. $\int_0^\infty \exp(-xt)\, dt = 1/x, x > 0$; the convergence is uniform for $a \leq x < \infty$ $(a > 0)$. Since $\exp(-xt) \leq \exp(-at)$, we can take $M(t) = \exp(-at)$. Note that

$$\int_0^\infty \exp(-at)\, dt = 1/a.$$

17. Take $M(t) = \exp(-t)$.

20. **(c)** When $\alpha = 1\sqrt{2t}$, y attains its maximum value $\exp(-1/2)/\sqrt{2t}$. The function is an odd function.

CHAPTER 7

Section 7.1, p. 295

1. **(b)** The case $n = 0$ leads to a constant that is also periodic of period 2π.

2. Divide by ρ, and then make the substitution $v = dR/d\rho$.

7. One interpretation is the following: find the steady-state temperatures in the walls of an infinitely long pipe having inner radius b and outer radius c if the outside of the pipe is kept at temperature zero, while the inside surface has a temperature given by $f(\phi)$.

8. The pipe is infinitely long, that is, $-\infty < z < \infty$, so that there are no boundaries in the z-direction. Moreover, the boundary values on the inside and outside surfaces are independent of z.

10. Note that the differential equations are Cauchy–Euler equations (see Section 1.3).

14. $$u(\rho, \phi) = \frac{4u_0}{\pi} \sum_{n=1} \left(\frac{\rho^{2n-1} - \rho^{-(2n-1)}}{c^{2n-1} - c^{-(2n-1)}} \right) \frac{\sin(2n - 1)\phi}{2n - 1}$$

16. $$u(\rho, \phi) = \frac{\log \rho}{2 \log c} a_0 + \sum_{n=1} a_n \frac{\rho^{2n} - \rho^{-2n}}{c^{2n} - c^{-2n}} \cos(2n\phi),$$

where $a_0 = \frac{4}{\pi}\int_0^{\pi/2} f(s)\, ds$ and $a_n = \frac{4}{\pi}\int_0^{\pi/2} f(s) \cos 2ns\, ds$, $n = 1, 2, 3, \ldots$

18. **(a)** $u(\rho) = [100 \log(\rho/b)]/\log(a/b)$

19. $$u(\rho, \phi) = \frac{100}{\pi}\left[\frac{\pi}{8} + \sum_{n=1}^{\infty}\left(\frac{\rho}{c}\right)^n \frac{\sin n\left(\frac{\pi}{4} - \phi\right) + \sin n\phi}{n}\right]$$

22. **(a)** $z(\rho) = (z_0 \log \rho)/\log \rho_0,\ 1 \le \rho \le \rho_0$

Section 7.2, p. 312

7. **(a)** $y = c_1 J_0(x) + c_2 Y_0(x)$;
(c) $y = c_1 J_0(e^x) + c_2 Y_0(e^x)$

8. **(a)** Make the substitution $u = \lambda_j s$;

(c) evaluate $\int_0^b J_1(\lambda_j s)\, ds$ as in part (a), and then let $b \to \infty$.

9. **(a)** Use integration by parts with $u = J_0(s)$, $dv = J_1(s)\, ds$.

10. **(b)** $$\frac{2}{c}\sum_{j=1}^{\infty} \frac{(\lambda_j c)^2 - 4}{\lambda_j^3 J_1(\lambda_j c)} J_0(\lambda_j x)$$

11. **(b)** $$2\sum_{j=1}^{\infty} \frac{(8 - \lambda_j^2)}{\lambda_j^3 J_1'(\lambda_j)} J_1(\lambda_j x),\ 0 \le x < 1$$

12. **(b)** The transformed equation is

$$r^2 \frac{d^2Z}{dr^2} + r\frac{dZ}{dr} + [\lambda^2 r^2 - (n + 1/2)^2]Z = 0.$$

14. **(a)** $$\frac{1}{2}\sum_{j=1}^{\infty} \frac{J_0(\lambda_j) - J_0(2\lambda_j)}{\lambda_j [J_2(2\lambda_j)]^2} J_1(\lambda_j x)$$

16. $x^n J_n(x)$

17. **(b)** $-x^{-n} J_n(x)$

23. **(c)** 1.108, −0.140, 0.045, −0.021, 0.012
(d) 0.997

24. Begin by using integration by parts with $u = s^{n-1}$, $dv = sJ_0(s)\, ds$.

25. **(a)** $$2\sum_{j=1}^{\infty} \frac{\lambda_j J_2(\lambda_j)}{(\lambda_j^2 - 1)[J_1(\lambda_j)]^2} J_1(\lambda_j x)$$

26. See Eq. (7.2–23).

29. **(c)** Make the substitution $t = \sin\theta$ in part (a).

33. **(a)** By Rolle's theorem the derivative of $x^{-n} J_n(x)$ as given in Exercise 17(a) must vanish between its zeros.

Section 7.3, p. 334

3. **(b)** From Eq. (7.3–14) it follows that

$$P_n(-x) = (-1)^n P_n(x), \; n = 0, 1, 2, \ldots;$$

(d) Since $P'_{2n}(x)$ contains terms in $x^{2n-2k-1}$, the only way nonzero terms can arise when $x = 0$ is to have $2n - 2k - 1 = 0$. But this is impossible, since it implies that $2(n - k) = 1$. Hence $P'_{2n}(0) = 0$.

(e) From Eq. (7.4–14),

$$P_{2n}(x) = \frac{1}{2^{2n}} \sum_{k=0}^{N} \frac{(-1)^k (4n - 2k)!}{(2n - 2k)!(2n - k)!} x^{2n-k}.$$

When $x = 0$, the only nonzero term in this sum occurs when $k = n$, and then $P_{2n}(0)$ has the required value.

12. **(a)** $aP_1(x) + bP_0(x)$;

(c) $\frac{2}{5} aP_3(x) + \frac{2}{3} bP_2(x) + \left(c + \frac{3a}{5}\right) P_1(x) + \left(d + \frac{b}{3}\right) P_0(x)$

15. **(b)** $A_{2n+1} = \frac{(4n + 3)}{2} \int_0^1 x P_{2n+1}(x)\, dx$, which is zero for all n, except $n = 0$.

18. Use Rodrigues' formula.

29. A_0 is the average value of the even extension of the function on $(-1, 1)$.

Section 7.4, p. 350

12. $u(r, \theta) = (r \cos \theta)/b$

13. **(a)** $u(\rho, z) \doteq 28.68 J_0(2.405\rho) \cosh (2.405z) - 0.85 J_0(5.520\rho) \cosh (5.520z) + 0.03 J_0(8.654\rho) \cosh (8.654z) - + \cdots$

(b) $u(0, 0) \doteq 27.86$

14. $$u(\rho, z) = \frac{200}{c} \sum_{j=1} \frac{J_0(\lambda_j \rho) \sinh (\lambda_j z)}{\lambda_j J_1(\lambda_j c) \sinh (\lambda_j b)},$$

where the $\lambda_j c$ are roots of $J_0(\lambda) = 0$.

16. $$z(\rho, t) = \frac{2}{c} \sum_{j=1} \frac{J_0(\lambda_j \rho) \cos (\lambda_j a t)}{\lambda_j J_1(\lambda_j c)},$$

where the $\lambda_j c$ are roots of $J_0(\lambda) = 0$.

18. $$u(r, \theta) = \sum_{m=0} (4m + 3)\left(\frac{r}{b}\right)^{2m+1} P_{2m+1} (\cos \theta) A_{2m+1},$$

where $A_{2m+1} = \int_0^{\pi/2} f(\cos \theta) P_{2m+1} (\cos \theta) \sin \theta \, d\theta$

21. $u(r, \theta) = 100$; the result should be obvious, but work through the appropriate steps of Example 7.4–6.

23. **(a)** $$u(\rho, z) = \frac{2}{c^2} \sum_{j=1} \frac{\exp(-\lambda_j z) J_0(\lambda_j \rho)}{J_1^2(\lambda_j c)} \int_0^c x f(x) J_0(\lambda_j x)\, dx,$$

where $J_0(\lambda_j c) = 0$

24. $$v(r) = \frac{1}{r(b - a)} [v_1 a(b - r) + v_2 b(r - a)]$$

25. $$v(r, \theta) = \frac{-3E}{K + 2} r \cos\theta, \; r < b;$$

$$V(r, \theta) = -Er\cos\theta + Eb^3\left(\frac{K - 1}{K + 2}\right) r^{-2}\cos\theta, \; r > b$$

26. $$v(r, \theta) = -Er\cos\theta + \frac{Eb^3}{r^2}\cos\theta$$

28. $$\frac{\partial}{\partial z}\left(\frac{1}{r}\right) = -\frac{z}{r^3} = -\frac{1}{r^2} P_1\left(\frac{z}{r}\right) = -\frac{1}{r^2} P_1(\cos\theta),$$

since $z/r = \cos\theta$ and $P_1(z/r) = z/r$.

35. $$u(\rho, z) = \frac{200}{a} \sum_{j=1} \frac{J_0(\lambda_j \rho) \sinh[\lambda_j(L - z)]}{\lambda_j J_1(a\lambda_j) \sinh(\lambda_j L)},$$

where the $\lambda_j a$ are positive roots of $J_0(\lambda) = 0$.

36. $$f(z) = \frac{2}{z^2} J_2(z)$$

CHAPTER 8

Section 8.2, p. 358

1. $cm^2/°C$ sec

7. 0.07 is acceptable

Section 8.3, p. 365

3. The diffusion equation $u_t = k u_{xx}$ must be dimensionally correct.

5. **(b)** $u(5, 5) = 1.786$, $u(10, 5) = 7.143$, $u(15, 5) = 26.786$

6. 21

7. $u(10/3, 10/3) = 0.69$, $u(20/3, 10/3) = 2.08$, $u(10, 10/3) = 5.56$, $u(40/3, 10/3) = 14.58$, $u(50/3, 10/3) = 38.19$; $u(x, 20/3) = u(x, 10/3)$ by symmetry. Note that the resulting *five* equations can be solved by elimination.

11. **(a)** $y_i^1 = \frac{1}{2}(y_{i+1}^0 + y_{i-1}^0)$

t \ x	0	10	20	30	40	50	60	70	80
0	0	0.3	0.6	0.5	0.4	0.3	0.2	0.1	0
Δt	0	0.3	0.4	0.5	0.4	0.3	0.2	0.1	0
$2\Delta t$	0	0.1	0.2	0.3	0.4	0.3	0.2	0.1	0
$3\Delta t$	0	−0.1	0	0.1	0.2	0.3	0.2	0.1	0
$4\Delta t$	0	−0.1	−0.2	−0.1	0	0.1	0.2	0.1	0
$5\Delta t$	0	−0.1	−0.2	−0.3	−0.2	−0.1	0	0.1	0
$6\Delta t$	0	−0.1	−0.2	−0.3	−0.4	−0.3	−0.2	−0.1	0
$7\Delta t$	0	−0.1	−0.2	−0.3	−0.4	−0.5	−0.4	−0.3	0
$8\Delta t$	0	−0.1	−0.2	−0.3	−0.4	−0.5	−0.6	−0.3	0
$9\Delta t$	0	−0.1	−0.2	−0.3	−0.4	−0.5	−0.4	−0.3	0
$10\Delta t$	0	−0.1	−0.2	−0.3	−0.4	−0.3	−0.2	−0.1	0
$11\Delta t$	0	−0.1	−0.2	−0.3	−0.2	−0.1	0	0.1	0
$12\Delta t$	0	−0.1	−0.2	−0.1	0	0.1	0.2	0.1	0
$13\Delta t$	0	−0.1	0	0.1	0.2	0.3	0.2	0.1	0
$14\Delta t$	0	0.1	0.2	0.3	0.4	0.3	0.2	0.1	0
$15\Delta t$	0	0.3	0.4	0.5	0.4	0.3	0.2	0.1	0
$16\Delta t$	0	0.3	0.6	0.5	0.4	0.3	0.2	0.1	0

(b) $f = \dfrac{1}{(16)(0.000179)} = 350$ Hz; one-half of a cycle has been completed in 8 steps; hence it requires 16 steps for a complete cycle.

12. **(a)** with $\Delta x = 0.1$ and $\Delta t = 0.1$ (for $r = 1$) representative values are shown in the table

t \ x	0.1	0.3	0.5
0.1	0.2939	0.7694	0.9511
0.7	−0.1816	−0.4755	−0.5878
1.6	0.0955	0.2500	0.3090

(b) The results in part (a) are the same as the analytical results to four decimals.

17. $\Delta t = 0.0175$ sec for $r = 1/2$

18. $\Delta t = 3.846$ sec for $r = 1/2$

Index

The most important page reference for each entry is listed first.